AF361701

Migration and Global Health

Migration and Global Health

Editors

Heiko Becher
Volker Winkler
Hajo Zeeb

MDPI • Basel • Beijing • Wuhan • Barcelona • Belgrade • Manchester • Tokyo • Cluj • Tianjin

Editors
Heiko Becher
University Medical Center
Hamburg-Eppendorf
Germany

Volker Winkler
Heidelberg University Hospital
Germany

Hajo Zeeb
Leibniz Institute for Prevention
Research and Epidemiology –
BIPS
Germany

Editorial Office
MDPI
St. Alban-Anlage 66
4052 Basel, Switzerland

This is a reprint of articles from the Special Issue published online in the open access journal *International Journal of Environmental Research and Public Health* (ISSN 1660-4601) (available at: https://www.mdpi.com/journal/ijerph/special_issues/migration_global_health).

For citation purposes, cite each article independently as indicated on the article page online and as indicated below:

LastName, A.A.; LastName, B.B.; LastName, C.C. Article Title. *Journal Name* **Year**, *Volume Number*, Page Range.

ISBN 978-3-0365-3607-1 (Hbk)
ISBN 978-3-0365-3608-8 (PDF)

Cover image courtesy of Heiko Becher

Contents

About the Editors

Heiko Becher is director of the Institute of Medical Biometry and Epidemiology, University Medical Center Hamburg-Eppendorf, Hamburg, Germany. Until 2015 he was deputy director of the Institute of Public Health, University of Heidelberg, Heidelberg, Germany and head of the Unit of Epidemiology and Biostatistics. He is also affiliated with the Institute of Global Health, University of Heidelberg, Heidelberg, Germany and with the Berlin Institute of Health at Charité, Berlin, Germany. He studied statistics at the Universities of Dortmund, Germany, and Sheffield, England, and habilitated in Epidemiology and Medical Biometry at the medical Faculty, University of Heidelberg. He was founding president of the German Society of Epidemiology (DGEpi). His research interests include: statistical methods in epidemiology, cancer epidemiology, epidemiology of neurological diseases, social epidemiology/migrant research, infectious disease epidemiology including malaria and COVID-19, and descriptive epidemiology in low-income countries.

Volker Winkler After his studies in biology Volker Winkler did his Ph.D. within the DFG graduate program 793 'Epidemiology of communicable and chronic, non-communicable diseases and their interrelationships'. In 2015, he became Deputy Head of the unit 'Epidemiology and Biostatistics' within the Heidelberg Institute of Global Health and one year later, he received the venia legendi in epidemiology. Since 2018, he is co-leading the crosscutting research group on non-communicable diseases and in 2020, he became leader of the research group 'Epidemiology of Transition'. Considerations of the health and its determinants of population groups facing major changes are his main interests. He focuses on migrants, on non-communicable diseases and on the analysis of longitudinal data including the development of methods for the scientific analysis of register data. Generally, his research, as well as his teaching activities focus on topics using quantitative methods. Furthermore, he is PI and coordinator of the Heidelberg Graduate School of Global Health, which aims to provide medical students with an opportunity to engage in global health research.

Hajo Zeeb serves as head of the Department of Prevention and Evaluation at the Leibniz-Institute for Prevention Research and Epidemiology—BIPS in Bremen, Germany. He holds a professorship for epidemiology at the University of Bremen. His earlier posts include scientist positions at the German Cancer Research Center (DKFZ, Heidelberg) (1996–1999), the School of Public Health of Bielefeld University (1999–2005), the WHO in Geneva, Switzerland (2005–2006), and a professorship at the University Medical Center of Mainz University (2006–2009). After his medical studies, he obtained his medical doctorate from RWTH Aachen (1990) and an M.Sc. in Community Health from Heidelberg University (1996). His research interests include social and environmental epidemiology, as well as evidence-based public health and intervention research. He is speaker of the Leibniz Science Campus Digital Public Health Bremen and is a past president of the German Society for Epidemiology (DGEpi). During the coronavirus pandemic, he actively contributed to the coordination of the Competence Network Public health COVID-19 in Germany.

Preface to "Migration and Global Health"

The recent years have witnessed a substantial increase in scientific research on migration-related topics, including health. Numerous large population-based studies have been implemented, mainly in Europe and North America, and these studies usually include individuals with a migration background. Therefore, these studies allow a comparison to the autochthonous population with respect to chronic disease risk, genetic predispositions, epigenetic aspects, and preventive aspects, including health literacy, health service use, and others. Although these perspectives are critical to understand and improve the health situation of migrants in their host countries, the relationship between migration and health remains complex. Migration can increase health risks but also improve the health and wellbeing of the migrants themselves and that of their family members "left behind". Today's migration-related research often neglects the perspectives regarding the countries of origin, which frequently are lower- or middle-income countries (LMIC).

This Special Issue of the *International Journal of Environmental Research and Public Health* focuses on research and experiences related to the broad topic of migration and health while recognizing its complexity. It contains 14 articles, which clearly show the large variety of research topics in this challenging field. Naturally, since we as the editors of this Special Issue are located in Germany, many but not all of the results presented here stem from research in our country, which has received millions of migrants in its recent history.

The two largest migrant groups in Germany are (i) the so-called "Aussiedler"(resettler), who are descendants of Germans who settled in Eastern Europe, mainly in the 17th and 18th century, and who migrated from countries of the former Union of Soviet Socialist Republics (USSR) back to Germany, mainly after 1989 and (ii) migrants from Turkey and their descendants who initially came to Germany in the 1960s. About seven million individuals resident in Germany belong to these two groups and represent almost 10% of the total population. Six of the papers in this Special Issue are devoted to these two migrant populations.

Studies by Mahanani et al. and Lindblad et al. investigate issues around colorectal and gastric cancer among resettlers from the former USSR, using secondary data from a large database of migrants. Huebner et al. link individual environmental and genetic data in resettlers and a possible effect on cardiovascular diseases. In Arena et al., general health issues among possible future resettlers, i.e., individuals whose ancestors moved from Germany to Russia, and which still live in Russia, were investigated. Krist et al. and Anapa et al. focus on the population descending from Turkey. They study health-related quality of life and acculturation, as well as validity aspects of cognitive assessment across comparison groups.

The two papers by Grochtdreis et al. use data from the well-renowned socio-economic panel (SOEP) to investigate healthcare utilization and health-related quality of life in groups with varying degrees of migration history. The SOEP is an interesting resource for migration and health research in Germany and can be exploited even more in the future.

The number of migrants from Sub-Saharan Africa in Germany is comparatively small, however there are clusters of African migrants, mainly in large cities. The study reported by Amoah et al. deals with migrants from Ghana and it is an interesting example of a feasibility study to perform an intervention in this group. The qualitative study by Kidane et al. focusses on oral health among refugees from Eritrea who recently came to Germany. It is a big challenge to do research on this group, not only because it has a very heterogeneous composition, but also because it is very difficult to approach them for research and to generalize the results due to lack of representativeness even for select groups.

Oral health is also in the focus of the systematic review on dental caries among refugees in Europe, reported by Buhari et al. This is the second paper of this Special Issue drawing attention to oral health an often overlooked but highly relevant public health concern. The study by Vonneilich et al. also focusses on the European context and draws attention to differences in changing health pattern between aging migrants and non-migrants. Given the history of immigration to European countries, the importance of healthy ageing of migrants is further increasing.

The two final studies of this Special Issue broaden its focus beyond Europe. Martinez-Cardoso and Geronimus examine how migration affects the cardio-metabolic health profile of Mexicans living temporarily in the United States and the systematic literature review by Racaite et al. looks globally at physical health consequences among so-called left-behind children, e.g., children whose parents temporarily migrated for work.

In summary, the current collection attests to the ample research needs and opportunities around migration and health, with a focus on recent, as well as earlier migration to Europe. It sheds light on several issues ranging from non-communicable disease epidemiology and health services utilization to aspects of quality of life, and of some methodological challenges. We hope that the collection stimulates further research in this important field and supports mainstreaming of migrant health in health research in coming years.

Heiko Becher, Volker Winkler, Hajo Zeeb
Editors

International Journal of
Environmental Research and Public Health

MDPI

Article

Colorectal Cancer among Resettlers from the Former Soviet Union and in the General German Population: Clinical and Pathological Characteristics and Trends

Melani Ratih Mahanani [1], Simone Kaucher [1], Hiltraud Kajüter [2], Bernd Holleczek [3], Heiko Becher [4] and Volker Winkler [1,*]

[1]   Epidemiology of Transition, Heidelberg Institute of Global Health, University Hospital Heidelberg, 69120 Heidelberg, Germany; melani.mahanani@uni-heidelberg.de (M.R.M.); simone.kaucher@uni-heidelberg.de (S.K.)
[2]   Cancer Registry, 44801 Bochum, Germany; Hiltraud.Kajueter@krebsregister.nrw.de
[3]   Saarland Cancer Registry, 66119 Saarbrücken, Germany; b.holleczek@krebsregister.saarland.de
[4]   Institute for Medical Biometry and Epidemiology, University Medical Center Hamburg-Eppendorf, 20246 Hamburg, Germany; h.becher@uke.de
*   Correspondence: volker.winkler@uni-heidelberg.de

Citation: Mahanani, M.R.; Kaucher, S.; Kajüter, H.; Holleczek, B.; Becher, H.; Winkler, V. Colorectal Cancer among Resettlers from the Former Soviet Union and in the General German Population: Clinical and Pathological Characteristics and Trends. *Int. J. Environ. Res. Public Health* **2021**, *18*, 4547. https://doi.org/10.3390/ijerph18094547

Academic Editor: Matteo Goldoni

Received: 11 January 2021
Accepted: 23 April 2021
Published: 25 April 2021

Publisher's Note: MDPI stays neutral with regard to jurisdictional claims in published maps and institutional affiliations.

Abstract: This study examined time trends and clinical and pathological characteristics of colorectal cancer (CRC) among ethnic German migrants from the Former Soviet Union (resettlers) and the general German population. Incidence data from two population-based cancer registries were used to analyze CRC as age-standardized rates (ASRs) over time. The respective general populations and resettler cohorts were used to calculate standardized incidence ratios (SIRs) by time-period (before and after the introduction of screening colonoscopy in 2002), tumor location, histologic type, grade, and stage at diagnosis. Additionally, SIRs were modeled with Poisson regression to depict time trends. During the study period from 1990 to 2013, the general populations showed a yearly increase of ASR, but for age above 55, truncated ASR started to decline after 2002. Among resettlers, 229 CRC cases were observed, resulting in a lowered incidence for all clinical and pathological characteristics compared to the general population (overall SIR: 0.78, 95% CI 0.68–0.89). Regression analysis revealed an increasing SIR trend after 2002. Population-wide CRC incidence decreases after the introduction of screening colonoscopy. In contrast the lowered CRC incidence among resettlers is attenuating to the general population after 2002, suggesting that resettlers do not benefit equally from screening colonoscopy.

Keywords: incidence; colorectal cancer; young-onset; clinical characteristics; pathological characteristics; migrants; Former Soviet Union; Germany

1. Introduction

The term colorectal cancer (CRC) summarizes malignancies of the colon and the rectum. In 2016, its median age of diagnosis was 76 and 72 years among German women and men, respectively [1]. The incidence of CRC started to decrease after the introduction of colon cancer screening by colonoscopy in 2002 [2]. Recently, rising CRC incidence among adolescents and young adults has attracted increased attention [3,4]. It is known that these so-called young-onset CRC tumors present distinctive clinical and pathological characteristics with lower survival compared to non-young-onset CRC cases [4–6]. An increasing prevalence of well-known CRC risk factors, such as alcohol consumption [7], red meat intake, low physical activity, cigarette smoking [8], obesity [9,10], and diabetes mellitus [11] are discussed as the main reasons. Additionally, the recommended screening age may influence the observed differences between young-onset and non-young-onset CRC, respectively [12,13].

Int. J. Environ. Res. Public Health **2021**, *18*, 4547. https://doi.org/10.3390/ijerph18094547   https://www.mdpi.com/journal/ijerph

Colonoscopy is a secondary prevention method that is generally offered to populations with increased age, family history of CRC syndromes, and chronic inflammatory bowel diseases [4,14]. Unlike other cancer screening methods aiming for early diagnosis to improve patients' outcomes, colonoscopy additionally allows removing precancerous lesions during the examination [15]. In Germany, statutory health insurance (SHI) has offered fully covered colonoscopies to people aged 55 years and above since 2002, as an alternative to the fecal occult blood test, which has been offered since 1977 [16]. In 2019, Germany lowered the recommended age for screening colonoscopies for men to 50 years, while it remained unchanged for women [17].

Ethnic German resettlers from the Former Soviet Union are the second-largest migrant group in Germany, with about 2.5 million people immigrating in large numbers in the early 1990s after the fall of the iron curtain [18]. They receive German citizenship upon arrival and are entitled to fully utilize the German healthcare system [19,20]. Focusing on cancer incidence and mortality compared to the general population of Germany, Kaucher et al. found that incidence and mortality of colorectal (both sexes), lung (women), prostate, and female breast cancer were lower among resettlers [21].

Considering the discussion about the increasing incidence of CRC at younger ages and the offer of screening colonoscopy, this study aims to explore the incidence of CRC and its temporal trends and to compare clinical and pathological characteristics of CRC cases between resettlers and the general German population.

2. Materials and Methods

We used data of two resettler cohorts, one in the administrative district of Münster (North Rhine-Westphalian) and another one in the federal state of the Saarland with the observation periods 1994 to 2013 and 1990 to 2009, respectively. The combined cohort comprised 51,311 resettlers (Saarland: 18,619; Münster: 32,692), who immigrated between 1990 and 2001 (Münster) and between 1990 and 2005 (Saarland). More details on the study population and the follow-up procedures can be found elsewhere [21,22]. The study protocol was approved by the Ethics Committee of the Medical Faculty, University Hospital Heidelberg [22].

In brief, the vital status of resettlers was derived from local population registries through record linkage or manually. The accumulated person-time was estimated for each sex, 5-year age group (up to 85+), and calendar year. For the general populations of the Saarland and Münster, person-time was ascertained from the mid-year populations provided by the federal statistic office of the Saarland and the federal cancer registry of North Rhine-Westphalian, respectively. Both population-based cancer registries provided data on CRC cases in the respective general population and the respective resettler cohort through record linkage. Incidence data included date of diagnosis, age at diagnosis, sex, tumor location (International Classification of Diseases 10th Revision (ICD-10)), tumor morphology (International Classification of Diseases for Oncology third revision (ICD O-3)), tumor grade, and stage at diagnosis. We restricted all analysis to histologically confirmed primary CRC cases (ICD-10 C18-C20) and categorized patients according to age (<55 years: young-onset CRC, ≥55 years: non-young-onset CRC).

The condensed stage at diagnosis coding system developed by the European Network of Cancer Registries was used to categorize tumors into a local, advanced, or unknown stage based on the status of lymph nodes (N) and the existence of metastasis (M) [23], due to expected missing values in population-wide registry data with varying versions of the TNM classification. Furthermore, we combined well and moderately differentiated tumors (grades 1 and 2) as low grade and poorly differentiated and undifferentiated ones (grades 3 and 4) as high grade. For three-year calendar periods, we calculated truncated age-standardized incidence rates (ASRs) separately for young-onset and non-young-onset CRC using the 1976 European standard population [24].

Afterward, we modeled the truncated ASR over time by first estimating age-specific rates with Poisson regression using the number of observed cases as the dependent variable and the log of the mid-year population P as the offset. For young-onset incidence Y, calendar year T from 1990 to 2013 (continuous, coded from 0 to 23, starting in 1990) and age group A (categorical, 5-year age groups) were used as covariables (see formula 1).

For non-young-onset CRC incidence Z, an interrupted time-series approach [25] was used to detect changes due to colonoscopy screening introduced in 2002. In addition to calendar year T and age group A, "colonoscopy" X (binary; 0: years 1990–2001, 1: years 2002–2013) and the interaction term between the calendar year and "colonoscopy" were used as covariables (see Formula (2)). Using the modeled yearly age-specific rates Y and Z from Formulas (1) and (2), we calculated the ASR shown in Figure 1:

$$\log(Y_{t,\,a}) = \log(P) + \beta_0 + \beta_1 T + \beta_2' A \tag{1}$$

$$\log(Z_{t,\,a}) = \log(P) + \beta_0 + \beta_1 T + \beta_2' A + \beta_3 X_t + \beta_4 T X_t \tag{2}$$

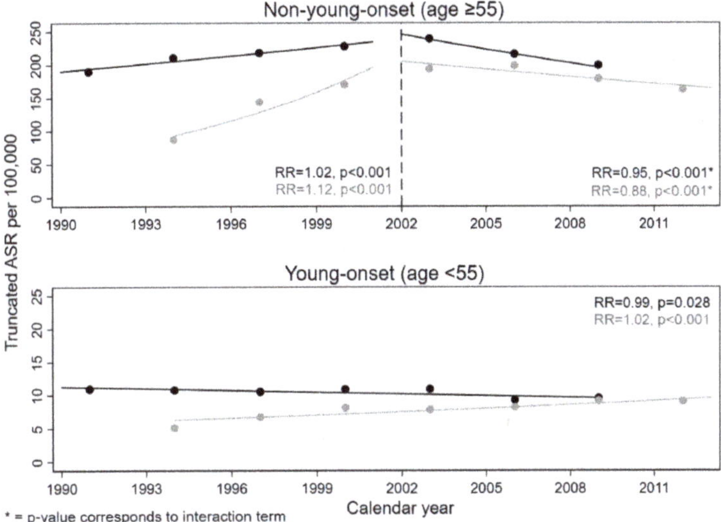

Figure 1. Observed and modeled young-onset and non-young-onset truncated ASRs of colorectal cancer incidence for the general population from 1990 to 2013; the rate ratio (RR) corresponds to the modeled calendar year effect (see Appendix A); the dashed line indicates the introduction of screening colonoscopy; black represents the Saarland population, gray the Münster population.

In the next step, we calculated standardized incidence ratios (SIRs) to compare observed CRC cases among resettlers to expected numbers in the respective host population using sex, age group, and calendar year-specific rates. SIRs were computed with exact 95% confidence intervals (95% CI) for all cases and before and after the introduction of colonoscopy in 2002, as well as for tumor location, grading, histology, and condensed stage at diagnosis.

We also modeled sex-specific SIRs with Poisson regression for young-onset and non-young onset cases separately using the number of observed cases among resettlers as the dependent variable and the log of the expected cases E as the offset. For young-onset cases U, we used sex S (binary; 0: male, 1: female) and calendar year T as covariables (see Formula (3)). For non-young-onset cases V, the model was again extended with the covariables colonoscopy X and the interaction term of the calendar year and colonoscopy (see Formula (4)):

$$\log(U_{t,\,s}) = \log(E) + \beta_0 + \beta_1 T + \beta_2 S \tag{3}$$

$$\log(V_{t,\,s}) = \log(E) + \beta_0 + \beta_1 T + \beta_2 S + \beta_3 X_t + \beta_4 T X_t \tag{4}$$

In all Poisson models, standard errors were controlled for overdispersion [26]. Statistical analyses were performed using Stata/IC 15.1 for Windows (64-bit x86-64) Revision 21 November 2017 (StataCorp LLC, 4905 Lakeway Drive, College Station, TX 77845, USA).

3. Results

Between 1990 and 2013, the combined resettler cohorts comprised 666,899 person-years and 238 diagnoses with primary colorectal cancer, of which 229 (96.2%) cases were histologically confirmed. In the host population, 48,980 (88.7%) CRC cases were histologically confirmed. Demographic characteristics of patients, as well as clinical and pathological features of the included tumors among the general population and the resettlers, are summarized in Table 1. Notably, there were more young-onset CRC cases among resettlers (22.3% vs. 10.0%). Rectal cancer was more frequently diagnosed for the general population, while among resettlers, CRC was more likely in the left colon. In both groups, most CRC cases were of other adenocarcinoma subtypes, low grade, and localized tumors.

Table 1. Demographic characteristics and clinical and pathological characteristics of CRC among resettlers and in the general population (Saarland, 1990–2009 and Münster, 1994–2013).

Characteristics		General Population		Resettler	
		N	**%**	**N**	**%**
Total		48,980	100.0	229	100.0
Region	Saarland	17,405	35.5	76	33.2
	Münster	31,575	64.5	153	66.8
Time period	1990–2001	19,466	39.7	52	22.7
	2002–2013 (colonoscopy)	29,514	60.3	177	77.3
Young-onset	yes (age < 55)	4906	10.0	51	22.3
	no (age ≥ 55)	44,074	90.0	178	77.7
Sex	Female	25,349	51.8	106	46.3
	Male	23,631	48.2	123	53.7
Anatomic location	Right colon	13,123	26.8	55	24.0
	Left colon	12,756	26.0	83	36.3
	Rectum	15,810	32.3	66	28.8
	Other/unknown	7291	14.9	25	10.9
Histologic Type	Mucinous adenocarcinoma	8463	17.3	34	14.9
	Signet-ring cell carcinoma	311	0.6	1	0.4
	Other adenocarcinoma subtypes	36,344	74.2	183	79.9
	Other/unknown	3862	7.9	11	4.8
Tumor grade	Low	34,663	70.8	174	76.0
	High	10,499	21.4	44	19.2
	Unknown	3818	7.8	11	4.8
Tumor stage	Local	18,469	37.7	93	40.6
	Advanced	17,404	35.5	85	37.1
	Unknown	13,107	26.8	51	22.3

Separated by age at onset, Figure 1 illustrates the observed and the modeled truncated ASR of the general populations. The underlying Poisson regression coefficients can be found in Appendix A.

Table 2 presents results of the SIR analyses of CRC among resettlers compared to the general population. Overall, the SIR was lower among resettlers in both cohorts and for both sexes. Resettlers showed a lower incidence of CRC according to all clinical and pathological characteristics.

Table 2. Standardized incidence ratios of resettlers compared to the general Saarland (1990–2009) and Münster (1994–2013) population with exact 95% confidence intervals.

Characteristics		Total		Saarland		Münster	
		Obs.	SIR (95% CI)	Obs.	SIR (95% CI)	Obs.	SIR (95% CI)
Total (1990–2013)		229	**0.78 (0.68–0.89)**	76	**0.73 (0.57–0.91)**	153	**0.81 (0.68–0.94)**
Time period	1990–2001	52	**0.61 (0.46–0.80)**	31	0.72 (0.49–1.02)	21	**0.50 (0.31–0.76)**
	2002–2013 (colonoscopy)	177	**0.85 (0.73–0.98)**	45	**0.74 (0.54–0.98)**	132	0.89 (0.75–1.06)
Young-onset	Yes (age < 55)	51	0.99 (0.74–1.31)	14	0.80 (0.44–1.34)	37	1.10 (0.77–1.51)
	No (age ≥ 55)	178	**0.73 (0.63–0.85)**	62	**0.72 (0.55–0.92)**	116	**0.74 (0.61–0.89)**
Sex	Female	123	0.85 (0.71–1.02)	47	0.97 (0.71–1.29)	76	**0.80 (0.63–0.99)**
	Male	106	**0.70 (0.58–0.85)**	29	**0.52 (0.35–0.75)**	77	0.81 (0.64–1.02)
Anatomical location	Right colon	55	**0.69 (0.52–0.91)**	15	0.65 (0.36–1.06)	40	**0.72 (0.51–0.98)**
	Left colon	83	1.08 (0.86–1.33)	29	1.08 (0.73–1.55)	54	1.07 (0.80–1.39)
	Rectum	66	**0.68 (0.53–0.87)**	16	**0.45 (0.26–0.73)**	50	0.82 (0.61–1.08)
	Others (incl. % unknown)	25 (20)	**0.60 (0.39–0.89)**	16 (31.3)	0.86 (0.49–1.40)	9 (0)	**0.39 (0.18–0.75)**
Histologic type	Mucinous adenocarcinoma	34	**0.67 (0.47–0.94)**	14	0.82 (0.45–1.38)	20	**0.60 (0.36–0.92)**
	Signet-ring cell carcinoma	1	0.45 (0.01–2.49)	1	1.13 (0.03–6.31)	0	0.00 (0.00–2.73)
	Other adenocarcinomas	183	**0.83 (0.71–0.96)**	59	**0.73 (0.56–0.94)**	124	0.88 (0.74–1.05)
	Others	11	**0.53 (0.26–0.95)**	2	0.35 (0.04–1.26)	9	0.60 (0.27–1.14)
Tumor grade	Low grade	174	**0.83 (0.71–0.96)**	61	0.82 (0.62–1.05)	113	**0.83 (0.68–0.99)**
	High grade	44	**0.68 (0.50–0.92)**	13	**0.58 (0.31–0.98)**	31	0.74 (0.50–1.05)
	Unknown	11	0.57 (0.29–1.03)	2	0.29 (0.03–1.04)	9	0.73 (0.34–1.40)
Tumor stage	Local stage	93	0.82 (0.66–1.01)	26	**0.67 (0.44–0.98)**	67	0.90 (0.70–1.15)
	Advanced stage	85	**0.79 (0.63–0.98)**	32	0.86 (0.59–1.22)	53	**0.76 (0.57–0.99)**
	Unknown	39	**0.64 (0.45–0.87)**	18	**0.49 (0.25–0.88)**	28	0.72 (0.48–1.05)

SIR, standardized incidence ratio; Obs., number of observations; CI, confidence interval. Significant results are bolded.

Modeled SIRs for CRC among resettlers are shown in Figure 2. Among non-young-onset CRC, an increasing SIR could be observed after the introduction of screening colonoscopy in 2002. Corresponding Poisson regression coefficients can be found in Appendix A.

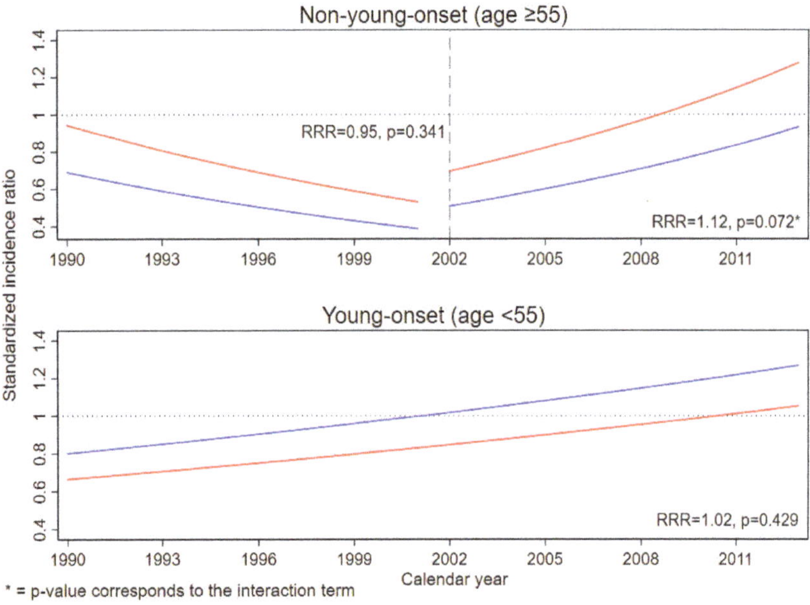

Figure 2. Modeled standardized incidence ratios for colorectal cancer among resettlers compared to the direct host populations using Poisson regression from 1990 to 2013; the relative SIR change (RRR) corresponds to the modeled calendar year effect (see Appendix A); the dashed line indicates the year of screening colonoscopy introduction; the blue line represents men, red line women.

4. Discussion

This study confirmed a declining incidence in the population for which screening colonoscopy is offered in Germany. Among resettlers, the CRC incidence in older age groups was lower compared to the general population; however, for young-onset CRC, there was no difference. The lower incidence of resettlers was increasing and, therefore, attenuating to the general population after the introduction of screening colonoscopy. With respect to clinical and pathological characteristics, there were hardly any differences to the general population except for a higher frequency of left colon tumors among resettlers.

The observed increase of CRC incidence among young individuals is consistent with a number of studies from different western countries [10,27–29]. This increase may to some extent be attributed to modifiable risk factors, such as obesity and physical inactivity [30,31]. Additionally, low awareness of young-onset CRC among both patients and physicians and that it also occurs in those who are not subjected to family history or apparent risk factors [32] might contribute.

Left-sided CRC diagnosis is associated with rectal bleeding and changes in bowel habits [33], which may generally lead to delayed diagnoses. Additionally, a previous study suggested that male resettlers were more likely to be diagnosed with advanced tumors when looking at the most frequent cancer-sites combined (stomach, colorectal, lung, breast, and prostate cancer) [21]. However, this study does not show delayed diagnoses for CRC among resettlers. In contrast, the general German population presented a higher incidence of mucinous adenocarcinoma, which is associated with poorer clinical and pathological characteristics, such as higher grade and advanced stage at diagnosis, leading to lower survival compared to other CRC types [34].

A possible explanation for the attenuating incidence between resettlers and the general population is that risk behaviors and lifestyle adjustments to the host population are likely among migrants, as well as improved screening and diagnostic accessibility [21,35]. Resettlers may gradually adjust their lifestyle and dietary habits due to greater availability and selection of food [21]. Another explanation for the attenuating incidence might be the overtime constant CRC incidence rate among resettlers, suggesting that resettlers do not benefit from screening colonoscopy equally to the German population, which experiences decreasing rates. If resettlers do not use screening colonoscopy, they also do not benefit from the possibility to remove precancerous lesions, which may result in a higher incidence of CRC. However, the constant CRC incidence rate among resettlers (analysis not shown) might also be explained by the limited number of observations.

Our study is the first population-based study looking at time trends and clinical and pathological characteristics of young-onset and non-young-onset CRC among resettlers from the Former Soviet Union compared to Germany's general population. It needs to be stated that the analysis relies only on secondary data without information on individual risk factors, such as lifestyle, family history of CRC, etc. Furthermore, the dataset was restricted to histologically confirmed CRC cases leavening out 11.0% (young-onset: 5.2%; non-young-onset: 11.9%) of all reported CRC cases. However, there was no time trend concerning histological confirmation, and the fraction of confirmed cases was close to or above 90% except for the years 1994 to 1996 when only about 70% of all cases were histologically confirmed. Therefore, the restriction to histologically confirmed cases does not introduce bias onto the time trend analysis of the general population. Concerning the resettler cohorts, selection bias was unlikely since all ethnic Germans were invited to migrate to Germany, and during the immigration process, they were allocated quasi-randomly to their first area of residence [22]. Due to data protection concerns, neither information on the date of immigration nor an individual mortality follow-up among individuals of the Münster cohort was available, which prevented us from analyzing the incidence among resettlers concerning lengths of stay in Germany. However, since most resettlers migrated to Germany in the first half of the 1990s, calendar time is highly correlated with length of stay. It should also be mentioned that the person-time of the Münster cohort had to be estimated due to an incomplete follow-up [36].

5. Conclusions

Similar to other countries, Germany is encountering a decreasing CRC incidence in the population eligible for screening colonoscopy. CRC incidence among ethnic German migrants from the Former Soviet Union is lower but continuously attenuates to the general population. This might hint towards less screening participation among resettlers, which may lead to increasing CRC incidence. However, the clinical and pathological characteristics of the resettler's tumor conditions were hardly different from the general population.

Author Contributions: H.B. and V.W. initiated the cohort studies, data were collected by V.W. and S.K. H.K. provided the incidence data for the Münster cohort. B.H. provided the incidence data for the Saarland cohort. M.R.M. did the statistical analyses. V.W. helped with the analyses. M.R.M. drafted the manuscript, and S.K., H.K., B.H., H.B., and V.W. contributed to writing and editing the manuscript and interpreting the results. All authors have read and agreed to the published version of the manuscript.

Funding: This research was funded by the German Cancer Aid (Grant Number 111232).

Institutional Review Board Statement: The study protocol was approved by the Ethics Committee of the Medical Faculty, University Hospital Heidelberg.

Informed Consent Statement: Not applicable.

Acknowledgments: We acknowledge financial support by the Deutscher Akademischer Austauschdienst—German Academic Exchange Service (DAAD) Research Grants—Doctoral Programs in Germany, 2019/20; grant number: 57440921.

Conflicts of Interest: The authors declare no conflict of interest.

Appendix A

Table A1. Poisson regression for modeled incidence rates displayed in Figure 1, non-young-onset CRC.

Variable	Saarland		Münster	
	Coefficient	*p*-Value	Coefficient	*p*-Value
Constant	−5.79	<0.001	−6.75	<0.001
Calendar year	0.02	<0.001	0.11	<0.001
Age group		<0.001		<0.001
55–59	−1.19		−1.44	
60–64	−0.73		−0.99	
65–69	−0.41		−0.62	
70–74	−0.16		−0.34	
75–79	0.03		−0.10	
80–84	0.18		0.00	
85+	Ref.		Ref.	
Colonoscopy		<0.001		<0.001
No (calendar year < 2002)	Ref.		Ref.	
Yes (calendar year ≥ 2002)	0.64		1.46	
Colonoscopy X calendar year		<0.001		<0.001
No	Ref.		Ref.	
Yes	−0.05		−0.13	

Table A2. Poisson regression for modeled incidence rates displayed in Figure 1, young-onset CRC.

Variable	Saarland		Münster	
	Coefficient	*p*-Value	Coefficient	*p*-Value
Constant	−27.42	0.976	−28.19	0.972
Calendar year	−0.01	0.028	0.02	<0.001
Age group		<0.001		<0.001
0–4	−20.09		−20.15	
5–9	−20.18		−6.42	
10–14	−20.24		−4.99	
15–19	−4.71		−4.21	
20–24	−4.13		−4.13	
25–29	−3.87		−3.58	
30–34	−3.14		−2.77	
35–39	−2.11		−2.10	
40–44	−1.45		−1.45	
45–49	−0.74		−0.70	
50–54	Ref.		Ref.	

Table A3. Poisson regression for standardized incidence ratios displayed in Figure 2.

Variable	Young-Onset CRC		Non-Young-Onset CRC	
	Coefficient	*p*-Value	Coefficient	*p*-Value
Constant	−0.22	0.593	−0.37	0.414
Calendar year	0.02	0.429	−0.05	0.341
Sex		0.455		0.024
Male	Ref.		Ref.	
Female	−0.19		0.31	
Colonoscopy				0.113
No (calendar year < 2002)			Ref.	
Yes (calendar year ≥ 2002)			−0.97	
Colonoscopy X calendar year				0.072
No			Ref.	
Yes			0.11	

References

1. Koch-Institut, R. *Krebs in Deutschland Für 2015/2016*; Robert Koch-Institut: Berlin, Germany, 2019.
2. Brenner, H.; Schrotz-King, P.; Holleczek, B.; Katalinic, A.; Hoffmeister, M. Declining bowel cancer incidence and mortality in Germany: An analysis of time trends in the first ten years after the introduction of screening colonoscopy. *Dtsch. Ärzteblatt Int.* **2016**, *113*, 101. [CrossRef]
3. Siegel, R.L.; Miller, K.D.; Fedewa, S.A.; Ahnen, D.J.; Meester, R.G.S.; Barzi, A.; Jemal, A. Colorectal cancer statistics, 2017. *CA Cancer J. Clin.* **2017**, *67*, 177–193. [CrossRef]
4. Ahnen, D.J.; Wade, S.W.; Jones, W.F.; Sifri, R.; Mendoza Silveiras, J.; Greenamyer, J.; Guiffre, S.; Axilbund, J.; Spiegel, A.; You, Y.N. The Increasing Incidence of Young-Onset Colorectal Cancer: A Call to Action. *Mayo Clin. Proc.* **2014**, *89*, 216–224. [CrossRef]
5. Ballester, V.; Rashtak, S.; Boardman, L. Clinical and molecular features of young-onset colorectal cancer. *World J. Gastroenterol.* **2016**, *22*, 1736–1744. [CrossRef] [PubMed]
6. Patel, S.G.; Ahnen, D.J. Colorectal Cancer in the Young. *Curr. Gastroenterol. Rep.* **2018**, *20*, 15. [CrossRef] [PubMed]
7. Fardet, A.; Druesne-Pecollo, N.; Touvier, M.; Latino-Martel, P. Do alcoholic beverages, obesity and other nutritional factors modify the risk of familial colorectal cancer? A systematic review. *Crit. Rev. Oncol. Hematol.* **2017**, *119*, 94–112. [CrossRef]
8. Johnson, C.M.; Wei, C.; Ensor, J.E.; Smolenski, D.J.; Amos, C.I.; Levin, B.; Berry, D.A. Meta-analyses of colorectal cancer risk factors. *Cancer Causes Control* **2013**, *24*, 1207–1222. [CrossRef] [PubMed]
9. Pan, S.Y.; DesMeules, M. Energy intake, physical activity, energy balance, and cancer: Epidemiologic evidence. *Methods Mol. Biol.* **2009**, *472*, 191–215. [CrossRef]
10. Edwards, B.K.; Ward, E.; Kohler, B.A.; Eheman, C.; Zauber, A.G.; Anderson, R.N.; Jemal, A.; Schymura, M.J.; Lansdorp-Vogelaar, I.; Seeff, L.C. Annual report to the nation on the status of cancer, 1975–2006, featuring colorectal cancer trends and impact of interventions (risk factors, screening, and treatment) to reduce future rates. *Cancer* **2010**, *116*, 544–573. [CrossRef]
11. Yuhara, H.; Steinmaus, C.; Cohen, S.E.; Corley, D.A.; Tei, Y.; Buffler, P.A. Is diabetes mellitus an independent risk factor for colon cancer and rectal cancer? *Am. J. Gastroenterol.* **2011**, *106*, 1911–1921, quiz 1922. [CrossRef] [PubMed]

12. Torre, L.A.; Siegel, R.L.; Ward, E.M.; Jemal, A. Global cancer incidence and mortality rates and trends—An update. *Cancer Epidemiol. Prev. Biomark.* **2015**, *25*, 16–27. [CrossRef] [PubMed]
13. Jung, Y.S. Is colorectal cancer screening necessary before 50 years of age? *Intest. Res.* **2017**, *15*, 550–551. [CrossRef]
14. Bibbins-Domingo, K.; Grossman, D.C.; Curry, S.J.; Davidson, K.W.; Epling, J.W.; García, F.A.; Gillman, M.W.; Harper, D.M.; Kemper, A.R.; Krist, A.H. Screening for colorectal cancer: US Preventive Services Task Force recommendation statement. *JAMA* **2016**, *315*, 2564–2575. [PubMed]
15. Austin, H.; Henley, S.J.; King, J.; Richardson, L.C.; Eheman, C. Changes in colorectal cancer incidence rates in young and older adults in the United States: What does it tell us about screening. *Cancer Causes Control* **2014**, *25*, 191–201. [CrossRef]
16. Brenner, H.; Altenhofen, L.; Stock, C.; Hoffmeister, M. Prevention, early detection, and overdiagnosis of colorectal cancer within 10 years of screening colonoscopy in Germany. *Clin. Gastroenterol. Hepatol.* **2015**, *13*, 717–723. [CrossRef]
17. Bundesausschuss, G. *Beschluss des Gemeinsamen Bundesausschusses über eine Richtlinie für Organisierte Krebsfrüherkennungsprogramme und eine Änderung der Krebsfrüherkennungs-Richtlinie. Decision of the Federal Joint Committee on a Directive of Organized Cancer Screening Programs and an Amendment to the Cancer Screening Directive*; Federal Joint Committee: Berlin, Germany, 2018.
18. Bundesverwaltungsamt. *(Spät-)Aussiedler und Ihre Angehörigen Zeitreihe 1950–2017*; Herkunftsstaaten: Cologne, Germany, 2017.
19. Bundesamt, S. *Bevölkerung und Erwerbstätigkeit*; Statistisches Bundesamt: Wiesbaden, Germany, 2018.
20. Kaucher, S.; Deckert, A.; Becher, H.; Winkler, V. Migration pattern and mortality of ethnic German migrants from the former Soviet Union: A cohort study in Germany. *J. BMJ Open* **2017**, *7*, e019213. [CrossRef] [PubMed]
21. Kaucher, S.; Kajüter, H.; Becher, H.; Winkler, V. Cancer Incidence and Mortality among Ethnic German Migrants From the Former Soviet Union. *Front. Oncol.* **2018**, *8*, 378. [CrossRef] [PubMed]
22. Winkler, V.; Kaucher, S.; Deckert, A.; Leier, V.; Holleczek, B.; Meisinger, C.; Razum, O.; Becher, H. Aussiedler Mortality (AMOR): Cohort studies on ethnic German migrants from the Former Soviet Union. *BMJ Open* **2019**, *9*, e024865. [CrossRef]
23. Berrino, F.; Brown, C.; Moller, T.; Sobin, L.; Faivre, J. ENCR Recommendations—Condensed TNM for Coding the Extent of Disease. *Lyon Eur. Netw. Cancer Regist.* **2002**, 2–5.
24. Pace, M.; Lanzieri, G.; Glickman, M.; Zupanič, T. *Revision of the European Standard Population: Report of Eurostat's Task Force*; Publications Office of the European Union: Luxembourg, 2013.
25. Bernal, J.L.; Cummins, S.; Gasparrini, A. Interrupted time series regression for the evaluation of public health inter-ventions: A tutorial. *Int. J. Epidemiol.* **2017**, *46*, 348–355, Corrigendum to **2020**, *49*, 1414.
26. McCullagh, P.; Nelder, J. *Generalized Linear Models*, 2nd ed.; Chapman and Hall: New York, NY, USA, 1989.
27. Liang, J.; Kalady, M.F.; Church, J. Young age of onset colorectal cancers. *Int. J. Colorectal Dis.* **2015**, *30*, 1653–1657. [CrossRef] [PubMed]
28. Young, J.P.; Win, A.K.; Rosty, C.; Flight, I.; Roder, D.; Young, G.P.; Frank, O.; Suthers, G.K.; Hewett, P.J.; Ruszkiewicz, A. Rising incidence of early-onset colorectal cancer in A ustralia over two decades: Report and review. *J. Gastroenterol. Hepatol.* **2015**, *30*, 6–13. [CrossRef]
29. Siegel, R.L.; Torre, L.A.; Soerjomataram, I.; Hayes, R.B.; Bray, F.; Weber, T.K.; Jemal, A. Global patterns and trends in colorectal cancer incidence in young adults. *Gut* **2019**, *68*, 2179–2185. [CrossRef]
30. Russo, A.; Franceschi, S.; La Vecchia, C.; Dal Maso, L.; Montella, M.; Conti, E.; Giacosa, A.; Falcini, F.; Negri, E. Body size and colorectal-cancer risk. *Int. J. Cancer* **1998**, *78*, 161–165. [CrossRef]
31. Aleksandrova, K.; Schlesinger, S.; Fedirko, V.; Jenab, M.; Bueno-de-Mesquita, B.; Freisling, H.; Romieu, I.; Pischon, T.; Kaaks, R.; Gunter, M.J. Metabolic mediators of the association between adult weight gain and colorectal cancer: Data from the European Prospective Investigation into Cancer and Nutrition (EPIC) cohort. *Am. J. Epidemiol.* **2017**, *185*, 751–764. [CrossRef]
32. O'Connell, J.B.; Maggard, M.A.; Livingston, E.H.; Cifford, K.Y. Colorectal cancer in the young. *Am. J. Surg.* **2004**, *187*, 343–348. [CrossRef]
33. Richman, S.; Adlard, J. Left and right sided large bowel cancer. *BMJ* **2002**, *324*, 931–932. [CrossRef]
34. Kang, H.; O'Connell, J.B.; Maggard, M.A.; Sack, J.; Ko, C.Y. A 10-year outcomes evaluation of mucinous and signet-ring cell carcinoma of the colon and rectum. *Dis. Colon Rectum* **2005**, *48*, 1161–1168. [CrossRef] [PubMed]
35. Arnold, M.; Razum, O.; Coebergh, J.-W. Cancer risk diversity in non-western migrants to Europe: An overview of the literature. *Eur. J. Cancer* **2010**, *46*, 2647–2659. [CrossRef] [PubMed]
36. Becher, H.; Winkler, V.J. Estimating the standardized incidence ratio (SIR) with incomplete follow-up data. *BMC Med. Res. Methodol.* **2017**, *17*, 55. [CrossRef]

International Journal of
*Environmental Research
and Public Health*

Article

The Incidence of Intestinal Gastric Cancer among Resettlers in Germany—Do Resettlers Remain at an Elevated Risk in Comparison to the General Population?

Anna Lindblad [1], Simone Kaucher [1], Philipp Jaehn [2], Hiltraud Kajüter [3], Bernd Holleczek [4], Lauren Lissner [5], Heiko Becher [6] and Volker Winkler [1,*]

[1] Heidelberg Institute of Global Health, University Hospital Heidelberg, 69120 Heidelberg, Germany; anna2.lindblad@gmail.com (A.L.); simone.kaucher@uni-heidelberg.de (S.K.)

[2] Institute of Social Medicine and Epidemiology, Brandenburg Medical School Theodor Fontane, 14770 Brandenburg an der Havel, Germany; Philipp.Jaehn@mhb-fontane.de

[3] Cancer Registry, North Rhine-Westphalia, 44801 Bochum, Germany; Hiltraud.Kajueter@krebsregister.nrw.de

[4] Saarland Cancer Registry, 66119 Saarbrücken, Germany; b.holleczek@krebsregister.saarland.de

[5] School of Public Health and Community Medicine, Institute of Medicine, Sahlgrenska Academy at University of Gothenburg, 41346 Gothenburg, Sweden; lauren.lissner@gu.se

[6] Institute for Medical Biometry and Epidemiology, University Medical Center Hamburg-Eppendorf, 20246 Hamburg, Germany; h.becher@uke.de

* Correspondence: volker.winkler@uni-heidelberg.de

Received: 4 November 2020; Accepted: 7 December 2020; Published: 9 December 2020

Abstract: Objective: Previous studies have shown that the incidence of gastric cancer (GC), and particularly intestinal GC, is higher among resettlers from the former Soviet Union (FSU) than in the general German population. Our aim was to investigate if the higher risk remains over time. Methods: GC cases between 1994 and 2013, in a cohort of 32,972 resettlers, were identified by the respective federal cancer registry. Age-standardized rates (ASRs) and standardized incidence ratios (SIRs) were analyzed in comparison to the general population for GC subtypes according to the Laurén classification. Additionally, the cohort was pooled with data from a second resettler cohort from Saarland to investigate time trends using negative binomial regression. Results: The incidence of intestinal GC was elevated among resettlers in comparison to the general population (SIR (men) 1.64, 95% CI: 1.09–2.37; SIR (women) 1.91, 95% CI: 1.15–2.98). The analysis with the pooled data confirmed an elevated SIR, which was stable over time. Conclusion: Resettlers' higher risk of developing intestinal GC does not attenuate towards the incidence in the general German population. Dietary and lifestyle patterns might amplify the risk of GC, and we believe that further investigation of risk behaviors is needed to better understand the development of disease pattern among migrants.

Keywords: incidence; stomach cancer; Laurén classification; migrants; former Soviet Union; cohort; Germany

---

1. Introduction

Migration is a growing phenomenon in the world. In 2019, 272 million people were international migrants according to the International Organization for Migration [1]. Given diverse push and pull factors behind migration, migrants are a heterogenous group of people [2]. Traditionally, the health of migrants is investigated by comparing it to the health of the host country's non-migrant population. Upon their arrival, migrants may suffer from less morbidity and mortality as a result of the "healthy migrant effect". In general, differences in health and disease are expected to attenuate over time as migrants adapt to exposure of risk factors and changing environments [3,4]. As exposures change over time, the importance of a life course approach has been emphasized for further understanding migrant health and its disparities [5].

In 2018, one out of four people living in Germany had a migration background [6]. To date, the second largest migrant group consists of ethnic German resettlers whose ancestors migrated to Russia in 18th and 19th centuries on invitation by the Russian empress to farm unsettled land. After the Second World War, ethnic Germans and their families were invited by the government of West Germany to return to Germany. They obtained German citizenship upon arrival and were quasi-randomly allocated to the federal states based on population density and economic conditions. Due to strict emigration regulations, it was not until after the collapse of the Soviet Union that a significant migration flow reached Germany peaking in 1994 with more than 200,000 people per year. Until 2019, about 2.4 million resettlers have migrated from the former Soviet Union to Germany [7–9].

Gastric cancer (GC) is the fifth most common cancer type worldwide, and it occurs twice as frequent among men than women [10]. The prognosis is poor: in Germany the relative five-year survival rate is 30–35% [11]. GC is a heterogeneous disorder, associated with both genetic and environmental factors. The most important risk factor is infection with *Helicobacter pylori* (*H. pylori*). Lifestyle and dietary risk factors associated to GC are alcohol, smoking, red and processed meat, salty foods, obesity, and low physical activity. Age, male gender, low socioeconomic status, and gastroesophageal reflux are other factors predisposing GC [12–16]. The incidence of GC has decreased significantly over the last decades, probably explained by higher standards of hygiene, better nutrition, and eradication of *H. pylori* infections [17,18]. According to the widely used Laurén classification, GC can be distinguished into two main histologic types: diffuse and intestinal. Briefly, the intestinal type is composed of well-differentiated polarized cells forming glandular structures, whereas the diffuse type consists of less differentiated and unpolarized cells in randomly ordered cell clusters [19]. Regarding epidemiology, intestinal GC occur predominantly in men and in older age groups, whereas diffuse GC is equally frequent in both sexes and is more common at younger ages [20]. Evidence suggests that the two subtypes are associated to risk factors such as lifestyle and dietary factors to varying degrees [21–23].

Previous studies have found an elevated incidence and mortality of GC among resettlers, and particularly a higher risk of developing intestinal GC, in comparison to the general population [24–27]. The aim of this study was to further investigate differences in intestinal GC among resettlers by (i) replicating previous findings in another cohort of resettlers and (ii) pooling all available data on resettlers' GC incidence for a joint analysis to provide insight into whether subtype specific risk differences attenuate or remain over time.

2. Methods

2.1. Münster Cohort

The study is mainly based on a registry-based cohort established in the administrative district of Münster (the AMIN cohort; Aussiedler in Münster—Incidence cohort study). Details about the Münster cohort and the follow up procedures can be found elsewhere [25,28]. In brief, the cohort comprised 32,962 resettlers who were assigned to the district between 1990 and 2001 and is a quasi-random sample of 53% of all resettlers in that district which is a part of the federal state North Rhine-Westphalia (NRW), Germany. Person-years (PY) of time under risk for developing a cancer condition were estimated from the beginning of 1994 or immigration (if later than 1994) to the diagnosis of GC or the end of follow-up (31st December 2013) taking into account deaths and out-migration of the study area [25,29]. The accumulated person-time of the general population of the Münster region was derived from the midyear population (from 1994 to 2013), provided by the federal cancer registry of NRW. The study was conducted in accordance with the Declaration of Helsinki, and the protocol was approved by the Ethics Committee of the Medical Faculty, University Hospital Heidelberg (S-319/2013).

Records of patients with incident GC cases diagnosed between 1994 and 2013 were provided by the NRW cancer registry (three-digit ICD-10 code: C16, ICD-9 code: 151). Cases among resettlers were identified by a pseudonymized two-step record linkage procedure [7,30]. GC cases were divided into three subgroups based on the tumor morphology according to ICD-O-3 (International Classification System of Diseases for Oncology, 3rd edition) and the Laurén classification: intestinal GC, diffuse GC, and other/missing GC. The latter group comprises carcinomas with other specified histology patterns or unspecified or missing tumor morphology. The classification and numbers of cases are summarized in Table 1 (morphology codes according to the second edition of the ICD-O used for cases diagnosed between 1992 and 2002 could be unambiguously translated into ICD-O-3 values) [31].

Table 1. Numbers of gastric cancer (GC) cases in the Münster population with corresponding ICD-O-3 [a,b] codes and classification by Laurén.

Laurén Classification	ICD-O-3 Codes	N [c]
Intestinal GC (44%)	8140/3 Adenocarcinoma, not otherwise specified	2486
	8144/3 Adenocarcinoma, intestinal type	1774
	8211/3 Tubular adenocarcinoma	283
	8260/3 Papillary adenocarcinoma, not otherwise specified	66
	8480/3 Mucinous adenocarcinoma	169
Diffuse GC (26%)	8490/3 Signet ring cell carcinoma	1540
	8142/3 Linitis plastica	21
	8145/3 Carcinoma, diffuse type	1321
Other/Missing GC (30%)	Sections 802–857: Other carcinoma	697
	Sections 804 and 824: Endocrine carcinoma	182
	Sections 880–914: Sarcoma	159
	Sections 917–971: Lymphoma	1
	Section 800: Neoplasm	498
	Section 801: Carcinoma, not otherwise specified	1676
Total		10,873

[a] International Classification System of Diseases for Oncology, 3rd edition; [b] Codes and terms can be unambiguously translated from ICD-O 2nd edition used 1992–2002; [c] N includes all gastric cancer cases among resettlers and the general population in Münster.

Statistical Analyses

Age-standardized incidence rates (ASR) for the Münster population were calculated by direct standardization using the old European standard population [32]. Calculations were performed by sex and two-year calendar periods (1: 1994–1995, 2: 1996–1997, ... , 10: 2012–2013) for the total group and for each histologic GC subgroup separately. Additionally, negative binomial regression was used to model ASR time trends by estimating age-specific rates, which were again standardized directly. The model used the number of cases as the dependent variable and sex (categorical), age group (categorical), and calendar year (continuous, calendar year—1990) as independent variables and log(PY) as the offset. Standardized incidence ratios (SIR) of resettlers in comparison to the general Münster population were calculated using indirect standardization. SIRs were calculated separated by sex and by the three subgroups of GC.

2.2. Pooled Data from Münster and Saarland Cohorts

To further investigate SIR and the secular trends in GC incidence among resettlers in comparison to the general population, data from the Münster cohort and the AMOR cohort, a second resettler cohort from the federal state of Saarland, were combined [24]. The characteristics of the pooled data are represented in Table 2, further description of the Saarland cohort can be found elsewhere [26,28]. In brief, the AMOR cohort comprises 18,619 resettlers who were assigned to the federal state of the Saarland, Germany, between 1990 and 2005. Incidence data from 1990 to 2009 were provided by the Saarland Cancer Registry, and vital status follow-up was performed by local registry offices. Data from the mid-year Saarland population were provided by the federal statistics offices. The combined observation time for the Münster cohort and the Saarland cohort was between 1990 and 2013.

Int. J. Environ. Res. Public Health **2020**, 17, 9215

Table 2. Characteristics of resettlers and the general population in Münster, and of resettlers and the general population in Münster and Saarland combined (pooled data).

	Münster (1994–2013)				Pooled Data (1990–2013)			
	Male [N (%)]		Female [N (%)]		Male [N (%)]		Female [N (%)]	
	Resettlers	Münster	Resettlers	Münster	Resettlers	Population	Resettlers	Population
Incident gastric cancer cases	51	6161	36	4712	87	8814	65	6914
Median age at diagnosis	66	70	72.5	76	68	70	71	76
(Interquartile range)	(52–75)	(61–78)	(57–78)	(65–83)	(58–75)	(61–77)	(51–77)	(66–83)
Laurén classification								
Intestinal GC	28 (55)	3033 (49)	19 (53)	1745 (37)	53 (61)	4400 (50)	32 (49)	2634 (38)
Diffuse GC	14 (27)	1442 (23)	6 (17)	1440 (31)	17 (19)	2204 (25)	16 (25)	2198 (32)
Other/missing GC	9 (18)	1686 (28)	11 (30)	1527 (32)	17 (20)	2210 (25)	17 (26)	2082 (30)

GC, Gastric cancer; N, Number of cases.

Statistical Analyses

The expected cases for the pooled SIR analysis were calculated for each cohort separately using incidence and population data of the respective host populations. SIRs were calculated for the whole study time and for the calendar periods 1990–2001 and 2002–2013. Additionally, time trends of SIRs were modeled using negative binomial regression. The model used the number of observed events among resettlers as the dependent variable and sex (categorical), subtype (categorical), calendar year (continuous, calendar year—1990), and cohort (categorical) as independent variables and the number of expected cases as the offset. ASRs and SIRs were calculated with exact 95% confidence intervals (95% CIs) and the significance level was set to 0.05. Statistical analyses were performed using Stata IC (version 14) (StataCorp LLC, Texas, USA).

3. Results

3.1. Descriptive Results

Table 2 summarizes characteristics of the study population in Münster and the pooled data. The Münster cohort comprised 16,033 men and 16,939 women, and a total of 462,823 PY with a mean follow-up time for 13.4 years. The mid-year population in Münster was 2,558,285 in 1994 and 2,574,148 in 2013. During the study period, 10,873 gastric cancer cases were registered in the Münster region. Intestinal GC represented the most common subtype. For both sexes combined, the median age at diagnosis was 73 years for patients with intestinal GC and 69 years for patients with diffuse GC. The median age at diagnosis was 75 years for patients with other/missing GC. For resettlers, the median age at diagnosis was significantly lower than their counterparts in Münster for diffuse GC ($p = 0.015$) and other/missing GC ($p = 0.021$), but not for intestinal GC ($p = 0.222$). No associations were seen between resettler status and histologic type. The tumor was histologically confirmed in 79% of all cancer cases, and 10% of the cancer cases were notified by death certificate only (DCO).

The pooled cohort data accumulated 667,190 person-years (52% female years). The midyear population was on average 3,055,145. Of 152 reported GC diagnoses among resettlers, 52% were of intestinal type. Median age at diagnosis for both male and female resettlers was also significantly lower than their counterparts in the general population (intestinal GC $p = 0.014$, diffuse GC $p < 0.001$, other/missing GC $p = 0.003$), and again there was no association between resettler status and histologic type ($p = 0.124$ for men and $p = 0.176$ for women).

3.2. Münster Population

Figure 1 illustrates ASRs for GC subtypes in the general population of Münster. The ASR for total GC decreased from 24.8 (95% CI: 22.7–26.8) to 16.4 (95% CI: 15.1–17.8) for men and from 13.1 (95% CI: 12.0–14.3) to 8.9 (95% CI: 7.9–9.8) for women from 1994–1995 to 2012–2013. Mean ASR for intestinal GC was more than twice as high among men than women (9.9 (95% CI: 9.5–10.2) and 3.6 (95% CI: 3.4–3.8), respectively) and did not change over time. There was strong evidence for a declining trend for other/missing GC for both sexes ($p < 0.001$). The group of missing/other GC consisted of 73% missing GC cases in 1994 and 54% in 2013.

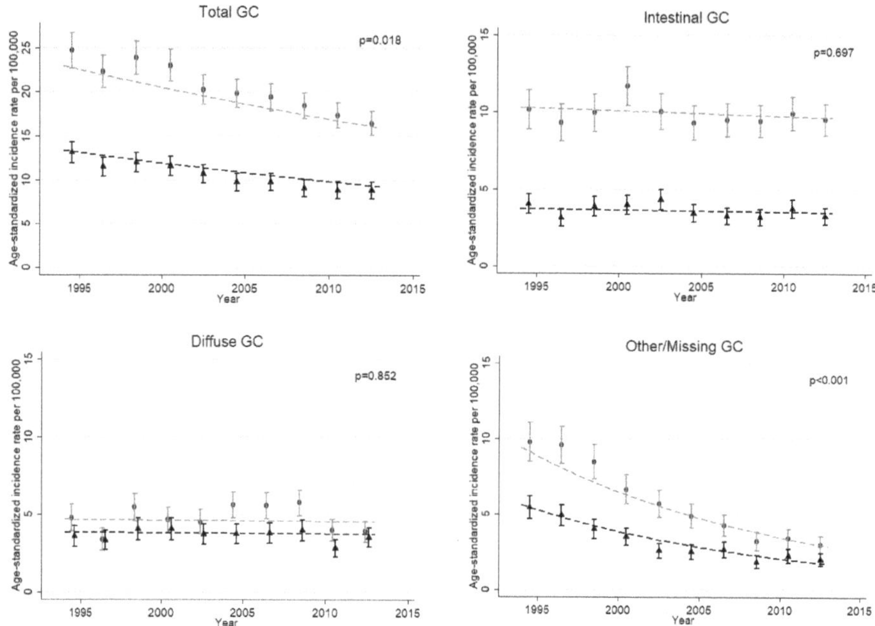

Figure 1. Age-standardized incidence rates (ASR) for each subtype of gastric cancer (GC) according to Laurén classification in the general Münster population. Plotted for two-year periods, from 1994–1995 to 2012–2013, with 95% confidence intervals. Males marked in gray, females in black. Dashed lines represent time trends modeled using negative binominal regression, with corresponding p-values of linear calendar year effect (rates are standardized with respect to the old European standard population).

3.3. Resettlers in Münster

Compared to the general Münster population, the SIR of total GC was elevated for male but not for female resettlers (Table 3). Intestinal GC was elevated among both sexes (men: 1.64 (95% CI: 1.09–2.37); women: 1.91 (95% CI: 1.15–2.98)). SIR of diffuse GC varied and was 1.61 (95% CI: 0.88–2.70) for men and 0.64 (95% CI: 0.23–1.39) for women.

Table 3. Standardized incidence ratios (SIR) for subtypes of gastric cancer of resettlers compared to the general Münster population.

	Male		Females		Both Sexes	
	Observed	SIR (95% CI)	Observed	SIR (95% CI)	Observed	SIR (95% CI)
Total GC	51	1.50 (1.12–1.98)	36	1.32 (0.93–1.83)	87	1.42 (1.14–1.76)
Intestinal GC	28	1.64 (1.09–2.37)	19	1.91 (1.15–2.98)	47	1.73 (1.28–2.31)
Diffuse GC	14	1.61 (0.88–2.70)	6	0.64 (0.23–1.39)	20	1.11 (0.68–1.71)
Other/Missing GC	9	1.10 (0.50–2.09)	11	1.40 (0.70–2.50)	20	1.25 (0.76–1.92)

GC, gastric cancer; SIR, standardized incidence ratio; CI, confidence interval.

3.4. Pooled Cohorts Analyses

Between 1990 and 2013, the SIRs of total GC and intestinal GC were elevated for both women and men in the Saarland cohort and Münster cohort combined (Table 4). When dividing the observation time into two periods (1990–2001 and 2002–2013), the incidence was elevated among resettlers in both periods. No difference in incidence was seen for diffuse GC. Regarding other/missing GC, SIR was elevated for female resettlers in the first period, with a major decrease in the second period. In Appendix A, SIRs are presented for each period and cohort separately. The incidence of intestinal GC was elevated in the second period for both sexes in both cohorts. Regarding diffuse GC, the incidence was only elevated among women in Saarland cohort during the second period.

Table 4. Standardized incidence ratios for subtypes of gastric cancer among resettlers in comparison to the general population in Münster and Saarland.

	1990–2001 [a]		2002–2013 [b]		Total	
	Observed	SIR (95% CI)	Observed	SIR (95% CI)	Observed	SIR (95% CI)
Total GC						
Male	31	1.87 (1.27–2.65)	56	1.72 (1.30–2.23)	87	1.77 (1.42–2.18)
Female	28	2.09 (1.39–3.03)	37	1.46 (1.03–2.01)	65	1.68 (1.29–2.14)
Total	59	1.97 (1.50–2.54)	93	1.60 (1.29–1.96)	152	1.73 (1.46–2.03)
Intestinal GC						
Male	19	2.45 (1.47–3.82)	34	1.98 (1.37–2.76)	53	2.12 (1.59–2.77)
Female	10	2.18 (1.05–4.01)	22	2.25 (1.41–3.41)	32	2.23 (1.53–3.15)
Total	29	2.35 (1.57–3.37)	56	2.08 (1.57–2.70)	85	2.16 (1.73–2.67)
Diffuse GC						
Male	5	1.15 (0.37–2.69)	12	1.31 (0.68–2.28)	17	1.26 (0.73–2.01)
Female	5	1.07 (0.35–2.49)	11	1.13 (0.57–2.03)	16	1.11 (0.64–1.81)
Total	10	1.11 (0.53–2.04)	23	1.22 (0.77–1.83)	33	1.18 (0.81–1.66)
Other/Missing GC						
Male	7	1.55 (0.62–3.20)	10	1.61 (0.77–2.96)	17	1.59 (0.92–2.54)
Female	13	3.17 (1.69–5.41)	4	0.67 (0.18–1.73)	17	1.69 (0.99–2.71)
Total	20	2.32 (1.42–3.59)	14	1.15 (0.63–1.93)	34	1.64 (1.13–2.29)

GC, gastric cancer; SIR, standardized incidence ratio; CI, confidence interval. [a] 256,089 person-years. [b] 411,101 person-years.

No time trend could be observed (calendar year coefficient $= 0.002$, $p = 0.889$) when modeling SIR (Table 5). Likewise, no difference in SIR for sex could be seen (female coefficient $= 0.025$, $p = 0.886$). The SIR for diffuse GC was lower than for intestinal GC (diffuse type coefficient $= -0.639$, $p = 0.013$), and one could see an effect for cohort (Saarland coefficient 0.558, $p < 0.001$). The estimated SIR for a covariable combination can be obtained; for example, for the year 2000, intestinal GC, female, Münster cohort, the estimate is $\exp(0.535 + 0.002 \times (2000 - 1990) + 0.025) = 1.79$.

Table 5. Negative binomial regression modeling standardized incidence rate ratios for gastric cancer among resettlers in comparison to the direct host populations in the Münster and Saarland cohorts.

Variable	Coefficient [a]	95% CI	*p*-Value
Year (calendar year–1990)	0.002	−0.030, 0.035	0.889
Subtype			0.013
Intestinal GC	Ref.		
Diffuse GC	−0.639	−1.062, −0.215	
Other/Missing GC	−0.224	−0.648, 0.200	
Sex			0.886
Male	Ref.		
Female	0.025	−0.318, 0.368	
Cohort			0.003
Münster	Ref.		
Saarland	0.558	0.195, 0.921	
Constant	0.535	−0.053, 1.123	0.075

GC, gastric cancer; Ref, reference; CI, confidence interval. [a] The coefficients represent the change in the log of the SIR by one unit change of the calendar year—1990 or relative to the reference group.

4. Discussion and Conclusions

4.1. Key Findings

Our results are largely consistent with previous findings in the AMOR cohort from the federal state of Saarland: we found that the incidence of intestinal GC was higher among resettlers than in the general population in Münster [24]. In the analysis of the Saarland and Münster cohort combined as well as separate, the incidence of intestinal GC among resettlers remained elevated in the second period of the observation time and no secular trend could be observed. Regarding the incidence in the general population in Münster, ASRs illustrated a decreasing trend of total GC and other/missing type of GC, whereas no significant trend for intestinal or diffuse GC could be seen.

Over time, risk behaviors of migrants are expected to adapt to the ones in the host population, resulting in converging rates of morbidity and mortality [5]. This was the case for several cancer types among resettlers in Germany according to Kaucher et al. [25]. Initially, the incidence of all malignant cancer types was lower among resettlers than the general German population. Over time, the risk for cancer types such as colorectal, breast, and prostate cancer increased over time, towards the risk in the general population. However, for the elevated incidence of gastric cancer, no converging trend could be seen, possibly due to the small sample size and low number of cases. These findings were consistent with previous research, where migrants from non-Western countries were less prone to develop colorectal, prostate, and breast cancer but more likely to develop infectious related cancer types such as gastric cancer [4,26]. The converging trend for the former cancer types among resettlers may be explained by the adaptation to Western lifestyle (poor diet, sedentary lifestyle, etc.) and/or improved screening/diagnostic possibilities [4,25,33].

Like other infectious-related cancer types, gastric cancer was found to be more prevalent in non-Western countries [4]. In 2018, the age-adjusted incidence was 29.4 for men and 8.8 for women in Russia, in contrast to 9.4 and 4.5 in Germany [34]. As discussed in the article of Jaehn et al. in 2016 [24], the strongest risk factor for GC, *H. pylori* infection, cannot alone explain the higher risk among resettlers for only intestinal GC given its equal association to both subtypes. The development of GC is described as a complex multifactorial process, and several studies claim that *H. pylori* plays a role in early transformation steps, causing chronic inflammation, but the transitions that follow are determined by environmental, bacterial, and host factors [20,35–38]. While host and bacterial factors are less likely to change due to migration, in this case from FSU to Germany, it is more likely to expect environmental factors to be influenced by lifestyle and dietary patterns in the new country of residence.

The persistent elevated risk for intestinal GC among resettlers indicates a remaining gap between the migrant group and the host population. The differential risk could be explained by dietary patterns and lifestyle factors, which overall seem to be more strongly correlated to intestinal than diffuse GC. Studies show that heavy alcohol consumption is a risk factor, while intake of a diet rich with fruit and vegetables is a protective factor more strongly correlated to intestinal than diffuse type [21,22]. According to comparative studies [39,40], resettlers were more likely to be obese and less likely to take part in cancer screening. Smoking habits differed by sex, with female resettlers smoking less than the German population whereas male smoking more. Furthermore, resettlers consumed smaller amounts of alcohol but stronger alcohol. Resettlers ate more meat and potatoes, but no difference in vegetable intake was seen. Based on these findings, a higher consumption of stronger alcohol could increase the risk for developing intestinal GC, whereas obesity could increase the risk of GC in general. Moreover, resettlers have been shown to have lower socioeconomic status (SES) in comparison to the native German population, but with a decreasing gap over time [41]. Lower SES is known have a negative impact on health status through pathways and mechanisms understood to a certain extent [42]. A less beneficial lifestyle and dietary patterns could have an adverse effect on the risk of developing intestinal GC. However, generalization can be misleading as resettlers are a relatively heterogenous group, originating from different states and regions and possibly bringing divergent lifestyle and dietary habits with them.

4.2. Shortcomings and Limitations

The study is based on data that do not provide information on dietary pattern, lifestyle factors, etc. Therefore, it does not allow analyzing the association between different risk behavior and cancer diagnosis among resettlers. As cancer registration in the Münster region was not sufficiently complete before 1994, we decided to limit the analysis to the calendar period 1994–2013. Regarding the classification of GC cases, there was no validation study available on how to apply the Laurén scheme on the ICD-O-3 morphology classification. We used the same criteria as the Saarland study which was based on previous literature [24,43,44] and believe that a possible misclassification of histologic types of GC among resettlers and the general population would be non-differential, leading to a bias towards the null of SIRs.

The decreasing trend of other/missing GC in the general Münster population is most likely explained by both increased completeness of data and improved diagnostics techniques to classify histologic patterns of GC. The cases with missing histologic type are likely one of the two main subtypes and considering the median age at diagnosis for this group (77 years in general population, 70.5 for resettlers), these cases are more likely to be of intestinal type which is more common in older ages. The number of missing GC is larger in the beginning than the end of the observation time. If assuming that most of these cases would be of intestinal type, there could be a decreasing time trend hidden.

Selection bias is unlikely as all ethnic Germans were invited to Germany and a majority of them immigrated back and were allocated quasi-randomly into federal states [45]. This fact also makes bias caused by healthy migrant effect unlikely. The proportions of resettlers with ethnic German background decreased from 78% in 1993 to 19% in 2004 due to larger immigration of non-ethnic German family members and relatives [46]. The altered proportions may have increased the heterogeneity regarding characteristics and risk factor patterns in the migrant group. The reference population comprises both resettlers included and excluded in the cohort who are resident in the district of Münster. Due to the contamination of the group, real differences are less likely to be observed as ratios will move toward zero.

Due to data protection concerns, neither information on date of immigration nor mortality among individuals of the Münster cohort was available, which prevented us from analyzing the incidence among resettlers with respect to lengths of stay in Germany. However, as the majority of resettlers migrated to Germany in the first half of the 1990s, calendar time is highly correlated with length of stay. It should also be mentioned that person-time of the Münster cohort had to be estimated due to an incomplete follow-up in method thoroughly explained elsewhere [29].

4.3. Implications for Future Research

In conclusion, resettlers remain at higher risk of developing intestinal GC than the general population in Germany. Lifestyle and dietary patterns may be likely to explain the discrepancy as intestinal GC is more strongly associated with these factors. Given the poor prognosis for GC, prevention and early detection might ease the individual's health burden as well as the society's economic burden. We believe that further investigation of risk behaviors is needed to understand the development of disease pattern among migrants.

Author Contributions: Conceptualization, P.J., H.B., and V.W.; Data curation, S.K., H.K., and B.H.; Formal analysis, A.L., S.K., P.J., and L.L.; Funding acquisition, H.B. and V.W.; Investigation, A.L. and P.J.; Methodology, P.J., H.B., and V.W.; Project administration, V.W.; Resources, H.K., B.H., and V.W.; Software, H.B. and V.W.; Supervision, V.W. and L.L.; Validation, H.B. and V.W.; Visualization, A.L.; Writing—original draft, A.L.; Writing—review and editing, S.K., P.J., H.K., B.H., L.L., H.B., and V.W. All authors have read and agreed to the published version of the manuscript.

Funding: This research was funded by the German Cancer Aid (Grant Number 111232).

Acknowledgments: We acknowledge financial support by the Else Kröner-Fresenius-Stiftung within the Heidelberg Graduate School of Global Health.

Conflicts of Interest: The authors declare no conflict of interest.

Appendix A

Table A1. Standardized incidence ratios for subtypes of gastric cancer among resettlers compared to the general population in Münster and Saarland separately.

Münster	1994–2003 [a]		2004–2013 [b]	
	Observed	SIR (95% CI)	Observed	SIR (95% CI)
Total GC				
Male	17	1.30 (0.76–2.09)	34	1.63 (1.13–2.27)
Female	16	1.48 (0.85–2.40)	20	1.22 (0.75–1.88)
Intestinal GC				
Male	8	1.33 (0.58–2.62)	20	1.81 (1.10–2.79)
Female	4	1.10 (0.30–2.81)	15	2.37 (1.33–3.92)
Diffuse GC				
Male	4	1.31 (0.36–3.36)	10	1.77 (0.85–3.26)
Female	3	0.84 (0.17–2.44)	3	0.52 (0.11–1.51)
Other/Missing GC				
Male	5	1.25 (0.41–2.00)	4	0.95 (0.26–2.44)
Female	9	2.51 (1.15–4.77)	2	0.47 (0.06–1.68)
Saarland	**1990–1999 [c]**		**2000–2009 [d]**	
	Observed	SIR (95% CI)	Observed	SIR (95% CI)
Total GC				
Male	15	2.88 (1.61–4.75)	21	2.08 (1.29–3.18)
Female	11	2.62 (1.31–4.68)	18	2.45 (1.45–3.87)
Intestinal GC				
Male	10	3.90 (1.87–7.18)	15	2.80 (1.57–4.62)
Female	6	3.78 (1.39–8.22)	7	2.50 (1.01–5.16)
Diffuse GC				
Male	2	1.32 (0.16–4.77)	1	0.30 (0.01–1.68)
Female	1	0.65 (0.02–3.63)	9	2.59 (1.19–4.92)
Other/Missing GC				
Male	3	2.65 (0.55–7.73)	5	3.55 (1.15–8.28)
Female	4	3.70 (1.01–9.48)	2	1.85 (0.22–6.67)

GC, gastric cancer; SIR, standardized incidence ratio; CI, confidence interval. [a] 215,008 person-years in Münster. [b] 247,816 person-years in Münster. [c] 70,450 person-years in Saarland. [d] 133,917 person-years in Saarland.

References

1. International Organization for Migration. Migration and Migrants: A Global Overview. Available online: https://www.un-ilibrary.org/content/component/b8cf9ec2-en (accessed on 5 June 2020).
2. European Commission Eurostat. Push and Pull Factors of International Migration: A Comparative Report. Available online: https://op.europa.eu/en/publication-detail/-/publication/90913700-dbbb-40b8-8f85-bec0a6e29e83 (accessed on 5 May 2020).
3. Rechel, B.; Mladovsky, P.; Ingleby, D.; Mackenbach, J.P.; McKee, M. Migration and health in an increasingly diverse Europe. *Lancet* **2013**, *381*, 1235–1245. [CrossRef]
4. Arnold, M.; Razum, O.; Coebergh, J.-W. Cancer risk diversity in non-western migrants to Europe: An overview of the literature. *Eur. J. Cancer* **2010**, *46*, 2647–2659. [CrossRef] [PubMed]
5. Spallek, J.; Zeeb, H.; Razum, O. What do we have to know from migrants' past exposures to understand their health status? A life course approach. *Emerg. Themes Epidemiol.* **2011**, *8*, 6. [CrossRef] [PubMed]
6. Federal Office for Migration and Refugees. *Migration Report 2018: Key Results*; Bundesamt für Migration und Flüchtlinge: Nürnberg, Germany, 2019.
7. Kaucher, S.; Deckert, A.; Becher, H.; Winkler, V. Migration pattern and mortality of ethnic German migrants from the former Soviet Union: A cohort study in Germany. *BMJ Open* **2017**, *7*, e019213. [CrossRef] [PubMed]
8. Bundeszentrale für politische Bildung. Zuzug Von (Spät-)Aussiedlern und Ihren Familienangehörigen. Available online: http://www.bpb.de/nachschlagen/zahlen-und-fakten/soziale-situation-in-deutschland/61643/aussiedler (accessed on 5 June 2020).
9. Bundesverwaltungsamt. Spätaussiedler und ihre Angehörigen, Zeitreihe 1950–2019. Available online: https://www.bva.bund.de/SharedDocs/Downloads/DE/Buerger/Migration-Integration/Spaetaussiedler/Statistik/Zeitreihe_1950_2019.html (accessed on 8 December 2020).
10. Bray, F.; Ferlay, J.; Soerjomataram, I.; Siegel, R.L.; Torre, L.A.; Jemal, A. Global cancer statistics 2018: GLOBOCAN estimates of incidence and mortality worldwide for 36 cancers in 185 countries. *CA A Cancer J. Clin.* **2018**, *68*, 394–424. [CrossRef] [PubMed]
11. Robert-Koch-Institut-Zentrum für Krebsregisterdaten. Available online: www.krebsdaten.de/Krebs/DE/Content/Krebsarten/Magenkrebs/magenkrebs_node.html (accessed on 3 July 2020).
12. Sitarz, R.; Skierucha, M.; Mielko, J.; Offerhaus, G.J.A.; Maciejewski, R.; Polkowski, W.P. Gastric cancer: Epidemiology, prevention, classification, and treatment. *Cancer Manag. Res.* **2018**, *10*, 239–248. [CrossRef]
13. Karimi, P.; Islami, F.; Anandasabapathy, S.; Freedman, N.D.; Kamangar, F. Gastric cancer: Descriptive epidemiology, risk factors, screening, and prevention. *Cancer Epidemiol. Biomark. Prev.* **2014**, *23*, 700–713. [CrossRef]
14. Ladeiras-Lopes, R.; Pereira, A.K.; Nogueira, A.; Pinheiro-Torres, T.; Pinto, I.; Santos-Pereira, R.; Lunet, N. Smoking and gastric cancer: Systematic review and meta-analysis of cohort studies. *Cancer Causes Control* **2008**, *19*, 689–701. [CrossRef]
15. Gonzalez, C.A.; Jakszyn, P.; Pera, G.; Agudo, A.; Bingham, S.; Palli, D.; Ferrari, P.; Boeing, H.H.; del Giudice, G.; Plebani, M.; et al. Meat intake and risk of stomach and esophageal adenocarcinoma within the European Prospective Investigation into Cancer and Nutrition (EPIC). *J. Natl. Cancer Inst.* **2006**, *98*, 345–8874. [CrossRef]
16. Wang, X.-Q.; Terry, P.-D.; Yan, H. Review of salt consumption and stomach cancer risk: Epidemiological and biological evidence. *World J. Gastroenterol.* **2009**, *15*, 2204–2213. [CrossRef] [PubMed]
17. Muñoz, N.; Franceschi, S. Epidemiology of gastric cancer and perspectives for prevention. *Salud Publica de Mex.* **1997**, *39*, 318. [CrossRef]
18. Jemal, A.; Bray, F.; Center, M.M.; Ferlay, J.; Ward, E.; Forman, D. Global cancer statistics. *CA A Cancer J. Clin.* **2011**, *61*, 69–90. [CrossRef] [PubMed]
19. Lauren, P. The two histological main types of gastric carcinoma: Diffuse and so-called intestinal-type carcinoma. An attempt at a histo-clinical classification. *Acta Pathol. Microbiol. Scand.* **1965**, *64*, 31–49. [CrossRef] [PubMed]
20. Vauhkonen, M.; Vauhkonen, H.; Sipponen, P. Pathology and molecular biology of gastric cancer. *Best Pract. Res. Clin. Gastroenterol.* **2006**, *20*, 651–674. [CrossRef] [PubMed]

21. Lunet, N.; Valbuena, C.; Vieira, A.L.; Lopes, C.; David, L.; Carneiro, F.; Barros, H. Fruit and vegetable consumption and gastric cancer by location and histological type: Case-control and meta-analysis. *Eur. J. Cancer Prev. Off. J. Eur. Cancer Prev. Organ. (ECP)* **2007**, *16*, 312–327. [CrossRef] [PubMed]

22. Duell, E.J.; Travier, N.; Lujan-Barroso, L.; Clavel-Chapelon, F.; Boutron-Ruault, M.-C.; Morois, S.; Palli, D.; Krogh, V.; Panico, S.; Tumino, R.; et al. Alcohol consumption and gastric cancer risk in the European Prospective Investigation into Cancer and Nutrition (EPIC) cohort. *Am. J. Clin. Nutr.* **2011**, *94*, 1266–1275. [CrossRef] [PubMed]

23. Manish, A.S.; David, P.K. Gastric Cancer: A Primer on the Epidemiology and Biology of the Disease and an Overview of the Medical Management of Advanced Disease. *J. Natl. Compr. Cancer Netw.* **2010**, *8*, 437–447. [CrossRef]

24. Jaehn, P.; Holleczek, B.; Becher, H.; Winkler, V. Histologic types of gastric cancer among migrants from the former Soviet Union and the general population in Germany: What kind of prevention do we need? *Eur. J. Gastroenterol. Hepatol.* **2016**, *28*, 863–870. [CrossRef]

25. Kaucher, S.; Kajüter, H.; Becher, H.; Winkler, V. Cancer Incidence and Mortality Among Ethnic German Migrants From the Former Soviet Union. *Front. Oncol.* **2018**, *8*. [CrossRef]

26. Winkler, V.; Holleczek, B.; Stegmaier, C.; Becher, H. Cancer incidence in ethnic German migrants from the Former Soviet Union in comparison to the host population. *Cancer Epidemiol.* **2014**, *38*, 22–27. [CrossRef]

27. Kaucher, S.; Leier, V.; Deckert, A.; Holleczek, B.; Meisinger, C.; Winkler, V.; Becher, H. Time trends of cause-specific mortality among resettlers in Germany, 1990 through 2009. *Eur. J. Epidemiol.* **2017**, *32*, 289–298. [CrossRef] [PubMed]

28. Winkler, V.; Kaucher, S.; Deckert, A.; Leier, V.; Holleczek, B.; Meisinger, C.; Razum, O.; Becher, H. Aussiedler Mortality (AMOR): Cohort studies on ethnic German migrants from the Former Soviet Union. *BMJ Open* **2019**, *9*, e024865. [CrossRef]

29. Becher, H.; Winkler, V. Estimating the standardized incidence ratio (SIR) with incomplete follow-up data. *BMC Med. Res. Methodol.* **2017**, *17*, 55. [CrossRef] [PubMed]

30. Kajüter, H.; Batzler, W.U.; Krieg, V.; Heidinger, O.; Hense, H.W. Abgleich von Sekundärdaten mit einem epidemiologischen Krebsregister auf der Basis verschlüsselter Personendaten-Ergebnisse einer Pilotstudie in Nordrhein-Westfalen. *Gesundheitswesen* **2012**, *74*, e84–e89. [CrossRef]

31. National Institutes of Health. N.C.I. ICD-O-2 to ICD-O-3 Neoplasms. Available online: https://seer.cancer.gov/tools/conversion/ICDO2-3manual.pdf (accessed on 20 May 2020).

32. European Commission Eurostat. Revision of the European Standard Population. Report of Eurostat's Task Force. 2013. Edition. Available online: https://ec.europa.eu/eurostat/documents/3859598/5926869/KS-RA-13-028-EN.PDF/e713fa79-1add-44e8-b23d-5e8fa09b3f8f (accessed on 2 July 2020).

33. John, E.M.; Phipps, A.I.; Davis, A.; Koo, J. Migration history, acculturation, and breast cancer risk in Hispanic women. *Cancer Epidemiol. Biomark. Prev.* **2005**, *14*, 2905. [CrossRef] [PubMed]

34. International Agency for Research on Cancer. Global Cancer Observatory: Cancer Today. Available online: https://gco.iarc.fr/today (accessed on 10 June 2020).

35. Correa, P. Human gastric carcinogenesis-a multistep and multifactorial process-1st american-cancer-society award lecture on cancer-epidemiology and prevention. *Cancer Res.* **1992**, *52*, 6735–6740.

36. Bornschein, J.; Selgrad, M.; Warnecke, M.; Kuester, D.; Wex, T.; Malfertheiner, P.H. Pylori Infection Is a Key Risk Factor for Proximal Gastric Cancer. *Dig. Dis. Sci.* **2010**, *55*, 3124–3131. [CrossRef]

37. Sipponen, P.; Riihelä, M.; Hyvärinen, H.; Seppälä, K. Chronic Nonatrophic ('Superficial') Gastritis Increases the Risk of Gastric Carcinoma: A Case-Control Study. *Scand. J. Gastroenterol.* **1994**, *29*, 336–340. [CrossRef]

38. Huang, J.-Q.; Sridhar, S.; Chen, Y.; Hunt, R.H. Meta-analysis of the relationship between Helicobacter pylori seropositivity and gastric cancer. *Gastroenterology* **1998**, *114*, 1169–1179. [CrossRef]

39. Aparicio, M.L.; Döring, A.; Mielck, A.; Holle, R.; KORA Studiengruppe. Unterschiede zwischen Aussiedlern und der übrigen deutschen Bevölkerung bezüglich Gesundheit, Gesundheitsversorgung und Gesundheitsverhalten: Eine vergleichende Analyse anhand des KORA-Surveys 2000. *Sozial-und Präventivmedizin/Soc. Prevent. Med. A* **2005**, *50*, 107–118. [CrossRef]

40. Kuhrs, E.; Winkler, V.; Becher, H. Risk factors for cardiovascular and cerebrovascular diseases among ethnic Germans from the former Soviet Union: Results of a nested case-control study. *BMC Public Health* **2012**, *12*, 190. [CrossRef] [PubMed]

41. Ronellenfitsch, U.; Razum, O. Deteriorating health satisfaction among immigrants from Eastern Europe to Germany. *Int. J. Equity Health* **2004**, *3*, 4. [CrossRef] [PubMed]
42. Adler, N.E.; Ostrove, J.M. Socioeconomic Status and Health: What We Know and What We Don't. *Ann. N. Y. Acad. Sci.* **1999**, *896*, 3–15. [CrossRef] [PubMed]
43. Berlth, F.; Bollschweiler, E.; Drebber, U.; Hoelscher, A.H.; Moenig, S. Pathohistological classification systems in gastric cancer: Diagnostic relevance and prognostic value. *World J. Gastroenterol.* **2014**, *20*, 5679–5684. [CrossRef]
44. Henson, D.E.; Dittus, C.; Younes, M.; Nguyen, H.; Albores-Saavedra, J. Differential trends in the intestinal and diffuse types of gastric carcinoma in the United States, 1973–2000: Increase in the signet ring cell type. *Arch. Pathol. Lab. Med.* **2004**, *128*, 765.
45. Becher, H.; Kyobutungi, C.; Laki, J.; Ott, J.J.; Razum, O.; Ronellenfitsch, U.; Winkler, V. Mortality of immigrants from the former Soviet Union: Results of a cohort study. *Dtsch. Arztebl. Int.* **2007**, *104*, 1655–1661.
46. Bund, E.; Kohls, M.; Worbs, S. Zuwanderung und Integration von (Spät-)Aussiedlern in Deutschland. *Z. Ausländerrecht und Ausländerpolitik (ZAR)* **2014**, *10*, 349–354.

Publisher's Note: MDPI stays neutral with regard to jurisdictional claims in published maps and institutional affiliations.

International Journal of
*Environmental Research
and Public Health*

Article

Genetic Variation and Cardiovascular Risk Factors: A Cohort Study on Migrants from the Former Soviet Union and a Native German Population

Marianne Huebner [1,2], Daniela Börnigen [3], Andreas Deckert [4], Rolf Holle [5], Christa Meisinger [6], Martina Müller-Nurasyid [7], Annette Peters [6], Wolfgang Rathmann [8,9] and Heiko Becher [1,*]

1 Institute for Biometry and Epidemiology, University Medical Center Hamburg-Eppendorf, Martinistraße 52, 20246 Hamburg, Germany; huebner@msu.edu
2 Department of Statistics and Probability, Michigan State University, East Lansing, MI 48864, USA
3 Bioinformatics Core Facility, University Medical Center Hamburg-Eppendorf, Martinistraße 52, 20246 Hamburg, Germany; d.boernigen@uke.de
4 Institute of Global Health, Epidemiology and Biostatistics, University Hospital Heidelberg, Im Neuenheimer Feld 324, 69120 Heidelberg, Germany; a.deckert@uni-heidelberg.de
5 Institute of Health Economics and Health Care Management, Helmholtz Zentrum München, GmbH, 85764 Neuherberg, Germany; rholle@ibe.med.uni-muenchen.de
6 German Research Center for Environmental Health, Institute of Epidemiology, Helmholtz Zentrum München, 85764 Neuherberg, Germany; c.meisinger@unika-t.de (C.M.); peters@helmholtz-muenchen.de (A.P.)
7 Institute of Genetic Epidemiology, Helmholtz Zentrum München—German Research Center for Environmental Health, 85764 Neuherberg, Germany; martina.mueller@helmholtz-muenchen.de
8 German Center for Diabetes Research (DZD), München-Neuherberg, 85764 Neuherberg, Germany; wolfgang.rathmann@ddz.de
9 German Diabetes Center, Institute for Biometrics and Epidemiology, 40225 Duesseldorf, Germany
* Correspondence: h.becher@uke.de; Tel.: +49-(0)40-7410-59550

Citation: Huebner, M.; Börnigen, D.; Deckert, A.; Holle, R.; Meisinger, C.; Müller-Nurasyid, M.; Peters, A.; Rathmann, W.; Becher, H. Genetic Variation and Cardiovascular Risk Factors: A Cohort Study on Migrants from the Former Soviet Union and a Native German Population. *Int. J. Environ. Res. Public Health* **2021**, *18*, 6215. https://doi.org/10.3390/ijerph18126215

Academic Editor: Paul B. Tchounwou

Received: 28 April 2021
Accepted: 5 June 2021
Published: 8 June 2021

Publisher's Note: MDPI stays neutral with regard to jurisdictional claims in published maps and institutional affiliations.

Abstract: Resettlers are a large migrant group of more than 2 million people in Germany who migrated mainly from the former Soviet Union to Germany after 1989. We sought to compare the distribution of the major risk factors for cardiovascular disease (CVD) and to investigate the overall genetic differences in a study population which consisted of resettlers and native (autochthone) Germans. This was a joint analysis of two cohort studies which were performed in the region of Augsburg, Bavaria, Germany, with 3363 native Germans and 363 resettlers. Data from questionnaires and physical examinations were used to compare the risk factors for cardiovascular diseases between the resettlers and native Germans. A population-based genome-wide association analysis was performed in order to identify the genetic differences between the two groups. The distribution of the major risk factors for CVD differed between the two groups. The resettlers lead a less active lifestyle. While female resettlers smoked less than their German counterparts, the men showed similar smoking behavior. SNPs from three genes (BTNL2, DGKB, TGFBR3) indicated a difference in the two populations. In other studies, these genes have been shown to be associated with CVD, rheumatoid arthritis and osteoporosis, respectively.

Keywords: migrants; resettlers; genetic differences; cardiovascular diseases; GWAS; lifestyle; Germany

1. Introduction

Germany is a country of immigration, and according to recent data about 25% of the German population has a migrant background. One large migrant group are the so-called Aussiedler (resettlers) from Eastern European countries, with a large subgroup immigrating from the former Soviet Union (FSU) to Germany. The ancestors of the resettlers emigrated to the Russian empire in the 18th and 19th century, by invitation of the government. They were privileged compared to the Russian population, but at the beginning of the 20th century they became victims of persecution and suffered increasing discrimination.

Many ethnic Germans were deported to Kazakhstan and Siberia in 1941. After the opening of the inner-German border in 1989, the majority of the resettlers from the former Soviet Union migrated to Germany. The highest number of migrants came to Germany between the years 1990 and 1995 [1,2].

There are about 3.2 Million resettlers living in Germany, which is approximately 3.5% of the total German population. Although the resettlers were quasi-randomly assigned to different regions in Germany, they are overrepresented in some regions. In the region of Augsburg, a city in Bavaria with about 300,000 inhabitants, the proportion of resettlers is about six percent (https://statistikinteraktiv.augsburg.de/, accessed on 6 March 2019).

In the FSU, the overall mortality—and in particular CVD mortality—was much higher in recent decades compared to that in Germany and other Western countries [3,4]. Due to the expectation that there could also be a high mortality among resettlers, register-based cohort studies on resettlers were performed [2]. However, these studies showed a significantly lower all-cause and cardiovascular disease mortality compared to the German population.

Recent data show that about 90% of the population in the former FSU with German ancestors migrated to Germany, of which very few migrated back after some time in Germany. To date, about 400,000 people who consider themselves to be ethnic Germans continue to live in Russia according to the 2010 census of the Russian Federation [5]. Therefore, common explanations for a low mortality in migrants, such as the healthy migrant effect and salmon bias (i.e., immigrant groups appear to be healthier than they are because less healthy individuals selectively return to their country of origin), are unlikely to explain the findings.

A few studies on risk factor prevalence for the migrant population [6,7] showed that the typical risk factors for CVD are high among resettlers. Thus, these observations do not explain the observed pattern of low CVD mortality. Because the history of this migrant group suggests that they are the descendants of a group of Germans who not only decided to migrate from their home country in the 18th century but also survived difficult living conditions, there may be differences in the genetic predisposition conveying a survival advantage. Genetic stratification may exist within racial/ethnic groups. In a population-based genome-wide association analysis between five European populations, several genes were reported to be stratified within European populations [8]. A genetic cluster analysis in a US study with whites, African Americans, East Asians and Hispanics of 326 microsatellite markers produced four major clusters, which showed near-perfect correspondence with the four self-reported race/ethnicity categories [9]. It is unknown whether there is such a clustering for the resettlers and the native (autochthone) German population. Phenotypes controlled by a dozen or fewer loci can be expected to show substantial overlap between human populations [10].

This is a joint analysis of two prospective cohort studies which were sampled from the same total population. One is the so-called KORA S4 cohort, which was recruited in the years 1999–2001 as an age- and sex-stratified sample [7,11]. The other is a cohort of resettlers [12]. Both cohorts are described in more detail below. In this study, we investigated whether the observed differences in CVD mortality from previous large register-based studies can partly be explained by a differing distribution of genetic factors that contribute to this disease group. In addition, we analysed the distribution of other risk factors for CVD. The underlying hypothesis is that the ancestors of the resettlers may have been selected because they were particularly healthy and physically advantaged, and therefore may have been genetically advantaged compared to the average population. Our objectives were to compare the distribution of major risk factors for cardiovascular diseases and to investigate the overall genetic differences in a study population which consists of resettlers and native Germans from the region of Augsburg. We sought to examine the magnitude of the differences in the SNP allele frequencies between the two populations.

2. Materials and Methods

2.1. Study Population

The study sample originates from two cohort studies which were both performed in the region of Augsburg, Bavaria, Germany.

2.2. KORA Cohort

The KORA study is a series of population-based surveys conducted in the city of Augsburg in Southern Germany and its two adjacent counties. The cohort KORA S4 was recruited in the years 1999–2001 as an age- and sex-stratified sample of German residents based on information from local registry offices. The study participants underwent intensive clinical examinations and answered health questionnaires. Overall, the cohort consisted of 3788 individuals. It included 233 resettlers (Figure 1).

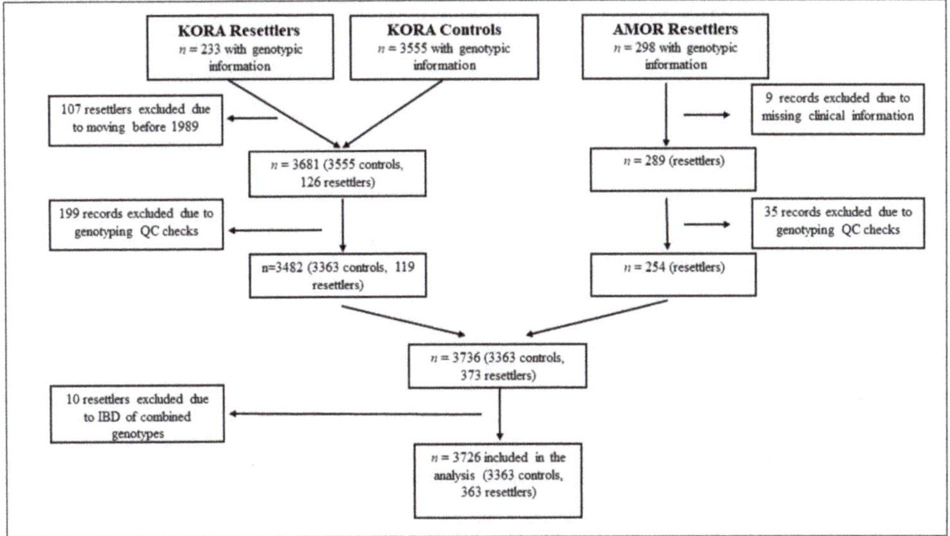

Figure 1. STROBE flow chart.

After excluding resettlers who moved before 1989 and excluding records due to genotyping QC checks, the final sub-cohort contained 3363 native Germans and 119 resettlers. Genotyping was performed on the Affymetrix Axiom. The details of the cohort are given in [7,11].

2.3. AMOR Cohort

Resettlers living in the greater Augsburg region in 2010 (n = 3718) were identified in the population registry. They were asked by written letter to provide a self-administered questionnaire (based on the KORA study questionnaire). The respondents then were invited to take part in a detailed physical examination (anthropometric measures and blood pressure measurements) and a personal interview, and to provide a blood sample. The data collection took place from 2011 to 2013. From the cohort of 673 individuals who provided questionnaire data, 298 provided blood samples for genetic testing. Of these, both the clinical examination and interview data were available for 180 subjects (63 males, 117 females). The genotyping was performed on the Illumina Infinium Global Screening Array GSA v.1.0 (Illumina Inc., San Diego, CA, USA). A comparison between the Illumina and Affymetrix platforms (Thermo Fisher Scientific, Waltham, MA, USA) indicated very high agreement [13]. More details of the AMOR cohort are given in [12].

Both cohorts combined thus included 531 resettlers (233 from KORA and 298 from the AMOR cohorts) and 3555 individuals as a comparison group (the KORA native German population) with genotypic information and questionnaire data. Of these, clinical examination data were available from 413 resettlers (233 from KORA and 180 from AMOR). People who participated in both studies could be detected through genetic identity, and were excluded from the analysis.

2.4. Data Preprocessing

For the genetics data, we implemented quality control measures, namely the QC pipeline by the eMERGE network [14]. The SNPs were excluded if: (i) the SNP genotyping call rate was < 0.99; (ii) the minor allele frequency (MAF) was < 0.01; (iii) the linkage disequilibrium (LD) was < 0.2; and (iv) the Hardy–Weinberg Equilibrium (HWE) $p < 10 \times 10^{-6}$. At the sample level, the (v) heterozygosity cutoff was 0.1 and (vi) the kinship coefficient was < 0.1. A principle component analysis (PCA) was performed in order to detect the outliers.

Variables from the questionnaires used in the KORA and AMOR cohorts were harmonized across the surveys. The hypertension categories were determined from the systolic and diastolic blood pressure, measured according to the National Joint Committee [15]. The measurement units were standardized for the serum measurements. The estimated glomular filtration rate (eGFR) was calculated using the CKD-EPI equation [16].

2.5. Statistical Methods

The continuous variables were summarized for the resettlers and native Germans with medians and quartiles, categorical variables as frequencies, and percentages. The self-reported comorbidity variables were myocardial infarction (MI), stroke, diabetes and cancer. The comorbidities derived from the laboratory measurements were hypertension and chronic kidney disease (CKD). The laboratory measurements were available for the KORA cohort and for a subset of the AMOR cohort, including systolic and diastolic blood pressure, triglycerides, creatinine, cholesterol, waist and hip measurements, height, and weight. The demographic variables and lifestyle factors included age, sex, sport activity, family status, education level, smoking status, and the family history of MI and stroke. These variables were described for the resettlers and native Germans, and were stratified by sex, as there are known sex differences. A conditional logistic regression analysis was used to analyze the differences in the risk factor distribution between resettlers and native Germans in a multivariable model. The strata were formed by 5-year age groups.

In order to investigate a possible general genetic difference between the resettlers and Germans, we first performed a principal component analysis (PCA) based on all of the SNPs available for both cohorts. A genome-wide association analysis was conducted with 'resettler' as the outcome, adjusted for age and sex. The p-values were adjusted for the Bonferroni criterion, and the false discovery rate (FDR) threshold was set at 0.1. The GWAS catalogue [17] was searched in order to observe whether the resulting SNPs and genes were identified in an association analysis for cardiovascular diseases, diabetes, or cancer. The SNPs thus identified were entered into unadjusted and adjusted logistic regression models with cardiovascular disease (MI or stroke) or diabetes as the outcomes for the entire population. The SNPs were coded as additive with the homozygous genotype for the minor allele coded as 2, the heterozygous genotype coded as 1, and otherwise as 0. Due to the possibility that there is an overrepresentation of subjects with cardiovascular disease among the resettlers, the GWAS was repeated after removing such cases.

The statistical analyses were performed using R version 3.5.1 (R Foundation for Statistical Computing: Vienna, Austria), including the packages SNPRelate v.1.12.2 [18] and snpStats v 1.28.0 [19] from the Comprehensive R Archive Network [20]. This study was reported according the guidelines of the STROBE Statement [21] and the STREGA checklist [22].

3. Results

Both cohorts combined included 531 resettlers (233 from KORA and 298 from the AMOR cohort) and 3555 individuals as a comparison group (the KORA native German population). The flow diagram (Figure 1) describes the reasons for various exclusions. Ten resettlers participated in both studies and were detected through their genetic identity. The final study group for the analysis consisted of 3726 individuals (363 resettlers and 3363 in the comparison group) for which both the genetic data and the data from the interview and examination were available. After the preprocessing, 50,340 SNPs were in common for the KORA and AMOR SNP arrays. The age range for the resettlers was 20 to 86 years, and 24 to 75 years for the native Germans. The resettlers were on average 5 years older for both women and men. The distribution of the demographic factors, CVD risk factors, anthropometric measurements and laboratory measurements are described in Table 1.

Table 1. Population characteristics of the resettlers and native Germans, stratified by sex.

	Women		Men	
	Resettlers (*n* = 214)	**Native Germans** (*n* = 1727)	**Resettlers** (*n* = 149)	**Native Germans** (*n* = 1636)
Age (years)	54 (44, 62)	49 (37, 60)	55 (43, 66)	51 (37, 62)
missing	4	0	0	0
Family Status				
single	161 (9.3%)	19 (9.0%)	5 (3.4%)	213 (13.0%)
cohabit	145 (68.4%)	1243 (72.0%)	126 (84.6%)	1255 (76.8%)
separated	19 (9.0%)	177 (10.3%)	13 (8.7%)	130 (8.0%)
widowed	145 (8.4%)	29 (13.7%)	5 (3.4%)	36 (2.2%)
missing	1	1	0	2
Anthropometrics				
Height (cm)	159.5 (156.2, 163.6)	162 (157.7, 166.5)	172.5 (168.2, 177.5)	175.1 (170.3, 179.8)
missing	56 (26.2%)	7 (0.4%)	34 (22.8%)	4 (0.2%)
Weight (kg)	76 (66.1, 85.0)	68.3 (60.7, 77.6)	84.4 (74.7, 93.6)	82.7 (75.4, 91.0)
missing	56 (26.2%)	18 (1.0%)	34 (22.8%)	6 (0.4%)
BMI (kg.m^{-2})	29.8 (25.6, 33.4)	25.9 (22.9, 29.6)	28.3 (25.6, 30.9)	27.0 (24.9, 29.6)
missing	56 (26.2%)	19 (1.1%)	34 (22.8%)	6 (0.4%)
Waist-to-hip ratio	0.84 (0.78, 0.88)	0.80 (0.76, 0.85)	0.93 (0.89, 0.97)	0.95 (0.90, 1.0)
missing	56 (26.2%)	18 (1.0%)	34 (22.8%)	4 (0.2%)
Laboratory measurements				
Cholesterol (mg/dl)	219 (196, 248)	222 (196, 255)	215 (191, 243)	224 (198, 256)
missing	56 (26.2%)	7 (0.4%)	35 (24.6%)	8 (0.5%)
Triglycerides (mg/dl)	149 (105, 223)	109 (83, 154)	161 (110, 265)	131 (91, 196)
missing	102 (47.7%)	1045 (60.5%)	72 (50.7%)	910 (55.6%)
Creatinine (mg/dl)	0.77 (0.70, 0.87)	0.75 (0.68, 0.82)	0.98 (0.87, 1.10)	0.93 (0.84, 1.02)
missing	57 (26.6%)	18 (1.0%)	35 (24.6%)	19 (1.2%)
EGFR	86.8 (73.2, 100.7)	93.9 (81.3, 105.4)	88.0 (71.2, 102.2)	93.3 (83.0, 104.5)
missing	59 (27.6%)	18 (1.0%)	35 (24.6%)	19 (1.2%)
Lifestyle factors				
Smoking				
Never	174 (82.9%)	911 (52.8%)	48 (32.7%)	530 (32.5%)
Previous	13 (6.2%)	451 (26.1%)	60 (40.8%)	628 (38.5%)
Current	23 (11.0%)	364 (21.1%)	39 (26.5%)	475 (29.1%)
missing	3	1	2	3
Physical activity				
Regular	41 (20.1%)	872 (50.6%)	40 (28.2%)	816 (50.1%)
Irregular	50 (24.5%)	297 (17.2%)	23 (16.2%)	292 (17.9%)
Inactive	113 (55.4%)	555 (32.2%)	79 (55.6%)	522 (32.0%)
missing	10	3	7	6
Hypertension				
normal	61 (39.4%)	800 (46.4%)	19 (16.5%)	323 (19.8%)
pre	61 (39.4%)	558 (32.4%)	57 (49.6%)	708 (43.4%)
Stage 1	26 (16.8%)	284 (16.5%)	26 (22.6%)	418 (25.6%)
Stage 2	7 (4.5%)	81 (4.7%)	13 (11.3%)	181 (11.1%)
missing	58 (27.1%)	4 (0.2%)	34 (22.8%)	6 (0.4%)

The data are presented as the median (quartiles) or as *n* (%) for the categorical variables. Abbreviations: EGFR—estimated glomular filtration rate.

The results of the univariable and the multivariable conditional logistic regression models with age in 5 year increments, in order to compare the resettlers and the native Germans stratified by gender, are shown in Table 2. The resettlers showed lower physical activity and a higher BMI. While the smoking prevalence in males was similar, with an OR for current male smokers of 1.02 (95% CI: 0.65, 1.61), $p = 0.940$ in comparison to native Germans, female resettlers smoke considerably less than the native German women, with an OR of 0.35 (95% CI: 0.22, 0.55), $p < 0.001$. Hypertension (normal, pre-hypertension, stage I and stage II) did not show a statistically significant different distribution in either gender. The cholesterol levels in the resettlers were lower for female resettlers (OR for women 0.94 (95% CI: 0.90, 0.98), $p = 0.006$; OR for men 0.97 (95% CI: 0.93, 1.02); $p = 0.207$), and the triglyceride levels were higher (OR for women 1.05 (95% CI: 1.03, 1.07), $p < 0.001$; OR for men 1.02 (95% CI: 1.00, 1.04); $p = 0.021$). Diabetes was more frequent in resettlers for both sexes (OR for women 2.82 (95% CI: 1.63, 4.86); $p < 0.001$; OR for men 2.34 (95% CI: 1.28, 4.23); $p = 0.006$).

Table 2. Univariable and multivariable logistic regression analysis to identify the differences in the risk factors for cardiovascular diseases between resettlers and native Germans.

	Women $n = 1941$		Men $n = 1785$	
	OR * (95% CI)	OR ** (95% CI)	OR * (95% CI)	OR ** (95% CI)
Number of resettlers (total) in model	146 (1830)		107 (1713)	
BMI	1.10 (1.07,1.14); $p < 0.001$	1.08 (1.05, 1.12); $p < 0.001$	1.04 (1.00,1.09); $p = 0.051$	1.02 (0.98, 1.08): $p = 0.329$
Cholesterol (10 mg/dl)	0.94 (0.90, 0.98); $p = 0.006$	0.94 (0.89, 0.99); $p = 0.015$	0.97 (0.93, 1.02); $p = 0.207$	0.98 (0.94, 1.03); $p = 0.524$
Triglycerides (10 mg/dl)	1.05 (1.03, 1.07); $p < 0.001$		1.02 (1.00,1.04); $p = 0.021$	
EGFR	0.98 (0.97, 0.99); $p < 0.001$	0.97 (0.96, 0.98); $p < 0.001$	0.98 (0.97, 0.99); $p = 0.006$	0.98 (0.97, 0.99); $p = 0.007$
Smoking never	1	1		1
previous	0.16 (0.09, 0.28); $p < 0.001$	0.22 (0.12, 0.41); $p < 0.001$	0.99 (0.66, 1.45); $p = 0.949$	1.12 (0.68, 2.05); $p = 0.665$
current	0.35 (0.22, 0.55); $p < 0.001$	0.37 (0.21, 0.65); $p < 0.001$	1.02 (0.65, 1.61); $p = 0.940$	1.19 (0.69,2.05); $p = 0.537$
Physical activity regular	1	1	1	1
irregular	3.87 (2.49, 6.03); $p < 0.001$	4.24 (2.48, 7.25); $p < 0.001$	1.63 (0.96, 2.78); $p = 0.075$	1.32 (0.70, 2.50); $p = 0.388$
inactive	4.26 (2.90, 6.28); $p < 0.001$	4.53 (2.79, 7.36); $p < 0.001$	2.92 (1.94, 4.40); $p < 0.001$	2.57 (1.62, 4.10); $p < 0.001$
Hypertension normal	1	1	1	1
Pre-hypertension	1.16 (0.78, 1.73); $p = 0.456$	0.83 (0.54, 1.29); $p = 0.416$	1.41 (0.81, 2.44); $p = 0.226$	1.32 (0.74, 2.35); $p = 0.346$
Stage 1	(0.97 (0.58, 1.62); $p = 0.913$	0.46 (0.25, 0.83); $p = 0.011$	1.01 (0.53, 1.92); $p = 0.970$	0.89 (0.45, 1.76); $p = 0.736$
Stage 2	0.95 (0.41, 2.23); $p = 0.914$	0.40 (0.15, 1.05); $p = 0.062$	1.07 (0.50, 23.0); $p = 0.860$	1.02 (0.45,2.30); $p = 0.956$

* Conditional logistic regression, with age in 5 yearage groups; ** conditional logistic regression, with age in 5 yearage groups; all of the risk factors were included in the model.

SNPs in the regions of 10 genes that have been shown to stratify European populations [8] were in the set of the 50K SNPs. However, none of these SNPs were shown to differentiate resettlers and native Germans in our dataset. A principal component analysis on all of the SNPs also did not indicate a separation between the resettlers and native Germans (Figure 2).

In a GWAS with resettler status as the outcome, three loci reached genome-wide significance to indicate a difference in the two populations after adjusting for sex and age, and correcting for multiple testing (Figure 3, Table 3).

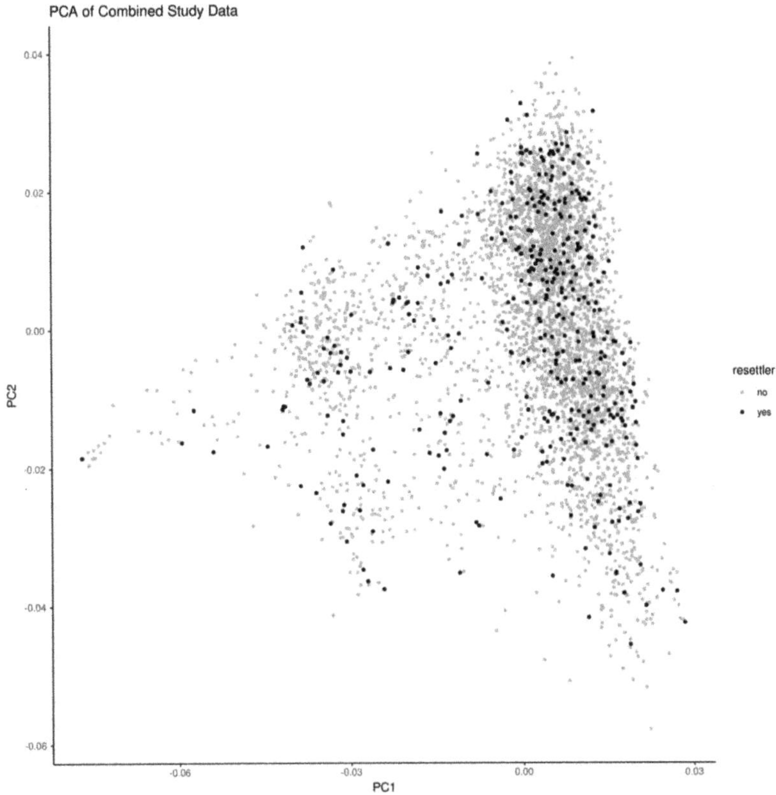

Figure 2. Principle component analysis (PCA) of the combined KORA and AMOR single nucleotide polymorphisms, with an indication of resettler status.

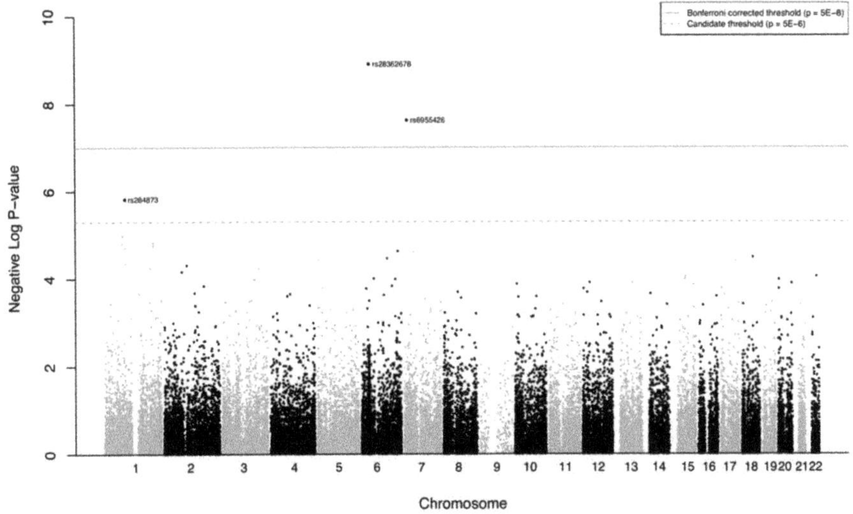

Figure 3. Manhattan plot for the genome-wide analysis of resettler status.

Table 3. SNPs differentiating resettlers and native Germans, and the distribution of the genotypes.

	rs28362678	rs6955426	rs284873
Chromosome: Gene	6: BTNL2	7: DGKB	1: TGFBR3
Gene function	Immunoregulators	Cellular processes	Cell surface receptor
Diseases *	Rheumatoid arthritis, sarcoidosis	Myocardial infarction	Bone mass and osteoporosis
Alleles	C > T	G > A	T > C
MAF (resettlers/controls)	5.8%/14.4%	19.5%/12.35%	13.46%/8.2%
AA	7(1.9%)/52 (1.5%)	8(2.2%)/39 (1.2%)	8 (2.2%)/20 (0.6%)
AB	28(7.7%)/864 (25.7%)	126 (34.6%) /750 (22.3%)	82 (22.5%)/510 (15.2%)
BB	326 (89.8%)/2443 (72.6%)	229 (63.2%)/2562 (76.2%)	273 (75.3%)/2829 (84.1%)
missing	2 (0.5%)/4 (0.1%)	0/12 (0.4%)	0/4 (0.1%)
p-value	1.24×10^{-9}	2.37×10^{-8}	1.51×10^{-6}
FDR	6.26×10^{-5}	0.000598	0.025374

Abbreviations: MAF—minor allele frequency; FDR—false discovery rate. 'A' denotes the minor allele and 'B' denotes the major allele. * BTNL2 [23]; DGKB [24]; TGFBR3 [25].

These three SNPs belong to the genes DGKB, BTNL2, and TGFBR3 that were indicated in previous genome wide association studies with CVD (DGKB), rheumatoid arthritis and sarcoidosis (BTNL2), and bone mass and osteoporosis (TGFBR3), respectively.

In a sensitivity analysis, we performed a GWAS excluding self-reported cases of previous myocardial infarction (20 resettlers, 66 native Germans); the same SNPs from BTNL2 and TGFBR3 marked a genetic difference between the resettlers and native Germans. An additional SNP rs11579207 from TGFBR3 was also associated with resettler status.

4. Discussion

This is the first study in which resettlers were investigated regarding a general genetic difference to the native German populations and their genetic predisposition for cardiovascular diseases. In addition, the study contributed to the previously limited information on risk factors in this migrant group. Because previous studies showed a lower CVD mortality in resettlers and, contrary to this finding, a higher prevalence of CVD risk factors, we hypothesized that genetic factors may partly explain this observation.

4.1. Gender Differences of Cardiovascular Disease Risk Factors in Resettlers

In order to explain the observed differences in disease incidence or mortality between different populations, the first natural analysis is a comparison of the known medical or lifestyle risk factors. The previous studies were small; however, they pointed towards a higher prevalence of CVD risk factors in resettlers [6,7]. Several of the main risk factors for CVD showed a different distribution in resettlers; however, it must be noted that the proportion of missing information was higher in resettlers, and a clear missing at random assumption is not justified. In both men and women, BMI and physical activity had a less favorable distribution in resettlers, which are indicators for a potentially higher CVD mortality. In men, smoking and hypertension, two other major risk factors for CVD, showed a similar distribution in resettlers as in native Germans. In women, smoking is much less common in female resettlers compared to native German women, but normal blood pressure has a lower prevalence. Overall, the results are partly in line with the prior data and, in our view, would indicate a higher CVD risk in resettlers.

4.2. Genetic Differences between Resettlers and Native Germans

The PCA analysis showed no general genetic difference between both groups; however, the GWAS analysis provided some interesting results. Genetic association studies with a large number of tests carry the risk of spurious results. In the present study, significance was retained at the genome-wide level after correction for multiple testing. Furthermore, we explored the functional relevance of the genes indicated in other association studies.

BTNL2 belongs to the butyrophilin-like B7 family of immunoregulators. Direct interaction between the BTN and its receptor on activated T cells leads to the suppression of the T cell response. TGFBR3 is a major mediator of TGF-beta signaling pathways and also functions as a BMP cell-surface receptor. Diacylglycerol kinases (DGKs) are regulators of the intracellular concentration of diacylglycerol and thus play a key role in cellular processes [24]. In genome-wide association studies, these genes have been identified in connection with myocardial infarction (DGKB [24]); immune function, rheumatoid arthritis, and sarcoidosis (BTNL2 [23]); and bone mass in different ethnic groups (TGFBR3 [25]). While these studies examined different SNPs of the same genes not included in our dataset, the SNP rs28362678 (BTNL2) identified in our study as having different variants for resettlers and native Germans has been shown to be associated with rheumatoid arthritis [23].

4.3. Limitations

There are several limitations. First, while the previous studies on the mortality of the resettlers were based on registries in which selection bias can be excluded, this study was based on the voluntary participation of individuals. The response rate was low, especially in the AMOR cohort. This may, in particular, have had an effect on the observed risk factor distribution and on self-reported disease. The resettler cohort consisted of older subjects than the native Germans, and this contributed to the comorbidity load. Second, due to our use of two different genotype platforms, the GWAS was limited to about 50K SNPs in common on both arrays. Third, the phenotypic data are missing for 22 percent of all of the resettlers because some individuals from the AMOR subcohort only provided a blood sample and did not undertake the physical examinations. This was because the blood samples were provided through the home physician, while all of the examinations were performed in the study center, which these individuals did not attend due to time constraints. We consider it unlikely, however, that the genetic information is correlated with participation, and thus we believe that our results are unbiased. Fourth, it was difficult to motivate the resettlers in the AMOR study to participate. The original study protocol assumed a much larger sample size.

Another set of risk factors for CVD are environmental exposures, such as air pollution [26]. The resettlers originated from several states in the former Soviet Union, which would make an exposure assessment almost impossible. Therefore, it is difficult to make comparisons of the exposure levels to the Augsburg area. We do not think, however, that possible differences in air pollution levels could explain the lower CVD mortality in the resettlers.

Our cohorts were not selected for specific diseases, and it would be desirable to follow up our results with further studies such as the German National Cohort NAKO [27], which recently finished its recruitment phase. It includes a random sample from 18 study centers widely distributed in Germany, with a study size of 205,000 participants, of whom 3500 are resettlers. A detailed assessment of disease risk factors, an extensive medical examination program and a collection of various biomaterials was performed. The data cleaning processes are underway, and analyses will follow in the near future.

5. Conclusions

This is the first study in which resettlers were investigated regarding a general genetic difference to the native German population. The gene DGKB was shown to be associated with CVD, but the SNP rs6955426 is a novel locus on this gene and should be considered further in future studies. The minor allele is a risk allele, and resettlers have a higher minor allele frequency. The SNP rs28362678 on BTNL2 has previously been shown to be associated with rheumatoid arthritis, and gene function related to diabetes. TGFBR3 has been shown to be associated with osteoporosis. The distribution of the CVD risk factors, such as BMI or physical activity, is less favorable in resettlers compared to the native Germans. Further ongoing studies, in particular the German National Cohort (NAKO) which includes about 3500 resettlers, with detailed data on physical examinations, questionnaire data and biologic specimens, will provide more insights on the genetics and the mortality and morbidity pattern of this migrant group.

Author Contributions: Conceptualization, H.B., C.M. and A.P.; methodology, M.H., M.M.-N. and H.B.; software, D.B., M.H.; validation, M.H., D.B.; formal analysis, M.H., D.B., H.B.; resources, H.B., A.P., C.M. data curation, M.H., A.D., H.B.; writing—original draft preparation, M.H. and H.B.; writing—review and editing, D.B., A.D., R.H., C.M., M.M.-N., A.P. and W.R.; visualization, M.H.; supervision, H.B.; project administration, H.B.; funding acquisition, H.B., C.M. and A.P. All authors have read and agreed to the published version of the manuscript.

Funding: The research was funded by the BMBF, German Federal Ministry of Education and Research (Grant Number 01ER1306 PERGOLA). The KORA study was initiated and financed by the Helmholtz Zentrum München, the German Research Center for Environmental Health, which is funded by the German Federal Ministry of Education and Research (BMBF) and by the State of Bavaria. Furthermore, the KORA research was supported within the Munich Center of Health Sciences (MC-Health), Ludwig-Maximilian-Universität, as part of LMUinnovativ.

Institutional Review Board Statement: The study was conducted according to the guidelines of the Declaration of Helsinki, and approved by the Ethics Committee of the Bavarian Medical Association.

Informed Consent Statement: Informed consent was obtained from all of the subjects involved in the study.

Data Availability Statement: The data are subject to national data protection laws; restrictions were imposed by the Ethics Committee of the Bavarian Medical Association to ensure the data privacy of the study participants and therefore the data cannot be made freely available in a public repository. The data are third party and belong to the KORA research platform, but can be accessed for specific research projects through individual project agreements. Interested researchers can request the data from KORA via the KORA.passt online tool (https://epi.helmholtz-muenchen.de/ accessed on 1 June 2021). In a data request, one has to briefly describe the intended scientific question and then select the variables of interest within the KORA.passt tool. We confirm that interested researchers who agree to the general terms and conditions of the KORA data user agreement can access the data of KORA in the same way we did. All other relevant data are within the manuscript and its Supporting Information files.

Acknowledgments: The authors acknowledge the support of the Bioinformatics Core Facility at the Medical University Hamburg-Eppendorf, Hamburg, Germany. Special thanks go to the Institute of Medical Biometry and Epidemiology, Medical University Hamburg-Eppendorf, who hosted one of the authors [M.H.] during her sabbatical.

Conflicts of Interest: The authors R.H., C.M., M.M.-N., A.P. have an affiliation with the Helmholtz Zentrum Muenchen. The funders had no role in the design of the study; in the collection, analyses, or interpretation of data; in the writing of the manuscript, or in the decision to publish the results.

References

1. Worbs, S.; Bund, E.; Kohls, M. Forschungsbericht 20... Nürnberg: Bundesamt für Migration und Flüchtlinge. Available online: https://publikationen.uni-tuebingen.de/xmlui/bitstream/handle/10900/60163/spaetaussiedler-in-deutschland.pdf?sequence=1&isAllowed=y (accessed on 1 June 2021).
2. Winkler, V.; Kaucher, S.; Deckert, A.; Leier, V.; Holleczek, B.; Meisinger, C.; Razum, O.; Becher, H. Aussiedler Mortality (AMOR): Cohort Studies on Ethnic German Migrants from the Former Soviet Union. *BMJ Open* **2019**, *9*, e024865. [CrossRef] [PubMed]
3. Deckert, A.; Winkler, V.; Meisinger, C.; Heier, M.; Becher, H. Myocardial Infarction Incidence and Ischemic Heart Disease Mortality: Overall and Trend Results in Repatriates, Germany. *Eur. J. Public Health* **2014**, *24*, 127–133. [CrossRef] [PubMed]
4. WHO|WHO Mortality Database. Available online: https://www.who.int/healthinfo/mortality_data/en/ (accessed on 14 September 2019).
5. Russian Federal Service of State Statistics (Rosstat) Том4. Национальный состав и владение языками, гражданство. In 10. НАСЕЛЕНИЕ НАИБОЛЕЕ МНОГОЧИСЛЕННЫХ НАЦИОНАЛЬНОСТЕЙПОВОЗРАСТНЫМГРУППАМ ИПОЛУПОСУБЪЕКТАМ РОССИЙСКОЙФЕДЕРАЦИИ; Russian Statistical Office: Moscow, Russia, 2010.
6. Kuhrs, E.; Winkler, V.; Becher, H. Risk Factors for Cardiovascular and Cerebrovascular Diseases among Ethnic Germans from the Former Soviet Union: Results of a Nested Case-Control Study. *BMC Public Health* **2012**, *12*, 190. [CrossRef]

7. Aparicio, M.; Doering, A.; Mielck, A.; Holle, R.; KORA Studiengruppe. Unterschiede zwischen Aussiedlern und der übrigen deutschen Bevölkerung bezüglich Gesundheit, Gesundheitsversorgung und Gesundheitsverhalten: Eine vergleichende Analyse anhand des KORA-Surveys 2000. *Sozial-und Präventivmedizin/Soc. Prev. Med.* **2005**, *50*, 107–118. [CrossRef] [PubMed]
8. Moskvina, V.; Smith, M.; Ivanov, D.; Blackwood, D.; StClair, D.; Hultman, C.; Toncheva, D.; Gill, M.; Corvin, A.; O'Dushlaine, C.; et al. Genetic Differences between Five European Populations. *Hum. Hered.* **2010**, *70*, 141–149. [CrossRef]
9. Tang, H.; Quertermous, T.; Rodriguez, B.; Kardia, S.L.R.; Zhu, X.; Brown, A.; Pankow, J.S.; Province, M.A.; Hunt, S.C.; Boerwinkle, E.; et al. Genetic Structure, Self-Identified Race/Ethnicity, and Confounding in Case-Control Association Studies. *Am. J. Hum. Genet.* **2005**, *76*, 268–275. [CrossRef]
10. Witherspoon, D.J.; Wooding, S.; Rogers, A.R.; Marchani, E.E.; Watkins, W.S.; Batzer, M.A.; Jorde, L.B. Genetic Similarities Within and Between Human Populations. *Genetics* **2007**, *176*, 351–359. [CrossRef]
11. Rabel, M.; Meisinger, C.; Peters, A.; Holle, R.; Laxy, M. The Longitudinal Association between Change in Physical Activity, Weight, and Health-Related Quality of Life: Results from the Population-Based KORA S4/F4/FF4 Cohort Study. *PLoS ONE* **2017**, *12*, e0185205. [CrossRef]
12. Stolpe, S.; Ouma, M.; Winkler, V.; Meisinger, C.; Becher, H.; Deckert, A. Self-Rated Health among Migrants from the Former Soviet Union in Germany: A Cross-Sectional Study. *BMJ Open* **2018**, *8*, e022947. [CrossRef] [PubMed]
13. Barnes, M.; Freudenberg, J.; Thompson, S.; Aronow, B.; Pavlidis, P. Experimental Comparison and Cross-Validation of the Affymetrix and Illumina Gene Expression Analysis Platforms. *Nucleic Acids Res.* **2005**, *33*, 5914–5923. [CrossRef] [PubMed]
14. Turner, S.; Armstrong, L.L.; Bradford, Y.; Carlson, C.S.; Crawford, D.C.; Crenshaw, A.T.; de Andrade, M.; Doheny, K.F.; Haines, J.L.; Hayes, G.; et al. Quality Control Procedures for Genome-Wide Association Studies. *Curr. Protoc. Hum. Genet.* **2011**, *68*, 1–19. [CrossRef] [PubMed]
15. National Joint Committee. The Seventh Report of the Joint National Committee on Prevention, Detection, Evaluation, and Treatment of High Blood Pressure. *Hypertension* **2003**, *42*, 1206–1252. [CrossRef]
16. Levey, A.S.; Stevens, L.A.; Schmid, C.H.; Zhang, Y.L.; Castro, A.F.; Feldman, H.I.; Kusek, J.W.; Eggers, P.; Van Lente, F.; Greene, T.; et al. A New Equation to Estimate Glomerular Filtration Rate. *Ann. Intern. Med.* **2009**, *150*, 604–612. [CrossRef]
17. GWAS Catalog. Available online: https://www.ebi.ac.uk/gwas/ (accessed on 14 September 2019).
18. Zheng, X.; Levine, D.; Shen, J.; Gogarten, S.M.; Laurie, C.; Weir, B.S. A High-Performance Computing Toolset for Relatedness and Principal Component Analysis of SNP Data. *Bioinforma. Oxf. Engl.* **2012**, *28*, 3326–3328. [CrossRef] [PubMed]
19. Clayton, D. SnpStats; snpStats: SnpMatrix and XSnpMatrix Classes and Methods. R package version 1.24.0; 2015. Available online: https://bioconductor.riken.jp/packages/3.4/bioc/manuals/snpStats/man/snpStats.pdf (accessed on 7 January 2019).
20. R Core Team. *R: A Language and Environment for Statistical Computing*; R Foundation for Statistical Computing: Vienna, Austria, 2017.
21. von Elm, E.; Altman, D.G.; Egger, M.; Pocock, S.J.; Gøtzsche, P.C.; Vandenbroucke, J.P.; STROBE Initiative. The Strengthening the Reporting of Observational Studies in Epidemiology (STROBE) Statement: Guidelines for Reporting Observational Studies. *J. Clin. Epidemiol.* **2008**, *61*, 344–349. [CrossRef] [PubMed]
22. Little, J.; Higgins, J.P.T.; Ioannidis, J.P.A.; Moher, D.; Gagnon, F.; von Elm, E.; Khoury, M.J.; Cohen, B.; Davey-Smith, G.; Grimshaw, J.; et al. STrengthening the REporting of Genetic Association Studies (STREGA)—An Extension of the STROBE Statement. *Genet. Epidemiol.* **2009**, *33*, 581–598. [CrossRef]
23. Mitsunaga, S.; Hosomichi, K.; Okudaira, Y.; Nakaoka, H.; Kunii, N.; Suzuki, Y.; Kuwana, M.; Sato, S.; Kaneko, Y.; Homma, Y.; et al. Exome Sequencing Identifies Novel Rheumatoid Arthritis-Susceptible Variants in the BTNL2. *J. Hum. Genet.* **2013**, *58*, 210–215. [CrossRef]
24. Dehghan, A.; Bis, J.C.; White, C.C.; Smith, A.V.; Morrison, A.C.; Cupples, L.A.; Trompet, S.; Chasman, D.I.; Lumley, T.; Völker, U.; et al. Genome-Wide Association Study for Incident Myocardial Infarction and Coronary Heart Disease in Prospective Cohort Studies: The CHARGE Consortium. *PLoS ONE* **2016**, *11*, e0144997. [CrossRef]
25. Xiong, D.-H.; Liu, X.-G.; Guo, Y.-F.; Tan, L.-J.; Wang, L.; Sha, B.-Y.; Tang, Z.-H.; Pan, F.; Yang, T.-L.; Chen, X.-D.; et al. Genome-Wide Association and Follow-up Replication Studies Identified ADAMTS18 and TGFBR3 as Bone Mass Candidate Genes in Different Ethnic Groups. *Am. J. Hum. Genet.* **2009**, *84*, 388–398. [CrossRef]
26. Kim, J.B.; Prunicki, M.; Haddad, F.; Dant, C.; Sampath, V.; Patel, R.; Smith, E.; Akdis, C.; Balmes, J.; Snyder, M.P.; et al. Cumulative Lifetime Burden of Cardiovascular Disease From Early Exposure to Air Pollution. *J. Am. Heart Assoc.* **2020**, *9*, e014944. [CrossRef] [PubMed]
27. German National Cohort (GNC) Consortium The German National Cohort: Aims, Study Design and Organization. *Eur. J. Epidemiol.* **2014**, *29*, 371–382. [CrossRef] [PubMed]

International Journal of
Environmental Research and Public Health

MDPI

Article

The Health Status and Healthcare Utilization of Ethnic Germans in Russia

Charlotte Arena [1], Christine Holmberg [1,2], Volker Winkler [3] and Philipp Jaehn [1,2,*]

[1] Institute of Social Medicine and Epidemiology, Brandenburg Medical School Theodor Fontane, 14770 Brandenburg an der Havel, Germany; charlotte.arena@mhb-fontane.de (C.A.); christine.holmberg@mhb-fontane.de (C.H.)

[2] Faculty of Health Sciences Brandenburg, Brandenburg Medical School Theodor Fontane, 14770 Brandenburg an der Havel, Germany

[3] Heidelberg Institute of Global Health, Heidelberg University Hospital, 69120 Heidelberg, Germany; volker.winkler@uni-heidelberg.de

* Correspondence: philipp.jaehn@mhb-fontane.de; Tel.: +49-3381-41-1283

Abstract: Ethnic German resettlers from the former Soviet Union are one of the largest migrant groups in Germany. In comparison with the majority of the German population, resettlers exhibit worse subjective health and utilize fewer preventive measures. However, there is little evidence on health among ethnic Germans who remained in Russia. Hence, the objective of this study was to determine the differences in subjective health, diabetes, smoking, and utilization of health check-ups between ethnic Germans and the majority population in Russia. We used data from the Russian Longitudinal Monitoring Survey II from 1994 to 2018 (general population of Russia n = 41,675, ethnic Germans n = 158). Multilevel logistic regression was used to calculate odds ratios (ORs) adjusted for age, sex, period, and place of residence. Analyses were furthermore stratified by the periods 1994–2005 and 2006–2018. Ethnic Germans in Russia rated their health less often as good compared with the Russian majority population (OR = 0.67, CI = 0.48–0.92). Furthermore, ethnic Germans were more likely to smoke after 2006 (OR = 1.91, CI = 1.09–3.37). Lower subjective health among ethnic Germans in Russia is in line with findings among minority populations in Europe. Increased odds of smoking after 2006 may indicate the deteriorating risk behavior of ethnic Germans in Russia.

Keywords: resettlers; migration; subjective health; smoking; diabetes; healthcare utilization

Citation: Arena, C.; Holmberg, C.; Winkler, V.; Jaehn, P. The Health Status and Healthcare Utilization of Ethnic Germans in Russia. *Int. J. Environ. Res. Public Health* **2022**, *19*, 166. https://doi.org/10.3390/ijerph19010166

Academic Editor: Paul B. Tchounwou

Received: 16 October 2021
Accepted: 22 December 2021
Published: 24 December 2021

Publisher's Note: MDPI stays neutral with regard to jurisdictional claims in published maps and institutional affiliations.

1. Introduction

Non-communicable diseases account for a large proportion of today's illnesses and are responsible for a major burden on the healthcare system [1]. Smoking behavior, diabetes, and the low utilization of health check-ups are related to an increased prevalence of non-communicable disease [2–4]. In particular, smoking behavior and diabetes represent important risk factors for the development of cardiovascular disease [5,6], which is one of the leading causes of death worldwide [7]. All-cause mortality can be predicted by subjective health, which is frequently operationalized using information about a person's self-rated general health [8]. In recent years, there has been increasing interest in social determinants of health (SDH) as distinguished from biomedical risk factors. It has become clear that social and socioeconomic factors shape the risk of non-communicable diseases in meaningful ways [9]. Minority status, migration background and the impact of a migration process on subsequent generations represent social determinants of increasing importance. A growing body of literature indicates inequalities of both risk factors and frequency of non-communicable diseases between ethnic minorities or migrants and the respective majority populations [10–12]. Immigrants represent an example of an ethnic minority. An ethnic minority is a societal subgroup with unique social and cultural characteristics that differ from the majority population. Ethnic minorities are often faced with oppression, whether or not the group is a numerical minority [13]. The importance of a minority status for health can be observed when considering the example of people who have migrated to Germany. On the one hand, a lower prevalence of physical activity, lower self-assessed health, and lower participation in healthcare and prevention programs has been observed among migrant groups in Germany compared with the general population [14–18]. On the other hand, immigrants to Germany may have specific health resources compared with the general population. For example, a lower all-cause mortality rate was described among immigrants compared with the general population [15]. Associations, thereby, may be heterogeneous within ethnic minority populations. For example, use of the hemoccult test is lower among people born in Eastern Europe, but not among people born in Southern Europe, compared with the autochthonous German population [19]. The autochthonous population is the stationary population that has not migrated to or from the respective country in recent times.

People who have migrated to another country are exposed to countless influencing factors, both in their countries of origin and arrival (further referred to as sending and receiving countries), as well as through the migration process itself. The impact of migration is often studied in the receiving countries, although rarely in the country of origin. Completing the picture of migrant's health by also studying the population of origin corresponds to a cross-national perspective. Such a perspective yields a more comprehensive picture of health behaviors, health risks, and resources among migrant populations by contrasting their health in the receiving countries with health in the sending population [20].

Resettlers are a relevant subgroup of people with an immigrant background in Germany. The term "resettlers" (in German, Aussiedler or Spätaussiedler) refers to ethnic Germans who were granted the right to migrate to Germany from Eastern European, Northern and Central Asian countries by a unique legal framework ("Bundesvertriebenengesetz"). Since 1990, they have immigrated almost exclusively from the former Soviet Union (FSU). Their populations of origin are Germans who moved to Russia in the 18th century after a call from Tsarina Catherine II to farm underpopulated regions [21,22]. They represented a minority group in the Russian Empire and the Soviet Union, as well as in modern successor states. Most resettlers migrated to Germany between 1990 and 1995. During this time, the Bundesvertriebenengesetz defined few restrictions for immigration from the FSU to Germany [18]. To this day, a total of about 500,000 ethnic Germans still reside in Russia, representing a largely recognized minority [23].

A cross-national perspective is especially important for resettlers because they represent a minority in their sending country. The health status of resettlers, therefore, cannot simply be compared with the health of the general populations of sending countries such as Russia, because it is unclear whether the health of ethnic Germans in Russia is comparable to the national average. Studies on health and healthcare utilization are only available for resettlers in Germany and have revealed significant differences compared with the autochthonous German population. Subjective perceptions of health and physical conditions were rated worse among resettlers compared with the German population [24]. Furthermore, lower participation in early detection hemoccult screening, but higher participation in mammography screening was observed [19,24,25]. However, resettlers exhibit lower all-cause mortality and a lower mortality from cardiovascular disease compared with the general population [15]. Research on the health of ethnic Germans in Russia is limited to one study that compared cancer incidence between ethnic Germans and the general population in the district of Tomsk, Russia [26].

These considerations imply the extension of studies of health and health behavior among resettlers from the FSU in Germany to ethnic Germans remaining in Russia; Russia is the second major country of origin of resettlers after Kazakhstan [27]. In this study, the question to be answered is whether there is a difference between ethnic Germans living in Russia and the general population of Russia in terms of their health status regarding smoking, diabetes, and subjective health, as well as their utilization of health check-ups.

2. Materials and Methods

2.1. Study Design and Population

Publicly available data of the Russia Longitudinal Monitoring Survey (RLMS) I and II, from 1994 to 2018, were used. The RLMS was originally designed by the G-7 countries to acquire objective and nationally representative data on social, health, and economic conditions in Russia. The RLMS employed a three-stage stratified clustered sampling design. First, 1850 pooled Raions (administrative-territorial districts) were created in a sampling frame. They contained 95.6% of the population and were considered primary sampling units (PSUs) [28].

Of the 98 PSUs selected, 63 were located in three metropolitan regions (Moscow City, Moscow Oblast and St. Petersburg City, Russia), whereas 35 were from the rest of Russia. Although intended to be conducted annually, due to lack of financing, the years 1997 and 1999 were omitted. Interviewers visited each selected household up to three times to obtain complete data. Any group of people sharing accommodation, income and expenses was defined as a "household", including unmarried children up to 18 years old who were living outside the home for a short period of time. As many household members as possible above age 13 were interviewed about their health and activities. Information about younger household members was provided by their parents or guardians [28].

The response rate of the fifth round (first round of the second phase) was 87.6%. More than half of all households completed 10 rounds of the RLMS [28]. We based our analysis on the first observation of each individual within the entire period 1994–2018, and therefore conducted a cross-sectional study.

For our analysis, we used the first observation per individual from the panel data (n = 55,660). We excluded 13,242 observations that included legitimate missing information. The RLMS defines missing observations as legitimate if information was missing due to instructions to skip certain questions. The remaining data contained missing observations that were due to the participant's refusal to answer. The variable ethnicity contained 585 (1.1%) missing observations; these were also excluded from the analyses. The total number of included observations was 41,833.

2.2. Variables

Subjective health, health check-ups visited, tobacco smoking and diabetes were the chosen outcomes. All these variables were assessed in every year of the survey, with little or no missing observations. With the question *"How would you evaluate your health?"*, the participants were queried about their subjective health. Answers to this question were provided on a 5-point Likert scale (very good, good, average, bad, very bad). For logistic regression, the five possible answers were trichotomized in the categories good (very good or good), average, and bad (bad or very bad). The following question asked about health check-ups: *"In the last three months have you seen a doctor for a medical check up, not because you were sick?"*. Moreover, participants were asked whether they currently smoked tobacco and whether they had been diagnosed with diabetes. The last three questions could only be answered as yes or no.

Self-identified German nationality was considered the exposure and was assessed by the following open question: *"What nationality do you consider yourself? I don't necessarily have in mind the nationality in your passport."* The reported nationalities "German" and "German-Jew" were considered as German nationality. All other nationalities, including no reported nationality, were contained in the comparison group.

A conceptual framework was established with confounders and mediators. The ascertained confounders are described below. To have an even age distribution that nevertheless represented individual stages of life in a meaningful way, age groups were coded as follows: ≤19 years, 20–39 years, 40–59 years, 60–79 years, and ≥80 years. Four categories for place of residency were available from the RLMS (regional center, big cities, small town, village). For descriptive analyses, a binary classification was used in which regional centers and big cities were defined as urban and all other categories as rural. Regarding sex, the possible answers were either male or female. For descriptive analyses, the year of survey was categorized into 4 periods of equal length (1994–1999, 2000–2005, 2006–2011, and 2012–2018).

The ascertained mediators are depicted hereafter. We selected education, employment, and marital status as mediators because a minority status can be regarded as an influencing factor for these social determinants. A study in Sweden found that first- and second-generation immigrants experienced discrimination in regard to employment, describing how they were less likely to be invited for a job interview [29]. Additionally, children from immigrant families received high education degrees less frequently compared with native children [30], and it has been shown that a lower level of educational attainment is linked to a higher prevalence of cardiovascular risk factors [10]. In addition to economic aspects, social support from family, friends, or the wider community might be an important resource for minority populations to preserve their health [31]. Proxies for social support, such as living in a relationship or being married, are associated with a lower risk of adverse cardiovascular events [32–34].

In the RLMS data, six categories were available for the variable level of education. The categories were *0–6 grades of comprehensive school; unfinished secondary education 7–8 grades of school]; unfinished secondary education [7–8 grades of school] plus something else, secondary school diploma; vocational secondary education diploma;* and *higher education diploma and more.* For descriptive statistics, the level of education was divided into three levels (non-completed secondary education, completed secondary education, and higher than secondary education) [35]. The employment status was measured by asking about the participants' primary work at present. Categories of the RLMS data were *currently working; on paid leave [maternity leave or taking care of a child under 3 years of age]; on another kind of paid leave; on unpaid leave;* and *not working.* For descriptive analyses, employment was recoded into a variable with three categories (unemployed, paid or unpaid leave, and employed). *Never married; in a registered marriage; living together and not registered; divorced and not remarried; widower or widow; registered but not living together;* and *married* were the seven categories of the variable marital status. To describe frequency distributions of marital status, the seven categories were condensed into four (living alone, divorced or widowed, living together but not married, and living together and married). Smoking, diabetes, and subjective health were only considered mediators for the association of self-reported nationality with healthcare utilization. Subjective health was used as variable with the original five categories.

2.3. Statistical Methods

For descriptive analysis, numbers of observations and weighted proportions together with 95% confidence intervals (CIs) were calculated. The clustered sampling design and survey weights were considered when calculating proportions and the 95% CI. All further analyses are from multilevel logistic regressions with primary sampling units as random effects to account for the clustered sampling design. In addition, survey weights were introduced in all regression models. Table 1 shows that missing data were below 2% in all variables. Therefore, regression analyses were performed on a complete dataset of 40,915 observations. Smoking, diabetes, subjective health, and the utilization of health check-ups were dependent variables. Subjective health was analyzed as a trichotomous variable. The category 'average' was used as a reference, and two logistic regression models were fitted which modelled: (1) odds of good versus average subjective health; and (2) odds of bad versus average subjective health. We chose this strategy because our analysis software did not allow the introduction of survey weights in multinomial multilevel regression. A stepwise modelling technique was employed for all outcomes, gradually introducing confounders (model 1) and mediators (social mediators: model 2, health-related mediators: model 3) in the models. For all confounders and mediators, we used the original categorizations of the variables provided by the RLMS to maintain maximum variability. One regression coefficient for each survey year was used. Finally, we investigated effect measure modifications of the association of self-reported German nationality with the four outcomes by period of the survey (1994–2005 vs. 2006–2018). These time periods were chosen to gain a high power when assessing the change in odds ratios over time. Studying effect modification by survey year is important, because our study covered a very long period in which important changes in associations might have occurred. In association with the utilization of health check-ups, we additionally investigated effect measure modifications by place of residency, because a rural infrastructure may represent a substantial barrier to healthcare in Russia [36]. Subsequently, the odds of all defined outcomes among ethnic Germans were compared with the odds among non-Germans (referred to as "other" in all tables) as the reference group using odds ratios (ORs) and 95% CIs. P-values were calculated using Wald tests. Analyses were conducted in R, version 4.0.2, and multilevel regression models were calculated calling MLwiN, version 3.05, from within R [37–39].

Table 1. Socio-demographic and health-related characteristics of the study population.

		Other (*n* = 41,675)		German (*n* = 158)	
		n	**Prop. (95% CI)**	*n*	**Prop. (95% CI)**
Sex	Female	23,080	52.9 (52.4–53.5)	77	46.5 (39.9–53.3)
	Male	18,595	47.1 (46.5–47.6)	81	53.5 (46.7–60.1)
Age	0–19	4112	11.0 (10.0–12.0)	12	8.79 (4.4–16.6)
	20–39	18,485	46.2 (44.9–47.5)	51	34.6 (26.8–43.3)
	40–59	11,630	27.9 (26.9–28.8)	54	35.6 (27.4–44.8)
	60–79	6437	12.9 (12.1–13.8)	36	18.5 (13.1–25.6)
	80+	1005	2.0 (1.8–2.3)	5	2.5 (0.8–7.6)
	Missing	6		0	
Place of residency	Urban	31,907	77.9 (64.7–87.1)	103	66.7 (39.2–86.2)
	Rural	9768	22.1 (12.9–35.3)	55	33.3 (13.8–60.8)
Year of survey	1994–1999	12,352	30.8 (27.5–34.2)	72	49.3 (40.7–57.9)
	2000–2005	6373	13.4 (9.7–18.3)	27	15.9 (9.3–25.7)
	2006–2011	14,788	36.0 (33.4–38.7)	36	21.1 (15.5–28.2)
	2012–2018	8162	19.8 (17.8–22.0)	23	13.7 (8.5–21.4)
Level of education	Unfinished secondary	9335	21.9 (19.3–24.8)	55	35.8 (26.2–46.7)
	Completed secondary	23,420	56.6 (55.1–58.0)	85	53.2 (44.8–61.5)
	Higher than secondary	8808	21.5 (18.4–25.0)	15	11.0 (6.7–17.6)
	Missing	112		3	
Employment	Unemployed	17,892	41.7 (39.9–43.4)	75	46.0 (35.5–56.9)
	Paid or unpaid leave	1273	3.1 (2.8–3.4)	3	2.2 (0.8–5.6)
	Employed	22,465	55.3 (53.5–57.0)	80	51.8 (40.6–62.9)
	Missing	45		0	
Marital status	Living alone	7581	20.8 (19.2–22.4)	17	11.7 (6.8–19.4)
	Divorced or widowed	7381	16.5 (15.7–17.4)	36	21.8 (13.5–33.3)
	Living together, not married	5178	11.2 (10.3–12.2)	24	13.5 (9.1–19.6)
	Living together, married	21,112	51.5 (49.4–53.5)	80	53.0 (42.5–63.3)
	Missing	423		1	
Smoking	No	27,142	64.2 (62.4–66.0)	104	66.1 (57.4–73.7)
	Yes	14,498	35.8 (34.0–37.6)	54	33.9 (26.3–42.6)
	Missing	423		0	
Diabetes	No	39,779	96.1 (95.6–96.5)	149	96.4 (91.5–98.5)
	Yes	1773	3.9 (3.5–4.4)	7	3.6 (1.5–8.5)
	Missing	123		2	

Table 1. *Cont.*

		Other (*n* = 41,675)		German (*n* = 158)	
		n	**Prop. (95% CI)**	*n*	**Prop. (95% CI)**
Subjective health	Very good	1105	2.9 (2.3–3.7)	3	2.0 (0.7–5.9)
	Good	14,189	35.7 (34.1–37.4)	30	20.6 (14.2–29.0)
	Average	20,747	49.3 (47.4–51.2)	99	62.4 (53.7–70.4)
	Bad	4709	10.5 (9.9–11.0)	19	11.8 (7.1–19.0)
	Very bad	744	1.6 (1.4–1.9)	7	3.1 (1.4–6.6)
	Missing	181		0	
Health check-up	No	33,482	80.3 (79.0–81.6)	132	83.1 (74.3–89.3)
	Yes	8147	19.7 (18.4–21.0)	26	16.9 (10.7–25.7)
	Missing	46		0	

Prop: weighted proportions (in %) using survey weights of the RLMS. 95% CI: 95% confidence interval.

2.4. Ethics Statement

The ethics committee of the Medical University of Brandenburg, Brandenburg an der Havel, Germany, approved the ethics application, with the number E-01-20191119.

3. Results

A total of 41,833 observations were analyzed, 158 of whom were ethnic Germans.

There was a higher proportion of ethnic Germans in the age groups 40–59, 60–79, and over 80, than the general population of Russia (Table 1). The proportion of males among ethnic Germans was higher than among non-Germans. A further difference was found in the place of residence of the respondents: 34.8% lived in small towns or villages. Among the non-Germans, this figure was 24.9%. Moreover, non-Germans had a higher level of education than ethnic Germans. Among non-Germans, 21.5% had an education classified as higher than secondary, whereas only 11.0% of ethnic Germans had this level of education. An unfinished secondary education was found among 35.8% of ethnic Germans and 21.9% of non-Germans. Among ethnic Germans, 46.0% were unemployed, compared with 41.7% among non-Germans. Finally, 11.7% of ethnic Germans were living alone compared with 20.8% among non-Germans.

3.1. Smoking

Briefly summarized, there were no noticeable difference between ethnic Germans and non-Germans concerning their smoking behavior (Table 1). After the multilevel logistic regression analysis, the unadjusted OR was 0.88 (CI 0.64–1.21) and hardly changed after adjusting for confounders as well as for social mediators (Table 2).

3.2. Diabetes

For self-reported diabetes, there was also no difference between both groups (Table 1). The crude OR equaled 1.12 (CI 0.56–2.25) and, after adjusting for confounding, decreased to 0.94 (CI 0.41–2.00) (Table 2). Subsequently adjusting for social mediators did not change the OR.

Table 2. Crude and adjusted associations of self-reported ethnicity with smoking, diabetes, subjective health, and the utilization of health check-ups.

		MODEL 1						MODEL 2			MODEL 3		
		Crude OR	95% CI	*p*-Value	Adjusted OR	95% CI	*p*-Value	Adjusted OR	95% CI	*p*-Value	Adjusted OR	95% CI	*p*-Value
Smoking	Other	1.00 (ref.)			1.00 (ref.)			1.00 (ref.)					
	German	0.88	0.64–1.21	0.43	0.82	0.55–1.21	0.31	0.75	0.53–1.08	0.12			
Diabetes	Other	1.00 (ref.)			1.00 (ref.)			1.00 (ref.)					
	German	1.12	0.56–2.25	0.74	0.94	0.41–2.00	0.89	0.94	0.39–2.25	0.89			
Subjective Health	(1) Models for odds of good vs. average subjective health (*n* = 35,537)												
	Other	1.00 (ref.)			1.00 (ref.)			1.00 (ref.)					
	German	0.49	0.34–0.72	<0.001	0.67	0.48–0.92	0.01	0.68	0.49–0.93	0.02			
	(2) Models for odds of bad vs. average subjective health (*n* = 25,880)												
	Other	1.00 (ref.)			1.00 (ref.)			1.00 (ref.)					
	German	0.97	0.63–1.47	0.87	0.77	0.47–1.25	0.29	0.69	0.42–1.14	0.15			
Health check-up	Other	1.00 (ref.)			1.00 (ref.)			1.00 (ref.)			1.00 (ref.)		
	German	0.76	0.48–1.19	0.23	0.84	0.54–1.30	0.44	0.86	0.56–1.31	0.48	0.85	0.55–1.31	0.46

Model 1: Odds ratio adjusted for year of survey, sex, place of residency, and age (confounders). Model 2: Odds ratio adjusted for year of survey, sex, place of residency, age, education, employment, and marital status (confounders and social mediators). Model 3: Odds ratio adjusted for year of survey, sex, place of residency, age, education, employment, marital status, smoking, diabetes, and subjective health (confounders, social mediators, and health mediators). OR, odds ratio. CI, 95% confidence interval.

3.3. Subjective Health

Of the non-Germans, 2.9% reported to have very good health, and 35.7% reported to have good health (Table 1). In comparison, the figures for ethnic Germans were 2.0% for very good and 20.6% for good health. In the multilevel logistic regression analysis modeling odds for good (very good or good) versus average subjective health, the unadjusted OR for German ethnicity was 0.49 (95% CI 0.34–0.72; p-value < 0.001) (Table 2). The OR adjusted for confounders yielded 0.67 (95% CI 0.48–0.92; p-value 0.01). After additional adjustment for the social mediators (education, employment, and marital status), the OR was 0.68 (95% CI 0.49–0.93; p-value 0.02). Considering regression analyses modeling the odds for bad (very bad or bad) versus average subjective health, the unadjusted OR for German ethnicity was 0.97 (95% CI 0.63–1.47, p-value 0.87). The OR after stepwise adjustment for confounders and mediators did not change substantially.

3.4. Health Check-Ups

Ethnic Germans reportedly made less frequent use of health check-ups than non-Germans (Table 1). Following the multilevel logistic regression analysis, the unadjusted odds ratio was 0.76 (CI 0.48–1.19; p-value 0.23) (Table 2). After adjusting for confounders, the OR came to 0.84 (CI 0.54-1.30) and the p-value equaled 0.44. There were no meaningful changes in the OR following adjustment for social mediators (education, employment, and marital status) and health-related mediators (smoking, diabetes, and subjective health).

3.5. Stratified Analyses

In the years before 2006, the odds for smoking were 0.48 times less for Germans than non-Germans (Table 3). After 2006, the OR was 1.91 with a p-value <0.001 for interaction. Concerning diabetes, a p-value of 0.65 did not confirm a significant interaction of ethnicity and year of survey. Finally, the odds of good versus average subjective health among ethnic Germans were lower than in the general population of Russia in both investigated periods. When modeling the odds of good versus average subjective health, there was no evidence for an effect measure modification of the association of ethnicity by year of survey. There was also no indication of effect measure modification when modeling odds of bad versus average subjective health.

Table 3. Effect modification of the association of self-reported ethnicity with smoking, having diabetes, and subjective health by year of survey.

	Year of Survey	OR	95% CI	*p*-Value
Smoking	<2006			
	Other	1.00 (ref.)		
	German	0.48	0.28–0.82	
	≥2006			
	Other	1.00 (ref.)		
	German	1.91	1.09–3.37	<0.001
Diabetes	<2006			
	Other	1.00 (ref.)		
	German	1.18	0.44–3.20	
	≥2006			
	Other	1 (ref.)		
	German	0.73	0.15–3.61	0.65
Subjective Health	(1) Model for odds of good vs. average subjective health (n = 35,537)			
	<2006			
	Other	1.00 (ref.)		
	German	0.71	0.45–1.12	
	≥2006			
	Other	1.00 (ref.)		
	German	0.51	0.30–0.86	0.39
	(2) Model for odds of bad vs. average subjective health (n = 25,880)			
	<2006			
	Other	1.00 (ref.)		
	German	0.66	0.36–1.20	
	≥2006			
	Other	1.00 (ref.)		
	German	1.03	0.41–2.59	0.46

OR, odds ratio. p-value: p-value for interaction. CI, 95% confidence interval. All models have been adjusted for confounding: sex, place of residency, and age.

Finally, the results of stratified analyses for the outcome health check-up utilization are presented. In regard to the year of survey, both ORs were <1, and the *p*-value for interaction by year of survey was 0.95, providing no evidence for effect measure modification (Table 4). When regarding the place of residency, there was some evidence of an effect measure modification: in urban areas, there was no evidence for a difference between ethnic Germans and non-Germans participating in health check-ups; however, in rural areas, the odds for utilizing health check-ups among German were 57% lower compared with the general Russian population (Table 4).

Table 4. Effect modification of the association of self-reported ethnicity with the utilization of health check-ups by year of survey and place of residency.

	Health Check-Up	OR	95% CI	*p*-Value
Year of survey [1]	<2006			
	Other	1.00 (ref.)		
	German	0.86	0.42–1.48	
	≥2006			
	Not German	1.00 (ref.)		
	German	0.83	0.38–1.82	0.95
Place of residency [2]	Urban			
	Other	1.00 (ref.)		
	German	1.03	0.65–1.66	
	Rural			
	Other	1.00 (ref.)		
	German	0.43	0.21–0.87	0.03

OR, odds ratio. *p*-value: *p*-value for interaction. CI, 95% confidence interval. [1] Model adjusted for confounding: sex, place of residency, and age. [2] Model adjusted for confounding: sex, year of survey, and age.

4. Discussion

Thus far, there have hardly been any studies on the health of ethnic Germans in Russia. This study represents an initial attempt to gain a better understanding of this minority in comparison with the majority population and enables a cross-national perspective on the health of this unique migrant group of resettlers. Ethnic Germans in Russia, the population of origin of resettlers, were less likely to evaluate their health as good compared with the general population of Russia. There seems to be no difference between ethnic Germans in Russia and the general Russian population regarding diabetes and utilizing health check-ups. Moreover, lower odds of smoking among ethnic Germans compared with non-Germans in Russia before 2006 seems to reverse to higher odds after 2006. Additionally, in rural areas of Russia, there is some evidence that ethnic Germans might be less likely to participate in health check-ups than the general population.

Self-rated health is regarded as a meaningful predictor for morbidity, the future use of healthcare services, and mortality [8]. Social networks and further social determinants, such as educational attainment, are related to subjective health; however, these and further mediators did not explain differences in subjective health between ethnic Germans and the autochthonous general population of Russia in our study [35]. Hence, further factors that might shape health of minorities, such as socio-structural racism, need to be considered. Immigrants and minorities tend to be "othered", and thus, face more exclusion from central social domains such as meaningful work or opportunities to increase household income. Social exclusion, in turn, can lead to barriers to healthcare and health-promoting resources [40]. At the same time the "othering" and excluding as such leads to more stress, fear, and experience of prejudice and violence, which might have a negative impact on health [11]. Even after adjusting for confounders and social mediators, ethnic Germans were less likely to evaluate their health as good compared with the general population of Russia. However, marital status, level of education, and employment status are only proxies, because we could not control for all possible socioeconomic influences and all forms of social support. Thus, the information in this case is limited. Several studies have found that immigrants in Germany rate their health worse than the autochthonous population: Rommel et al. found that even after adjusting for socioeconomic factors, women with a migration background rated their health significantly worse than women with no migration background [17]. Additionally, Ronellenfitsch and Razum showed that the effect of migration had the biggest effect on health satisfaction compared with other socioeconomic factors among Eastern European migrants, mainly resettlers [41]. These results coincide with studies in which resettlers from the FSU rated their health worse than the general German population [42]. This raises the question as to whether the subjective health of resettlers might have partly originated from their sending country and what role social exclusion and structural racism might play. Overall, regarding subjective health, migrants and ethnic minorities seem to be disadvantaged compared with the majority population after controlling for confounding, which is in line with the findings among migrants and ethnic minorities in Europe [12].

The smoking habits in ethnic Germans and the general population of Russia reversed after 2006. Considering that the overall level of smoking seems to have decreased worldwide, especially since the beginning of the 2000s [43,44], this is an unexpected finding. In Russia, smoking among men has decreased since the beginning of the 21st century, whereas smoking among women has increased, especially in older groups [45]. Overall, our findings suggest that smoking prevention among ethnic Germans in Russia should be carefully evaluated. Moreover, Reiss et al. described a higher smoking prevalence amongst male resettlers in Germany compared with the male German population, which adapts with increasing durations of stay, suggesting that smoking could be an imported behavior amongst resettlers [46]. In addition, lower smoking prevalence compared with the general Russian population before 2006 is in line with the finding of rather low cardiovascular mortality among resettlers in Germany [47]. Our finding would support the hypothesis that ethnic Germans are healthier compared with the Russian population at the time of migration to Germany, and that the high cardiovascular mortality which was expected among resettlers based on the large excess mortality in Russia in the 1990s was not observed [48].

Generally, there does not seem to be a difference in the utilization of health check-ups between Germans and non-Germans. In our results, however, place of residence seemed to be important. There was some evidence that Germans in rural areas participated less often in preventative examinations than non-Germans, whereas there was no difference in urban areas. However, the number of cases in rural areas was too low to yield a more precise estimate in this subgroup. Moreover, urbanized areas continue to benefit from the medical care infrastructure of the former Soviet Union [36]. At the same time, many medical facilities and small hospitals in rural areas were closed and a considerable number of beds were eliminated, which might have contributed to the observation that people in urbanized regions utilize preventative consultations by physicians more often than people in rural areas in Russia [36]. Furthermore, irregular access to public transport, prolonged traveling time, and lack of access to a car have frequently been described as a barrier to healthcare for ethnic minorities [49]. Finally, organized measures of secondary prevention have been available in Russia since the inclusion of periodic health check-ups (dispansertizatsiya) in the compulsory health insurance system in 2013 [50]. However, especially in rural areas with worse medical care [36], accessibility of the health check-up program might have been suboptimal for ethnic Germans.

Limitations and Strengths

In this study, there was only a small sample of ethnic Germans. Thus, the power to identify differences between Germans and non-Germans was limited. Another limitation is the fact that all information provided by the participants is based on self-reported data. This leaves a possibility for systematic distortions such as social desirability bias. Moreover, we only covered a snapshot of risk factors for non-communicable diseases. It would be desirable to gain a more comprehensive picture of risk factors and disease burden in this population to detect health inequalities. Finally, more research is needed in additional countries of origin of resettlers. Russia is one of the most important countries of origin; however, data from Kazakhstan and Ukraine are needed to derive a more complete picture.

At the same time, this study offers first insights into the health of ethnic Germans in Russia, and thus provides further information about the population of origin of resettlers in Germany. The RLMS survey was designed to be nationally representative of Russia. We used survey weights to adjust for non-responses; hence, our results can be considered representative. The fact that ethnicity was self-reported is an additional strength of the study. Self-reported ethnicity might give a more comprehensive picture of how persons identify themselves rather than asking for formal citizenship. In addition, the participants were interviewed face-to-face, which is expected to yield high data quality.

5. Conclusions

This study provides the first results on the health status and healthcare utilization among ethnic Germans in Russia compared with the autochthonous population. A higher smoking prevalence between 2006 and 2018 suggests that primary prevention strategies specifically targeting smoking are important among this minority in Russia. Further studies with oversampling of ethnic Germans could aid in further identifying specific prevention needs. In addition, the lower use of health check-ups among ethnic Germans in rural regions highlights the importance of improving the accessibility of healthcare for this minority, especially in sparsely populated areas.

Finally, our study facilitates comparisons of health between resettlers in Germany and parts of their population of origin in Russia. Subjective health seems to be low, both in resettlers in Germany and among their population of origin in Russia. This finding might highlight the importance of a better social inclusion of minority populations in order to alleviate psychosocial stressors. Finally, cross-national research should be extended to additional countries of origin of resettlers such as Kazakhstan or Ukraine.

Author Contributions: Conceptualization, P.J.; methodology, C.A., V.W. and P.J.; validation, C.A. and P.J.; formal analysis, C.A. and P.J.; investigation, C.A.; resources, C.H.; writing—original draft preparation, C.A.; writing—review and editing, C.A., V.W., C.H. and P.J.; visualization, C.A.; supervision, P.J. and C.H. All authors have read and agreed to the published version of the manuscript.

Funding: The project received no funding.

Institutional Review Board Statement: The study was conducted according to the guidelines of the Declaration of Helsinki, and approved by the Ethics Committee of the Medical University of Brandenburg (protocol code E-01-20191119 and 9 December 2019).

Informed Consent Statement: Informed consent was obtained from all subjects involved in the study.

Data Availability Statement: The data presented in this study are openly available in the Russia Longitudinal Monitoring Survey Dataverse (RLMS-HSE Longitudinal Data Files) at https://doi.org/10.15139/S3/12438.

Acknowledgments: A special thanks to all participants of the survey, making the analysis possible in the first place.

Conflicts of Interest: The authors declare no conflict of interest.

References

1. Benziger, C.P.; Roth, G.A.; Moran, A.E. The Global Burden of Disease Study and the Preventable Burden of NCD. *Glob. Heart* **2016**, *11*, 393–397. [CrossRef] [PubMed]
2. Di Angelantonio, E.; Kaptoge, S.; Wormser, D.; Willeit, P.; Butterworth, A.S.; Bansal, N.; O'Keeffe, L.M.; Gao, P.; Wood, A.M.; Burgess, S.; et al. Association of Cardiometabolic Multimorbidity With Mortality. *JAMA* **2015**, *314*, 52–60. [CrossRef] [PubMed]
3. Lee, J.T.; Hamid, F.; Pati, S.; Atun, R.; Millett, C. Impact of Noncommunicable Disease Multimorbidity on Healthcare Utilisation and Out-Of-Pocket Expenditures in Middle-Income Countries: Cross Sectional Analysis. *PLoS ONE* **2015**, *10*, e0127199. [CrossRef] [PubMed]

4. Petersen, J.; Kontsevaya, A.; McKee, M.; Richardson, E.; Cook, S.; Malyutina, S.; Kudryavtsev, A.V.; Leon, D.A. Primary care use and cardiovascular disease risk in Russian 40-69 year olds: A cross-sectional study. *J. Epidemiol. Community Health* **2020**, *74*, 692–967. [CrossRef] [PubMed]

5. Neaton, J.D.; Wentworth, D. Serum cholesterol, blood pressure, cigarette smoking, and death from coronary heart disease. Overall findings and differences by age for 316,099 white men. Multiple Risk Factor Intervention Trial Research Group. *Arch. Intern. Med.* **1992**, *152*, 56–64. [CrossRef] [PubMed]

6. Balakumar, P.; Maung-U, K.; Jagadeesh, G. Prevalence and prevention of cardiovascular disease and diabetes mellitus. *Pharm. Res.* **2016**, *113*, 600–609. [CrossRef] [PubMed]

7. Abubakar, I.I.; Tillmann, T.; Banerjee, A. Global, regional, and national age-sex specific all-cause and cause-specific mortality for 240 causes of death, 1990–2013: A systematic analysis for the Global Burden of Disease Study 2013. *Lancet* **2015**, *385*, 117–171. [CrossRef] [PubMed]

8. Mackenbach, J.P.; Simon, J.G.; Looman, C.W.N.; Joung, I.M.A. Self-assessed health and mortality: Could psychosocial factors explain the association? *Int. J. Epidemiol.* **2002**, *31*, 1162–1168. [CrossRef] [PubMed]

9. Braveman, P.; Gottlieb, L. The social determinants of health: It's time to consider the causes of the causes. *Public Health Rep.* **2014**, *129*, 19–31. [CrossRef] [PubMed]

10. Hughes, J.; Kabir, Z.; Kee, F.; Bennett, K. Cardiovascular risk factors-using repeated cross-sectional surveys to assess time trends in socioeconomic inequalities in neighbouring countries. *BMJ Open* **2017**, *7*, e013442. [CrossRef] [PubMed]

11. Hossin, M.Z. International migration and health: It is time to go beyond conventional theoretical frameworks. *BMJ Glob. Health* **2020**, *5*, e001938. [CrossRef] [PubMed]

12. Nielsen, S.S.; Krasnik, A. Poorer self-perceived health among migrants and ethnic minorities versus the majority population in Europe: A systematic review. *Int. J. Public Health* **2010**, *55*, 357–371. [CrossRef] [PubMed]

13. Perkins, K.; Wiley, S. Minorities. In *Encyclopedia of Critical Psychology*; Teo, T., Ed.; Springer: New York, NY, USA, 2014; pp. 1192–1195.

14. Brand, T.; Kleer, D.; Samkange-Zeeb, F.; Zeeb, H. Prevention among migrants: Participation, migrant sensitive strategies and programme characteristics. *Bundesgesundheitsblatt Gesundh. Gesundh.* **2015**, *58*, 584–592. [CrossRef] [PubMed]

15. Spallek, J.; Razum, O. Health of migrants: Deficiencies in the field of prevention. *Med. Klin.* **2007**, *102*, 451–456. [CrossRef] [PubMed]

16. Tillmann, J.; Puth, M.-T.; Frank, L.; Weckbecker, K.; Klaschik, M.; Münster, E. Determinants of having no general practitioner in Germany and the influence of a migration background: Results of the German health interview and examination survey for adults (DEGS1). *BMC Health Serv. Res.* **2018**, *18*, 755. [CrossRef] [PubMed]

17. Rommel, A.; Saß, A.C.; Born, S.; Ellert, U. Health status of people with a migrant background and impact of socio-economic factors: First results of the German Health Interview and Examination Survey for Adults (DEGS1). *Bundesgesundheitsblatt Gesundh. Gesundh.* **2015**, *58*, 543–552. [CrossRef] [PubMed]

18. Cho, A.B.; Jaehn, P.; Holleczek, B.; Becher, H.; Winkler, V. Stage of cancer diagnoses among migrants from the former Soviet Union in comparison to the German population—are diagnoses among migrants delayed? *BMC Public Health* **2018**, *18*, 148. [CrossRef] [PubMed]

19. Wiessner, C.; Keil, T.; Krist, L.; Zeeb, H.; Dragano, N.; Schmidt, B.; Ahrens, W.; Berger, K.; Castell, S.; Fricke, J.; et al. Persons with migration background in the German National Cohort (NAKO)-sociodemographic characteristics and comparisons with the German autochthonous population. *Bundesgesundheitsblatt Gesundh. Gesundh.* **2020**, *63*, 279–289. [CrossRef] [PubMed]

20. Acevedo-Garcia, D.; Sanchez-Vaznaugh, E.V.; Viruell-Fuentes, E.A.; Almeida, J. Integrating social epidemiology into immigrant health research: A cross-national framework. *Soc. Sci. Med.* **2012**, *75*, 2060–2068. [CrossRef] [PubMed]

21. Hensen, J. *Zur Geschichte der Aussiedler- und Spätaussiedleraufnahme*; Bergner, C., Weber, M., Eds.; Aussielder- und Minderheiten-politik in Deutschland Bilanz und Perspektiven; Oldenbourg Wissenschaftsverlag: München, Germany, 2009.

22. Eisfeld, A. *Vom Stolperstein zur Brücke—Die Deutschen in Russland*; Bergner, C., Weber, M., Eds.; Aussiedler- und Aussiedler-und Minderheitenpolitik in Deutschland Bilanz und Perpektiven; Oldenbourg Wissenschaftsverlag: München, Germany, 2009; pp. 79–89.

23. Bundesministerium des Inneren, für Bau und Heimat. Deutsche Minderheiten Stellen Sich vor: Deutsche Minderheiten in Europa und Zentralasien. 2018. Available online: https://www.aussiedlerbeauftragter.de/SharedDocs/downloads/Webs/AUSB/DE/deutsche-minderheiten-stellen-sich-vor-neuauflage-2018.pdf?__blob=publicationFile&v=1 (accessed on 11 October 2021).

24. Aparicio, M.L.; Döring, A.; Mielck, A.; Holle, R. Differences between Eastern European immigrants of German origin and the rest of the German population in health status, health care use and health behaviour: A comparative study using data from the KORA-Survey 2000. *Soz. Prav.* **2005**, *50*, 107–118. [CrossRef] [PubMed]

25. Kaucher, S.; Khil, L.; Kajüter, H.; Becher, H.; Reder, M.; Kolip, P.; Spallek, J.; Winkler, V.; Berens, E.-M. Breast cancer incidence and mammography screening among resettlers in Germany. *BMC Public Health* **2020**, *20*, 417. [CrossRef] [PubMed]

26. Jaehn, P.; Kaucher, S.; Pikalova, L.V.; Mazeina, S.; Kajüter, H.; Becher, H.; Valkov, M.; Winkler, V. A cross-national perspective of migration and cancer: Incidence of five major cancer types among resettlers from the former Soviet Union in Germany and ethnic Germans in Russia. *BMC Cancer* **2019**, *19*, 869. [CrossRef] [PubMed]

27. Worbs, S.; Bund, E.; Kohls, M.; Babka von Gostomski, C. *(Spät-)Aussiedler in Deutschland. Eine Analyse Aktueller Daten und Forschungsergebnisse*; Bundesamt für Migration und Flüchtlinge: Nürnberg, Germany, 2013.

28. Kozyreva, P.; Kosolapov, M.; Popkin, B.M. Data Resource Profile: The Russia Longitudinal Monitoring Survey-Higher School of Economics (RLMS-HSE) Phase II: Monitoring the Economic and Health Situation in Russia, 1994–2013. *Int. J. Epidemiol.* **2016**, *45*, 395–401. [CrossRef] [PubMed]

29. Carlsson, M. Experimental Evidence of Discrimination in the Hiring of First- and Second-generation Immigrants. *Labour* **2010**, *24*, 263–278. [CrossRef]

30. Dustmann, C.; Glitz, A. Migration and Education. 2011. Available online: http://globalleadershipineducation.com/wp-content/uploads/2020/01/Dustmann_Glitz-Migration-and-Education.pdf (accessed on 16 September 2021).

31. Hombrados-Mendieta, I.; Millán-Franco, M.; Gómez-Jacinto, L.; Gonzalez-Castro, F.; Martos-Méndez, M.J.; García-Cid, A. Positive Influences of Social Support on Sense of Community, Life Satisfaction and the Health of Immigrants in Spain. *Front. Psychol.* **2019**, *10*, 2555. [CrossRef] [PubMed]

32. Singh, G.K.; Daus, G.P.; Allender, M.; Ramey, C.T.; Martin, E.K.; Perry, C.; De Los Reyes, I.P. Social Determinants of Health in the United States: Addressing Major Health Inequality Trends for the Nation, 1935–2016. *Int. J. MCH AIDS* **2017**, *6*, 139–164. [CrossRef] [PubMed]

33. Dhindsa, D.S.; Khambhati, J.; Schultz, W.M.; Tahhan, A.S.; Quyyumi, A.A. Marital status and outcomes in patients with cardiovascular disease. *Trends Cardiovasc. Med.* **2020**, *30*, 215–220. [CrossRef] [PubMed]

34. Chen, R.; Zhan, Y.; Pedersen, N.; Fall, K.; Valdimarsdóttir, U.A.; Hägg, S.; Fang, F. Marital status, telomere length and cardiovascular disease risk in a Swedish prospective cohort. *Heart* **2020**, *106*, 267–272. [CrossRef] [PubMed]

35. Jasilionis, D.; Shkolnikov, V.M. Longevity and Education: A Demographic Perspective. *Gerontology* **2016**, *62*, 253–262. [CrossRef] [PubMed]

36. Popovich, L.; Potapchik, E.; Shishkin, S.; Richardson, E.; Vacroux, A.; Mathivet, B. Russian Federation. Health system review. *Health Syst. Transit.* **2011**, *13*, 1–190. [PubMed]

37. Zhang, Z.; Parker, R.M.; Charlton, C.M.; Leckie, G.; Browne, W.J. R2MLwiN: A Package to Run MLwiN from within. *J. Stat. Softw.* **2016**, *10*, 1–43.

38. *MLwiN Version 3.05: Centre for Multilevel Modelling*; University of Bristol: Bristol, UK, 2020.

39. R Core Team. *R: A Language and Environment for Statistical Computing*; R Foundation for Statistical Computing: Vienna, Austria, 2020.

40. Popay, J.; Escorel, S.; Hernandez, M.; Mathieson, J.; Rispel, L. Understanding and tacklingsocial exclusion. *J. Res. Nurs.* **2010**, *15*, 295–297. [CrossRef]

41. Ronellenfitsch, U.; Razum, O. Deteriorating health satisfaction among immigrants from Eastern Europe to Germany. *Int. J. Equity Health* **2004**, *3*, 4. [CrossRef] [PubMed]

42. Stolpe, S.; Ouma, M.; Winkler, V.; Meisinger, C.; Becher, H.; Deckert, A. Self-rated health among migrants from the former Soviet Union in Germany: A cross-sectional study. *BMJ Open* **2018**, *8*, e022947. [CrossRef] [PubMed]

43. Azagba, S.; Shan, L.; Latham, K. Assessing trends and healthy migrant paradox in cigarette smoking among U.S. immigrant adults. *Prev. Med.* **2019**, *129*, 105830. [CrossRef] [PubMed]

44. Kuntz, B.; Kroll, L.E.; Hoebel, J.; Schumann, M.; Zeiher, J.; Starker, A.; Lampert, T. Time trends of occupational differences in smoking behaviour of employed men and women in Germany: Results of the 1999–2013 microcensus. *Bundesgesundheitsblatt Gesundh. Gesundh.* **2018**, *61*, 1388–1398. [CrossRef] [PubMed]

45. Shkolnikov, V.M.; Churilova, E.; Jdanov, D.A.; Shalnova, S.A.; Nilssen, O.; Kudryavtsev, A.; Cook, S.; Malyutina, S.; Mckee, M.; Leon, D.A. Time trends in smoking in Russia in the light of recent tobacco control measures: Synthesis of evidence from multiple sources. *BMC Public Health* **2020**, *20*, 378. [CrossRef] [PubMed]

46. Reiss, K.; Spallek, J.; Razum, O. 'Imported risk' or 'health transition'? Smoking prevalence among ethnic German immigrants from the Former Soviet Union by duration of stay in Germany—analysis of microcensus data. *Int. J. Equity Health* **2010**, *9*, 15. [CrossRef] [PubMed]

47. Becher, H.; Razum, O.; Kyobutungi, C.; Laki, J.; Ott, J.J.; Ronellenfitsch, U.; Winkler, V. Mortality of Immigrants from the Former Soviet Union: Results of a Cohort Study. *Dtsch. Ärzteblatt* **2007**, *104*, 1655–1661.

48. Boytsov, S.A.; Deev, A.D.; Shalnova, S.A. Mortality and risk factors for non-communicable diseases in Russia: Specific features, trends, and prognosis. *Ter. Arkh.* **2017**, *89*, 5–13. [CrossRef] [PubMed]

49. Scheppers, E.; van Dongen, E.; Dekker, J.; Geertzen, J.; Dekker, J. Potential barriers to the use of health services among ethnic minorities: A review. *Fam. Pract.* **2006**, *23*, 325–348. [CrossRef] [PubMed]

50. European Observatory on Health Systems and Policies. *Trends in Health Systems in the Former Soviet Countries*; European Observatory Health Policy Series: Copenhagen, Denmark, 2014.

International Journal of
Environmental Research and Public Health

MDPI

Article

Association of Acculturation Status with Longitudinal Changes in Health-Related Quality of Life—Results from a Cohort Study of Adults with Turkish Origin in Germany

Lilian Krist [1,*], Christina Dornquast [1,2], Thomas Reinhold [1], Heiko Becher [3], Karl-Heinz Jöckel [4], Börge Schmidt [4], Sara Schramm [4], Katja Icke [1], Ina Danquah [1,5], Stefan N. Willich [1], Thomas Keil [1,6,7] and Tilman Brand [8]

[1] Institute for Social Medicine, Epidemiology and Health Economics, Charité-Universitätsmedizin Berlin, 10117 Berlin, Germany; christina.dornquast@dzne.de (C.D.); thomas.reinhold@charite.de (T.R.); katja.icke@charite.de (K.I.); ina.danquah@uni-heidelberg.de (I.D.); stefan.willich@charite.de (S.N.W.); thomas.keil@charite.de (T.K.)

[2] German Center for Neurodegenerative Diseases (DZNE), 17489 Greifswald, Germany

[3] Institute of Medical Biometry and Epidemiology, University Medical Center Hamburg-Eppendorf, 20246 Hamburg, Germany; h.becher@uke.de

[4] Institute for Medical Informatics, Biometry und Epidemiology, University Hospital Essen, University Duisburg-Essen, 45122 Essen, Germany; k-h.joeckel@uk-essen.de (K.-H.J.); boerge.schmidt@uk-essen.de (B.S.); sara.schramm@uk-essen.de (S.S.)

[5] Institute of Global Health (HIGH), Heidelberg University Hospital, 69120 Heidelberg, Germany

[6] Institute for Clinical Epidemiology and Biometry, University of Würzburg, 97070 Würzburg, Germany

[7] State Institute of Health, Bavarian Health and Food Safety Authority, 97688 Bad Kissingen, Germany

[8] Department of Prevention and Evaluation, Leibniz Institute for Prevention Research and Epidemiology—BIPS, 28359 Bremen, Germany; brand@leibniz-bips.de

* Correspondence: lilian.krist@charite.de

Citation: Krist, L.; Dornquast, C.; Reinhold, T.; Becher, H.; Jöckel, K.-H.; Schmidt, B.; Schramm, S.; Icke, K.; Danquah, I.; Willich, S.N.; et al. Association of Acculturation Status with Longitudinal Changes in Health-Related Quality of Life—Results from a Cohort Study of Adults with Turkish Origin in Germany. *Int. J. Environ. Res. Public Health* **2021**, *18*, 2827. https://doi.org/10.3390/ijerph18062827

Academic Editor: Paul Tchounwou

Received: 3 February 2021
Accepted: 5 March 2021
Published: 10 March 2021

Publisher's Note: MDPI stays neutral with regard to jurisdictional claims in published maps and institutional affiliations.

Abstract: Health-related quality of life (HRQL) among migrant populations can be associated with acculturation (i.e., the process of adopting, acquiring and adjusting to a new cultural environment). Since there is a lack of longitudinal studies, we aimed to describe HRQL changes among adults of Turkish descent living in Berlin and Essen, Germany, and their association with acculturation. Participants of a population-based study were recruited in 2012–2013 and reinvited six years later to complete a questionnaire. Acculturation was assessed at baseline using the Frankfurt acculturation scale (integration, assimilation, separation and marginalization). HRQL was assessed at baseline (SF-8) and at follow-up (SF-12) resulting in a physical (PCS) and mental (MCS) sum score. Associations with acculturation and HRQL were analyzed with linear regression models using a time-by-acculturation status interaction term. In the study 330 persons were included (65% women, mean age ± standard deviation 43.3 ± 11.8 years). Over the 6 years, MCS decreased, while PCS remained stable. While cross-sectional analyses showed associations of acculturation status with both MCS and PCS, temporal changes including the time interaction term did not reveal associations of baseline acculturation status with HRQL. When investigating HRQL in acculturation, more longitudinal studies are needed to take changes in both HRQL and acculturation status into account.

Keywords: health-related quality of life; HRQL; acculturation; Turkish; migrants

1. Introduction

Migrants are more often in poorer health than the population in the host country [1–8]. On the one hand this is due to migration-related factors such as the country of origin, reason for immigration, traumatic experiences or genetic dispositions to certain diseases [1,2,9,10]. On the other hand, low socioeconomic status, poorer education, cultural differences, language barriers and low health literacy can cause a poor health status by reduced access to health information and health services or a lower use of health screenings [1,11–13].

51

Many studies reported also lower subjective health-related quality of life (HRQL) among migrants than among the native population [14–18]. Reasons in addition to those mentioned may include environmental and migration-related factors, experienced discrimination, socioeconomic hardship, occupational stress and poor working conditions, but also acculturative stress [15,19–24].

Acculturation is an anthropological term that was first introduced in the late 19th century in the context of colonization, then, in the 1930s, defined for the use in studies including cultural terms such as attitudes, beliefs or values in the concept of acculturation [25,26]. In the 1960s, Gordon reconceptualized the term and described acculturation as a linear continuum ranging from not acculturated to acculturated [27]. Another concept was later developed by Berry, who described acculturation as a bidimensional construct with two coexisting components (adaption of the host culture and maintenance of the culture of origin) [28]. Berry's model differentiates four acculturation strategies: (1) marginalization (low affiliation with both cultures); (2) separation (high origin-culture affiliation, low new-culture affiliation); (3) assimilation (high new-culture affiliation, low origin-culture affiliation) and (4) integration (high affiliation with both cultures) [28,29].

In the last years, a variety of studies have investigated acculturation as a predictor for physical or mental health in first and second generation migrants indicating that separation and marginalization rather predict poorer health outcomes than integration or assimilation [7,25,30,31]. A meta-analysis concluded that the most favorable acculturation strategy was integration, where migrants adapt to the host country but maintain their home country culture at the same time [21,32]. However, most studies in this field are cross-sectional and thus cannot provide information on the direction of the effect. It is as well possible that the respective acculturation strategy is rather the result of the experiences made in the host country, poor health status or low HRQL, than predicting these factors. Yoon et al. proposed that the effect of acculturation was mediated by other factors such as social connectedness and social status [22]. Additionally, the cultural distance between host country and country of origin may play a role resulting in greater difficulties of acculturation in non-Western or non-European migrants compared to Western migrants [33]. As acculturation is a complex concept, studies that examined HRQL in the context of acculturation show heterogeneous results. A cross-sectional study conducted in Greece found no association between acculturation and HRQL, but orientation to the heritage culture was negatively associated with psychological wellbeing [34]. Urzua et al. found that integration and separation were associated with more favorable quality of life in different domains in a migrant sample in Chile [35]. A cross-sectional study from Singapore showed a correlation between higher acculturation levels and higher HRQL [30]. Only two longitudinal studies investigated HRQL and acculturation. A study from the Netherlands showed that certain dimensions of acculturation were associated with higher HRQL among migrants living in the Netherlands [36]. A German study did not focus on acculturation, but showed a more pronounced decline of the mental component of the HRQL among 1st generation immigrants than among the host population, while the physical component was only associated with age [37]. Although a considerable number of studies on acculturation and HRQL among migrants have been carried out in recent years, there is still a lack of longitudinal data on this topic. In the 1960s and 1970s Germany recruited so called "guest workers" from predominantly Southern Europe and the Mediterranean region. Since then, persons with Turkish background have been the largest migrant group (currently 2.82 million) in Germany [38].

The aim of our study was therefore to investigate whether the acculturation status among persons of Turkish descent living in Germany has long-term effects on their health-related quality of life.

2. Materials and Methods

2.1. Study Sample and Design

For the present cohort study, 1236 adults of Turkish descent were recruited in two large German cities, Essen and Berlin. A detailed description of the baseline recruitment has been provided by Reiss et al. [39]. Briefly, the baseline assessment was conducted between 2012 and 2013 during the pretest phase of the German National Cohort Study (NAKO) with the aim to evaluate different recruitment strategies (register-based versus network approach) among persons with a Turkish background. For the first recruitment method, random samples from residents' registration offices were drawn and an onomastic procedure was used to identify eligible persons. For the network approach, representatives of the Turkish community were contacted to spread information of the study and to support the recruitment. All recruited participants were invited to the study center where they completed a questionnaire and underwent some medical examinations (measurement of body height and weight, blood pressure and blood sample). In 2018–2019, all participants who had agreed to be recontacted were invited to the follow-up. Participants received a self-report questionnaire via mail asking for the health status, health behavior, HRQL and others. Baseline and follow-up recruitment were conducted using bilingual written invitations, telephone contacts, and home visits performed by bilingual study staff. A description of the follow-up recruitment and retention methods was published by Krist et al. [40]. The study was approved by the ethical review committee of the Charité—Universitätsmedizin Berlin (EA1/206/17), Germany, and registered at the German Clinical Trials Register under the registration number DRKS00013545. Written informed consent was obtained from all participants.

2.2. Measures

2.2.1. Health-Related Quality of Life

For the assessment of HRQL, the Short Form Health Surveys 8 (SF-8) [41] and 12 (SF-12) [42,43] were used at baseline and at follow-up, respectively. Both instruments have been used in numerous countries and validated in different populations [44–47]. The instruments are short versions of the SF-36 yielding an eight-scale profile of different domains of physical and mental health (physical functioning, role participation with physical health problems (role–physical), bodily pain, general health, vitality, social functioning, role participation with emotional health problems (role–emotional) and mental health). For example, one question regarding mental health is "During the past 4 weeks, how much did personal or emotional problems keep you from doing your usual work, school or other daily activities?". According to the SF-12 and SF-8 manual, single items are aggregated into a physical component summary score (PCS) and a mental component summary score (MCS). Summary scores were then transformed into T-scores with a mean of 50 and a standard deviation of 10 using US general population norms obtained from the QualityMetric 2009 Norming Study [43]. Internal consistency (Cronbach's alpha) in our sample was $\alpha = 0.90$ for the SF-8 and 0.89 for the SF-12. The correlation between SF-8 at baseline and SF-12 at follow-up was $r = 0.47$ ($p > 0.001$) for PCS and $r = 0.45$ ($p > 0.001$) for MCS.

2.2.2. Acculturation

Acculturation status was assessed at baseline using the Frankfurt acculturation scale (FRACC) developed by Bongard et al. [48,49]. The scale consists of two subscales (subjectively assessed orientation towards the culture of origin (CO) and towards the host culture (HC)). Each scale includes ten items rated on a seven point Likert-like scale (0 = absolutely not, 6 = absolutely) assessing mechanisms of social integration: cultural identification, cultural practices and interethnic social networks. On each scale, the possible range was 0–60 points, with higher values indicating a stronger orientation to CO and HC, respectively. Cronbach's alpha was acceptable, with $\alpha = 0.83$ for CO and $\alpha = 0.78$ for HC. Participants were categorized as having higher or lower orientation towards CO and HC, respectively, using the median of the subscales as cut-off. By combining the two subscales, four acculturation groups were created: integration (CO+, HC+), assimilation (CO−, HC+), separation (CO+, HC−) and marginalization (CO−, HC−) following the concept of Berry [28].

2.2.3. Sociodemographic Covariates

We included sex, age, educational level (assessed at baseline) and net household income (assessed at follow-up) as sociodemographic variables (all variables were assessed via questionnaire). Age was categorized into five groups: 20–29 years, 30–39, 40–49, 50–59 and 60–69 years. Educational level was assessed as years of education, school type and country. We harmonized these data taking the Turkish schooling reform in 1997 into account [50] and then categorized them into <10 years, 10–12 years and >12 years of attained formal education in Turkey and/or Germany. Monthly net household income was categorized into <1000 Euro, 1000 to <2500 Euro and ≥2500 Euro. As additional migration-related variable, we included country of birth.

2.3. Statistical Analyses

The characteristics of the study population were analyzed descriptively by using frequencies (n) and means (±standard deviations). The association of HRQL with acculturation status was assessed in two steps. First, we analyzed the association of acculturation status (assessed at baseline) with each HRQL measurement using the cross-sectional data at baseline and at follow-up in ordinary linear regression models, respectively. In the second step, we used hierarchical linear regression models to assess change over time in HRQL. The advantage of this method is that all available data is included in the analysis while other methods such as analysis of variance with repeated measure include only subjects with complete datasets [51]. In the hierarchical linear model, baseline and follow-up observations were clustered in individuals. The relationship of acculturation status and change in HRQL was assessed by including a time (survey wave)-by-acculturation status interaction term. Since interaction terms are often difficult to interpret, we visualized the change over time in the acculturation groups by calculating predictive margins. Age, sex, education and income were included as covariates in all analyses. Assimilation was the reference category for acculturation status because it constituted the largest group.

Research on baseline data of this study indicated sex differences in HRQL (Brand et al. 2017). Therefore, we additionally applied sex-stratified models. Furthermore, since a large proportion of the participants had missing values on the FRACC scale (27%), we imputed the missing values in this variable using multiple imputations (MI, 5 imputations). MI uses various estimates to account for the uncertainty in the estimation of missing values. Compared to single imputations, MI yields wider standard errors and confidence intervals, which are supposed to be closer to the "true values" [52]. To ensure the consistency of the findings, we ran both a complete case analysis and an analysis with MI. All analyses were conducted using Stata 15 (StataCorp College Station, TX, USA).

3. Results

3.1. Characteristics of the Study Sample

Out of 1236 baseline participants, 1193 agreed to be recontacted. Of those, 330 completed the follow-up questionnaire (249/557 in Berlin and 81/636 in Essen). In Berlin, 248 persons refused actively or passively, while for 60 persons no valid address could be retrieved. In Essen, 544 persons refused actively or passively or could never be contacted, six persons died since the baseline observation.

Finally, 330 persons of Turkish origin were included in the analysis, but only 291 provided complete information for the HRQL scale at baseline and 314 at follow up (278 with complete information on both occasions, missing values were not imputed). The average age ± standard deviation was 43.3 ± 11.8 years at baseline. More women (65%) than men participated in the study and women were slightly overrepresented in the younger age groups (Table 1). About 21% of the study participants were born in Germany.

Participants with their own migration experience lived for an average of 29.4 ± 10.6 years in Germany. A substantial share of the sample had a low level of education (37.4%) and a net household income of less than 1000 Euros (20.4%). While assimilation formed the largest category of the acculturation status in the total sample, this was only the case for men but not for women. There was a sex difference in distribution of the acculturation status with more women belonging to the separated group and more men being in the integrated group. Participants' characteristics stratified for the acculturation status are presented in Supplementary Table S1.

Table 1. Characteristics of study participants.

N	Men (*n* = 118)	Women (*n* = 211)	Total (*n* = 330) [§]
Baseline variables			
Age groups (%)			
20–29	9.3	19.0	15.5
30–39	20.3	26.1	24.0
40–49	39.0	28.0	31.9
50–59	19.5	14.7	16.4
60–69	11.9	12.3	12.2
Country of birth (%)			
Turkey	79.7	70.6	73.9
Germany	13.6	24.6	20.7
Missing	6.8	4.7	5.5
Educational level (%)			
Low	30.0	17.0	37.4
Medium	29.7	35.6	26.5
High	31.2	42.4	25.1
Missing	9.1	5.1	10.9
Monthly net income (%)			
<1000 Euro	16.1	22.8	20.4
1000–<2500 Euro	44.9	41.2	42.6
2500 Euro or more	29.7	20.9	24.0
Missing	9.3	15.2	13.1
Acculturation status (%)			
Integration	20.3	11.9	14.9
Assimilation	33.1	21.3	25.5
Separation	17.8	23.7	21.6
Marginalization	11.0	10.4	10.6
Missing	17.8	32.7	27.4
PCS (mean, SD)	47.6 (9.1)	45.2 (10.2)	46.1 (9.8)
MCS (mean, SD)	49.0 (10.4)	43.7 (11.0)	45.7 (11.0)
Follow-up variables			
PCS (mean, SD)	48.4 (9.3)	43.6 (10.5)	45.3 (10.3)
MCS (mean, SD)	45.8 (9.8)	40.5 (10.5)	42.3 (10.6)

Numbers are percentages or means (SD); SD: standard deviation; PCS: physical component summary score; MCS: mental component summary score; [§] missing information on sex in one case.

3.2. Changes in HRQL

HRQL declined from baseline to follow-up in the total sample. In case of PCS, the decline was small (less than one scale point) and only observed among women. PCS sores slightly increased among men. The decline in MCS scores was larger and found for both women and men. Overall, there were large sex differences at baseline and follow-up on both HRQL scales with women consistently reporting lower scores (Table 1).

3.3. Association of Acculturation with HRQL

3.3.1. Cross-Sectional Analysis

Analyzing the association between acculturation status and HRQL showed that while acculturation status was not related with PCS at baseline, PCS scores at follow-up were significantly lower in the integrated and the separated group when compared to the assimilated group. Conversely, MCS scores were significantly lower among the separated compared to the assimilated group at baseline, but no significant differences were observable at follow-up. Furthermore, the cross-sectional analysis confirmed the sex differences in both scales and showed a strong association between HRQL and level of income (Table 2). As the complete case and the MI analysis provided similar findings, we reported only the results from the MI analysis in Table 2.

Table 2. Cross-sectional analysis of the association between acculturation status and health-related quality of life at baseline and follow-up (linear regression).

	PCS Baseline	PCS Follow-Up	MCS Baseline	MCS Follow-Up
N	287	289	287	289
Acculturation status (Ref. Assimilation)				
Integration	−1.86	−3.95 *	−2.96	−1.87
	[−5.69, 1.98]	[−7.37, −0.54]	[−6.59, 0.66]	[−5.89, 2.16]
Separation	−3.44	−3.40 *	−4.45 *	−2.87
	[−7.50, 0.61]	[−6.85, −0.04]	[−8.57, −0.34]	[−638, 0.63]
Marginalization	−0.88	−1.92	−2.90	−1.81
	[−4.65, 2.87]	[−5.27, 1.43]	[−7.98, 2.17]	[−6.62, 2.98]
Sex (Ref. Men)				
Women	−1.60	−3.89 **	−4.08 **	−3.51 **
	[−4.10, 0.89]	[−6.20, −1.57]	[−6.87, −1.30]	[−6.05, −0.97]
Monthly net income (Ref. <1000€)				
1000–<2500€	2.01	2.52	3.84 *	2.29
	[−1.10, 5.12]	[−0.42, 5.47]	[0.33, 7.36]	[0.03, 6.25]
≥2500€	4.65 **	6.31 ***	7.16 ***	6.97 ***
	[1.19, 8.12]	[3.01, 9.61]	[3.21, 11.1]	[3.30, 10.64]
Educational level (Ref. Low)				
Medium	0.06	1.30	−0.56	1.17
	[−2.89, 3.00]	[−1.47, 4.07]	[−3.84, 2.72]	[−1.93, 4.27]
High	2.52	1.97	0.74	2.51
	[−0.48, 5.48]	[−0.85, 4.78]	[−2.62, 4.11]	[−0.62, 5.65]

Adjusted for age; 95% confidence intervals in brackets; * $p < 0.05$, ** $p < 0.01$, *** $p < 0.001$; PCS: physical component summary score; MCS: mental component summary score.

3.3.2. Longitudinal Analysis

As already indicated by the descriptive analysis, the hierarchical linear model showed a significant decline in MCS over time, whereas no substantial change occurred in the average PCS scores (Table 3). The "main effect" of the acculturation status in this model refers to the differences in PCS and MCS at baseline. It showed significantly lower MCS scores in the separated group compared to the assimilated group as already indicated by the cross-sectional analysis. The coefficients of the time-by-acculturation status interaction term indicate the difference in change over time in HRQL between the assimilated group and the other three acculturation groups. As shown in Table 3, no significant interaction effect occurred. This was also the case in the sex-stratified analysis (Supplementary Table S2).

Table 3. Change over time in health-related quality of life and acculturation status (hierarchical linear regression).

N	PCS 585		MCS 585	
	Coef.	[95% CI]	Coef.	[95% CI]
Change over time	0.24	[−1.94, 2.41]	−3.86 **	[−6.21, −1.51]
Acculturation status (Ref. Assimilation)				
Integration	−1.91	[−5.65, 1.82]	−3.13	[−6.63, 0.35]
Separation	−3.02	[−6.87, 0.83]	−4.04 *	[−8.00, −0.08]
Marginalization	−1.03	[−4.71, 2.63]	−2.37	[−7.27, 2.51]
Time by acculturation (Ref. Time#Assimilation)				
Time#Integration	−2.04	[−6.68, 2.61]	1.44	[−3.24, 6.13]
Time#Separation	−0.86	[−4.16, 2.45]	1.06	[−2.41, 4.53]
Time#Marginalization	−0.93	[−5.06, 3.19]	0.27	[−4.11, 4.65]

Adjusted for age, sex, education and income; 95% confidence intervals in brackets; * $p < 0.05$, ** $p < 0.01$; PCS: physical component summary score; MCS: mental component summary score.

Figures 1 and 2 present a graphical depiction of the hierarchical linear model in Table 3. Figure 1 shows that PCS scores slightly increased among the assimilated while they decreased in the other groups. Thus, the difference in PCS scores increased, which corresponds to the negative coefficients in the interaction term (Table 3) and the significant differences in PCS at follow-up in the cross-sectional analysis (Table 2). However, the differences in MCS scores became smaller over time across the acculturation groups as can be seen in Figure 2. This corresponds with the positive coefficients in the interaction term for MCS in Table 3 and the lack of any significant difference in MCS at follow-up in the cross-sectional analysis (Table 2).

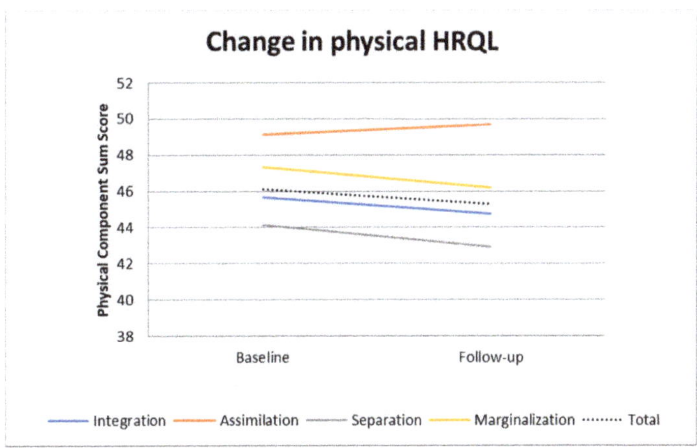

Figure 1. Change over time in physical component summary score (PCS) by acculturation status, adjusted for age, sex, education and income (predictive margins).

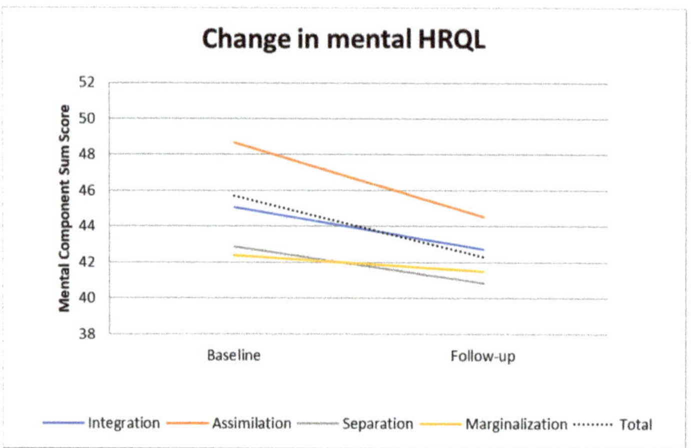

Figure 2. Change over time in mental component summary score (MCS) by acculturation status, adjusted for age, sex, education and income (predictive margins).

4. Discussion

4.1. Main Study Findings and Comparison with Other Studies

In the present cohort study among adults of Turkish descent living in Germany, HRQL was only partially associated with acculturation status. Separate cross-sectional analyses at baseline and follow-up revealed associations of acculturation with MCS only at baseline and with PCS only at follow-up. For both sum scores, the assimilated group had the best outcome, worst were separated and marginalized for PCS and MCS, respectively. When comparing our results to cross-sectional studies that used a similar definition of acculturation, it is noticeable that, in contrast, all of them found integration to be the most favorable acculturation strategy. However, only one focused on quality of life [35], while three investigated depression [7,8,32], and one anxiety [7]. Other studies focusing on quality of life but using other measures for acculturation showed that a higher acculturation score (measured with the "A Short Acculturation Scale for Filipino-Americans" (ASASFN)) [53], but also self-perceived integration [54], were associated with better quality of life. A reason for the better outcome in the assimilated group compared to the integrated group might be the higher percentage of second generation migrants among the assimilated. Although we controlled for age, education and income, we could not rule out that factors related to experiences and exposition before and shortly after migration (exposition to physically strenuous work and poor living condition) show a long-term effect here.

In our sample, MCS decreased from baseline to follow-up in all acculturation groups, while the differences between the groups remained relatively stable. Especially the marginalized and the separated had an MCS lower than the average indicating that those groups are particularly vulnerable. PCS decreased as well, except for the assimilated group that showed even an increase. The decline of MSC was, however, the steepest among the assimilated. At follow-up, both PCS and MCS were, however, still the highest in the assimilated group, followed by the marginalized, integrated and separated group. Considering the follow-up, the separated group had significantly lower MCS than the assimilated, while differences among the other groups were at a similar size, but were not statistically significant.

When taking the time as interaction term into account, our analyses revealed that no significant differences between the changes of HRQL in the acculturation groups occurred. The assimilated group with the highest baseline MCS score had even the most unfavorable trajectory compared to the other groups. A check for outliers did show a normal distribution, so there are rather other underlying reasons for that. One possible explanation could be that persons in that group were not successful with their acculturation strategy, e.g., they could have experienced discrimination, exclusion or did not have sufficiently social support, which can mitigate acculturative stress [15]. Reports of the Federal Antidiscrimination Agency confirm discrimination among persons with Turkish background [55,56]. One German study investigated time dependent changes in HRQL. Nesterko et al. showed a decline of MCS among migrants with Turkish background, and among second generation migrants compared to the German population and other migrant groups where no significant changes were observed [37]. This observation was confirmed by our study with the highest proportion of second generation migrants in the assimilated group (Supplementary Table S1). Similar to the results of Nesterko, the baseline MCS was better compared to the groups with a higher proportion of first generation migrants, but showed a stronger decrease over time. This might be explained with the difficult situation of this group torn between the culture of their parents (so that they are "stamped" as migrants by many Germans) and the German culture, their country of birth, which might create a distance to their Turkish community. This already challenging situation is often aggravated by discriminative experiences in the political, social or educational environment [57].

4.2. Strengths and Limitations

This study is to our knowledge the first longitudinal study examining trajectories of HRQL and their association with acculturation among a migrant sample in Germany. Second, the baseline recruitment was conducted very thoroughly and yielded a representative population-based sample of persons of Turkish descent covering different regions and a broad age range [39,58].

Some limitations have to be mentioned as well. Although great effort has been made to recruit participants for follow-up participation, the overall retention rate was rather low, which may have led to a selection bias. Second, our sample consists of first and second generation migrants, who were analyzed together, because the percentage of second generation migrants was too small (20.7%). Thus, the results may be valid rather for first than for second generation migrants. Lastly, acculturation status was assessed only at baseline so that changes over time could not be measured. However, first generation migrants of our sample lived on average for already almost 30 years in Germany. Therefore, the dynamic of acculturation might be negligible at this point, since significant changes have been reported only for the first years after migration [59,60].

4.3. Implications

The finding that acculturation patterns HRQL is also of importance for public health professionals, policy makers and researchers. Providing access to appropriate health care and preventative services on the one hand, but also inclusion in the host society as a whole is a task of the German state. Given the low HRQL among the separated and marginalized migrants, public health interventions should target these groups in order to improve their health. Peer to peer approaches such as the health initiative "With migrants for migrants (MiMi)" [61] are a promising way to improve reach and access to these groups. In addition, awareness by physicians and health professionals of the concept of acculturation should be enhanced in order to improve HRQL among migrants.

Although acculturation has been shown to be associated with HRQL in some way, more longitudinal studies should be conducted to assess trajectories of both acculturation and HRQL and their associations. Future research should also explore factors predicting different acculturation strategies in longitudinal studies in order to identify high risk profiles of migrants at an early stage. These risk profiles could then be used in the frame of screening programs or other health initiatives.

5. Conclusions

In a sample of adults of Turkish descent living in Germany, HRQL was partially associated with acculturation. While physical HRQL remained relatively stable, mental

HRQL decreased significantly over six years in all four acculturation groups. Persons who adapted towards the host culture (assimilated) had the best HRQL at baseline and follow-up. There was no association of acculturation with HRQL changes over time.

Further research should include longitudinal studies with larger study samples and assess acculturation status at each follow-up to take its dynamic nature into account. This may be especially useful among recently arrived migrants.

Supplementary Materials: The following are available online at https://www.mdpi.com/1660-460 1/18/6/2827/s1, Table S1: Participants' characteristics by acculturation status. Table S2: Change over time in health-related quality of life and acculturation status by sex (hierarchical linear regression).

Author Contributions: Conceptualization, L.K. and T.B.; methodology, L.K., T.B.; validation, H.B., T.R., I.D., and T.K.; formal analysis, T.B.; investigation, L.K.; resources, L.K., C.D., and K.I.; data curation, K.I., C.D., and S.S.; writing—original draft preparation, L.K. and T.B.; writing—review and editing, C.D., T.R., H.B., K.I., I.D., T.K., B.S., S.S., K.-H.J., and S.N.W.; visualization, L.K., and T.B.; supervision, T.K.; project administration, L.K., T.B., and T.K.; funding acquisition, T.R. All authors have read and agreed to the published version of the manuscript.

Funding: This project was funded by the Federal Ministry of Education and Research, Germany [grant number 01EH1604B].

Institutional Review Board Statement: The study was conducted according to the guidelines of the Declaration of Helsinki, and approved by the ethical review committee of the Charité— Universitätsmedizin Berlin, Germany (EA1/206/17), and registered at the German Clinical Trials Register under the registration number DRKS00013545.

Informed Consent Statement: Informed consent was obtained from all subjects involved in the study.

Data Availability Statement: The data presented in this study are available on request from the corresponding author.

Acknowledgments: The authors thank Saliha Solak, and Melike Durak for the recruitment of the study participants.

Conflicts of Interest: The authors declare no conflict of interest.

References

1. WHO. *Report on the Health of Refugees and Migrants in the WHO European Region*; WHO: Geneva, Switzerland, 2018.
2. Ujcic-Voortman, J.K.; Baan, C.A.; Seidell, J.C.; Verhoeff, A.P. Obesity and cardiovascular disease risk among Turkish and Moroccan migrant groups in Europe: A systematic review. *Obes. Rev.* **2012**, *13*, 2–16. [CrossRef] [PubMed]
3. Reeske, A.; Spallek, J.; Razum, O. Changes in smoking prevalence among first- and second-generation Turkish migrants in Germany - an analysis of the 2005 Microcensus. *Int. J. Equity Health* **2009**, *8*, 26. [CrossRef]
4. Hjörleifsdottir-Steiner, K.; Satman, I.; Sundquist, J.; Kaya, A.; Wändell, P. Diabetes and impaired glucose tolerance among Turkish immigrants in Sweden. *Diabetes Res. Clin. Pract.* **2011**, *92*, 118–123. [CrossRef] [PubMed]
5. Beutel, M.E.; Jünger, C.; Klein, E.M.; Wild, P.; Lackner, K.J.; Blettner, M.; Banerjee, M.; Michal, M.; Wiltink, J.; Brähler, E. Depression, anxiety and suicidal ideation among 1st and 2nd generation migrants-results from the Gutenberg health study. *BMC Psychiatry* **2016**, *16*, 1–10. [CrossRef]
6. Levecque, K.; Lodewyckx, I.; Vranken, J. Depression and generalised anxiety in the general population in Belgium: A comparison between native and immigrant groups. *J. Affect. Disord.* **2007**, *97*, 229–239. [CrossRef] [PubMed]
7. Ünlü Ince, B.; Fassaert, T.; de Wit, M.A.S.; Cuijpers, P.; Smit, J.; Ruwaard, J.; Riper, H. The relationship between acculturation strategies and depressive and anxiety disorders in Turkish migrants in the Netherlands. *BMC Psychiatry* **2014**, *14*, 1–11. [CrossRef] [PubMed]
8. Morawa, E.; Erim, Y. Acculturation and depressive symptoms among Turkish immigrants in Germany. *Int. J. Environ. Res. Public Health* **2014**, *11*, 9503–9521. [CrossRef]
9. International Organization for Migration. *World Migration Report 2020*; International Organization for Migration: Grand-Saconnex, Switzerland, 2020.
10. Solé-Auró, A.; Crimmins, E.M. Health of immigrants in European countries. *Int. Migr. Rev.* **2008**, *42*, 861–876. [CrossRef]
11. Walter, U.; Bisson, S.; Gerken, U.; Machleidt, W.; Krauth, C. Gesundheits- und Präventionsverhalten von Personen mit und ohne Migrationshintergrund. *Gesundheitswesen Suppl.* **2015**, *77*, S85–S86. [CrossRef]
12. Gazard, B.; Frissa, S.; Nellums, L.; Hotopf, M.; Hatch, S.L. Challenges in researching migration status, health and health service use: An intersectional analysis of a South London community. *Ethn. Health* **2015**, *20*, 564–593. [CrossRef]

13. Kristiansen, M.; Razum, O.; Tezcan-Güntekin, H.; Krasnik, A. Aging and health among migrants in a European perspective. *Public Health Rev.* **2016**, *37*, 1–14. [CrossRef]
14. Sand, G.; Gruber, S. Differences in Subjective Well-being Between Older Migrants and Natives in Europe. *J. Immigr. Minor. Health* **2018**, *20*, 83–90. [CrossRef] [PubMed]
15. Salgado, H.; Castañeda, S.F.; Talavera, G.A.; Lindsay, S.P. The role of social support and acculturative stress in health-related quality of life among day laborers in Northern San Diego. *J. Immigr. Minor. Health* **2012**, *14*, 379–385. [CrossRef] [PubMed]
16. Toselli, S.; Gualdi-Russo, E.; Marzouk, D.; Sundquist, J.; Sundquist, K. Psychosocial health among immigrants in central and southern Europe. *Eur. J. Public Health* **2014**, *24*, 26–30. [CrossRef]
17. Pantzer, K.; Rajmil, L.; Tebé, C.; Codina, F.; Serra-Sutton, V.; Ferrer, M.; Ravens-Sieberer, U.; Simeoni, M.C.; Alonso, J. Health related quality of life in immigrants and native school aged adolescents in Spain. *J. Epidemiol. Community Health* **2006**, *60*, 694–698. [CrossRef] [PubMed]
18. Brand, T.; Samkange-Zeeb, F.; Ellert, U.; Keil, T.; Krist, L.; Dragano, N.; Jöckel, K.-H.; Razum, O.; Reiss, K.; Greiser, K.H.; et al. Acculturation and health-related quality of life: Results from the German National Cohort migrant feasibility study. *Int. J. Public Health* **2017**, *62*, 521–529. [CrossRef] [PubMed]
19. Suzuki, P.T. Psychological problems of Turkish migrants in West Germany. *Am. J. Psychother.* **1981**, *35*, 187–194. [CrossRef]
20. Aichberger, M.C.; Bromand, Z.; Heredia Montesinos, A.; Temur-Erman, S.; Mundt, A.; Heinz, A.; Rapp, M.A.; Schouler-Ocak, M. Socio-economic status and emotional distress of female Turkish immigrants and native German women living in Berlin. *Eur. Psychiatry* **2012**, *27*, S10–S16. [CrossRef]
21. Yoon, E.; Chang, C.T.; Kim, S.; Clawson, A.; Cleary, S.E.; Hansen, M.; Bruner, J.P.; Chan, T.K.; Gomes, A.M. A meta-analysis of acculturation/enculturation and mental health. *J. Couns. Psychol.* **2013**, *60*, 15–30. [CrossRef]
22. Yoon, E.; Hacker, J.; Hewitt, A.; Abrams, M.; Cleary, S. Social connectedness, discrimination, and social status as mediators of acculturation/enculturation and well-being. *J. Couns. Psychol.* **2012**, *59*, 86–96. [CrossRef]
23. Verhagen, I.; Ros, W.J.G.; Steunenberg, B.; de Wit, N.J. Ethnicity does not account for differences in the health-related quality of life of Turkish, Moroccan, and Moluccan elderly in the Netherlands. *Health Qual. Life Outcomes* **2014**, *12*, 1–8. [CrossRef]
24. Cho, S.; Lee, H.; Oh, E.G.; Kim, G.S.; Kim, Y.C.; Park, C. gi Health-related quality of life among migrant workers: The impact of health-promoting behaviors. *Nurs. Health Sci.* **2020**, *22*, 318–327. [CrossRef]
25. Schumann, M.; Bug, M.; Kajikhina, K.; Koschollek, C.; Bartig, S.; Lampert, T.; Santos-Hövener, C. SSM-Population Health The concept of acculturation in epidemiological research among migrant populations: A systematic review. *SSM Popul. Health* **2020**, *10*, 1–10. [CrossRef]
26. Rudmin, F.W. Critical History of the Acculturation Psychology of Assimilation, Separation, Integration, and Marginalization. *Rev. Gen. Psychol.* **2003**, *7*, 3–37. [CrossRef]
27. Gordon, M.M. *Assimilation in Americal Life*; 1964; ISBN 9788578110796. Available online: https://research.vu.nl/en/publications/the-professional-teacher-educator-roles-behaviour-and-professiona (accessed on 31 December 2014).
28. Berry, J.W. Immigration, Acculturation, and Adaptation. *Appl. Psychol.* **1997**, *46*, 5–34. [CrossRef]
29. Berry, J.W. *Acculturation: A Personal Journey across Cultures*; Cambridge University Press: Cambridge, UK, 2019; ISBN 9781108589666.
30. Goh, Y.S.; Lopez, V. Acculturation, quality of life and work environment of international nurses in a multi-cultural society: A cross-sectional, correlational study. *Appl. Nurs. Res.* **2016**, *30*, 111–118. [CrossRef] [PubMed]
31. Berry, J.W.; Sabatier, C. Acculturation, discrimination, and adaptation among second generation immigrant youth in Montreal and Paris. *Int. J. Intercult. Relations* **2010**, *34*, 191–207. [CrossRef]
32. Behrens, K.; Del Pozo, M.A.; Großhennig, A.; Sieberer, M.; Graef-Calliess, I.T. How much orientation towards the host culture is healthy? Acculturation style as risk enhancement for depressive symptoms in immigrants. *Int. J. Soc. Psychiatry* **2015**, *61*, 498–505. [CrossRef] [PubMed]
33. Borraccino, A.; Charrier, L.; Berchialla, P.; Lazzeri, G.; Vieno, A.; Dalmasso, P.; Lemma, P. Perceived well-being in adolescent immigrants: It matters where they come from. *Int. J. Public Health* **2018**, *63*, 1037–1045. [CrossRef]
34. Prapas, C.; Mavreas, V. The Relationship Between Quality of Life, Psychological Wellbeing, Satisfaction with Life and Acculturation of Immigrants in Greece. *Cult. Med. Psychiatry* **2019**, *43*, 77–92. [CrossRef]
35. Urzúa, A.; Ferrer, R.; Canales Gaete, V.; Núñez Aragón, D.; Ravanal Labraña, I.; Tabilo Poblete, B. The influence of acculturation strategies in quality of life by immigrants in Northern Chile. *Qual. Life Res.* **2017**, *26*, 717–726. [CrossRef]
36. Nap, A.; Van Loon, A.; Peen, J.; Van Schaik, D.J.F.; Beekman, A.T.F.; Dekker, J.J.M. The influence of acculturation on mental health and specialized mental healthcare for non-western migrants. *Int. J. Soc. Psychiatry* **2015**, *61*, 530–538. [CrossRef]
37. Nesterko, Y.; Turrión, C.M.; Friedrich, M.; Glaesmer, H. Trajectories of health-related quality of life in immigrants and non-immigrants in Germany: A population-based longitudinal study. *Int. J. Public Health* **2019**, *64*, 49–58. [CrossRef]
38. Statistisches Bundesamt (Destatis). *Bevölkerung und Erwerbstätigkeit: Bevölkerung mit Migrationshintergrund–Ergebnisse des Mikrozensus 2019*; Statistisches Bundesamt (Destatis): Wiesbaden, Germany, 2020.
39. Reiss, K.; Dragano, N.; Ellert, U.; Fricke, J.; Greiser, K.H.; Keil, T.; Krist, L.; Moebus, S.; Pundt, N.; Schlaud, M.; et al. Comparing sampling strategies to recruit migrants for an epidemiological study. Results from a German feasibility study. *Eur. J. Public Health* **2014**, *24*, 721–726. [CrossRef]

40. Krist, L.; Dornquast, C.; Reinhold, T.; Solak, S.; Durak, M.; Keil, T. Strategies to enhance follow-up response in a cohort study with Berliners of Turkish descent. *Eur. J. Public Health* **2019**, *29*, 400–401. [CrossRef]
41. Ware, J.; Kosinski, M.; Dewey, J.; Gandek, B. *How to Score and Interpret Single-Item Health Status Measures: A Manual for Users of the SF-8TM Health Survey.–ScienceOpen*; QualyMetric: Lincoln, RI, USA, 2001.
42. Ware, J.; Kosinski, M.; Keller, S.D. A 12-Item Short-Form Health Survey: Construction of scales and preliminary tests of reliability and validity. *Med. Care* **1996**, *34*, 220–233. [CrossRef]
43. Maruish, M.E. (Ed.) *User's Manual for the SF-12v2 Health Survey*, 3rd ed.; ScienceOpen: Berlin, Germany, 2012; ISBN 1891810294.
44. Ellert, U.; Lampert, T.; Ravens-Sieberer, U. Messung der gesundheitsbezogenen Lebensqualität mit dem SF-8. Eine Normstichprobe für Deutschland. *Bundesgesundheitsblatt-Gesundheitsforsch.-Gesundheitsschutz* **2005**, *48*, 1330–1337. [CrossRef] [PubMed]
45. Lang, L.; Zhang, L.; Zhang, P.; Li, Q.; Bian, J.; Guo, Y. Evaluating the reliability and validity of SF-8 with a large representative sample of urban Chinese. *Health Qual. Life Outcomes* **2018**, *16*, 1–8. [CrossRef] [PubMed]
46. Schofield, M.J.; Mishra, G. Validity of the SF-12 compared with the SF-36 Health Survey in pilot studies of the Australian Longitudinal Study on Women's Health. *J. Health Psychol.* **1998**, *3*, 259–271. [CrossRef] [PubMed]
47. Gandek, B.; Ware, J.E.; Aaronson, N.K.; Apolone, G.; Bjorner, J.B.; Brazier, J.E.; Bullinger, M.; Kaasa, S.; Leplege, A.; Prieto, L.; et al. Cross-validation of item selection and scoring for the SF-12 Health Survey in nine countries: Results from the IQOLA Project. *J. Clin. Epidemiol.* **1998**, *51*, 1171–1178. [CrossRef]
48. Bongard, S.; Pogge, S.F.; Arslaner, H.; Rohrmann, S.; Hodapp, V. Acculturation and cardiovascular reactivity of second-generation Turkish migrants in Germany. *J. Psychosom. Res.* **2002**, *53*, 795–803. [CrossRef]
49. Bongard, S.; Etzler, S.F.E. *FRACC: Frankfurt Acculturation Scale*; Hogrefe: Göttingen, Germany, 2020.
50. Aksit, N. Educational reform in Turkey. *Int. J. Educ. Dev.* **2007**, *27*, 129–137. [CrossRef]
51. Twisk, J.W.R. *Applied Longitudinal Data Analysis for Epidemiology: A Practical Guide*, 2nd ed.; Cambridge University Press: Camebridge, UK, 2013.
52. Donders, A.R.T.; van der Heijden, G.J.M.G.; Stijnen, T.; Moons, K.G.M. Review: A gentle introduction to imputation of missing values. *J. Clin. Epidemiol.* **2006**, *59*, 1087–1091. [CrossRef] [PubMed]
53. de la Cruz, F.A.; Padilia, G.V.; Butts, E. Validating a Short Acculturation Scale For Filipino-Americans. *Search Res.* **1998**, *10*, 453–460.
54. Adedeji, A.; Bullinger, M. Subjective integration and quality of life of Sub-Saharan African migrants in Germany. *Public Health* **2019**, *174*, 134–144. [CrossRef] [PubMed]
55. Antidiskriminierungsstelle des Bundes. *Diskriminierungserfahrungen von Migrant_innen*; Antidiskriminierungsstelle des Bundes: Berlin, Germany, 2012.
56. Antidiskriminierungsstelle des Bundes. *Akzeptanz Religiöser und Weltanschaulicher Vielfalt in Deutschland Ergebnisse einer Repräsentativen Umfrage*; Antidiskriminierungsstelle des Bundes: Berlin, Germany, 2016.
57. Decker, O.; Kiess, J.; Brähler, E. *Die Enthemmte Mitte. Autoritäre und Rechtsextreme Einstellung in Deutschland*; Psychosozial-Verlag: Gießen, Germany, 2016; ISBN 9783837926309.
58. Brand, T.; Samkange-Zeeb, F.; Dragano, N.; Keil, T.; Krist, L.; Yesil-Jürgens, R.; Schlaud, M.; Jöckel, K.H.; Razum, O.; Reiss, K.; et al. Participation of Turkish Migrants in an Epidemiological Study: Does the Recruitment Strategy Affect the Sample Characteristics? *J. Immigr. Minor. Health* **2019**, *21*, 811–819. [CrossRef] [PubMed]
59. Murray, K.E.; Klonoff, E.A.; Garcini, L.M.; Ullman, J.B.; Wall, T.L.; Myers, M.G. Assessing Acculturation Over Time: A Four-year Prospective Study of Asian American Young Adults. *Asian Am. J. Psychol* **2014**, *5*, 252–261. [CrossRef] [PubMed]
60. Miller, A.M.; Wang, E.; Szalacha, L.A.; Sorokin, O. Longitudinal Changes in Acculturation for Immigrant Women from the Former Soviet Union. *J Cross Cult Psychol.* **2009**, *40*, 400–415. [CrossRef]
61. Salman, R. Gesundheit mit Migranten für Migranten - Die MiMi Präventionstechnologie als interkulturelles Health-Literacy-Programm. *Public Health Forum* **2015**, *23*, 109–112. [CrossRef]

International Journal of
Environmental Research and Public Health

Article

Effects of Sociodemographic Variables and Depressive Symptoms on MoCA Test Performance in Native Germans and Turkish Migrants in Germany

Görkem Anapa [1], Mandy Roheger [2], Ümran Sema Seven [1], Hannah Liebermann-Jordanidis [1], Oezguer A. Onur [3], Josef Kessler [3] and Elke Kalbe [1,*]

[1] Neuropsychology and Gender Studies & Center for Neuropsychological Diagnostics and Intervention (CeNDI), Department of Medical Psychology, Faculty of Medicine and University Hospital of Cologne, University of Cologne, 50937 Cologne, Germany; goerkem.anapa@gmail.com (G.A.); uemran.seven@uk-koeln.de (Ü.S.S.); hannah.liebermann-jordanidis1@uk-koeln.de (H.L.-J.)

[2] Department of Neurology, University Medicine Greifswald, 17489 Greifswald, Germany; mandy.roheger@med.uni-greifswald.de

[3] Department of Neurology, Faculty of Medicine and University Hospital of Cologne, University of Cologne, 50937 Cologne, Germany; oezguer.onur@uk-koeln.de (O.A.O.); josef.kessler@uk-koeln.de (J.K.)

* Correspondence: elke.kalbe@uk-koeln.de; Tel.: +49-221-478-6669; Fax: +49-221-478-3420

Abstract: The validity of the Montreal Cognitive Assessment (MoCA) in migrants is questionable, as sociodemographic factors and the migration process may influence performance. Our aim was to evaluate possible predictors (age, education, sex, depression, and migration) of MoCA results in Turkish migrants and Germans living in Germany. Linear regression models were conducted with a German ($n = 419$), a Turkish ($n = 133$), and an overall sample. All predictor analyses reached statistical significance. For the German sample, age, sex, education, and depression were significant predictors, whereas education was the only predictor for Turkish migrants. For the overall sample, having no migration background and higher education were significant predictors. Migration background and education had an impact on MoCA performance in a sample of German and Turkish individuals living in Germany. Thus, culture-specific normative data for the MoCA are needed, and the development of culture-sensitive cognitive screening tools is encouraged.

Keywords: Montreal Cognitive Assessment (MoCA); cognition; Turkish migrants; predictors

Citation: Anapa, G.; Roheger, M.; Seven, Ü.S.; Liebermann-Jordanidis, H.; Onur, O.A.; Kessler, J.; Kalbe, E. Effects of Sociodemographic Variables and Depressive Symptoms on MoCA Test Performance in Native Germans and Turkish Migrants in Germany. *Int. J. Environ. Res. Public Health* **2021**, *18*, 6335. https://doi.org/10.3390/ijerph18126335

Academic Editors: Heiko Becher, Volker Winkler and Hajo Zeeb

Received: 4 April 2021
Accepted: 7 June 2021
Published: 11 June 2021

Publisher's Note: MDPI stays neutral with regard to jurisdictional claims in published maps and institutional affiliations.

1. Introduction

Migration is an increasing phenomenon worldwide [1], and in Germany, Turkish migrants represent the largest population group with a migration background [2]. There is an increasing number of older people being diagnosed with dementia in general [3,4], and a high number of people with migration background living in Germany with dementia [5]. Therefore, valid screening of cognitive impairment and dementia in this population group is important and will become even more relevant with regard to the demographic change [6,7]. Migration is a critical life event and causes changes concerning different life areas (e.g., socioeconomic status, social chances and risks, and health chances and risks), which may have an influence on health and cognition [8,9]. For several countries, studies have shown that general health status and cognitive test performance can be lower in individuals with a migration background than in those without [8,9]. For example, in Germany, language barriers and low educational level of immigrated persons are known to negatively influence access to the German health care system [10]. Moreover, recent studies show correlations also between level of acculturation and health-related quality of life [11,12]. However, cognitive screening tools that provide normative values for migrants are lacking.

One of the most frequently used cognitive screening tools in both Germany and worldwide is the Montreal Cognitive Assessment (MoCA; [13]). Yet, limitations exist concerning the assessment and results of the MoCA in general and for migrants in particular, which affect its interpretation and decrease its diagnostic accuracy. For example, test scores are corrected for education, but not for age and sex, although all of these variables have been demonstrated to have an impact on test performance [14–17]. Moreover, depressive symptoms may influence cognitive performance and, in fact, have been shown to decrease MoCA performance [18,19]. Concerning Turkish people living in Germany (and migrants more generally), it is possible that factors associated with the process of migrating also influence MoCA outcomes. Notably, migration is known to be a stressful event with manifold challenges, and previous research indicates that stress is a possible etiological factor for cognitive impairment and Alzheimer's disease [20–22].

In terms of language barriers, a Turkish version of the MoCA already exists. However, it is questionable whether it is an appropriate and culture-sensitive tool to assess the cognitive status of Turkish migrants living in Germany, as it was validated in a Turkish population living in Turkey and did not consider the possible bias of specific characteristics of this population, including the migration processes and its consequences [23].

Taken together, there is a need to investigate possible influencing factors on MoCA test performance in individuals with a migration background, particularly sociodemographic variables (e.g., age, sex, and education), depression, and the migration process. To address this issue, we investigated MoCA test performance in native Germans and Turkish migrants aged 50 years or older living in Germany by testing them in their mother tongue (i.e., with the German and Turkish version of the screening tool). Next, the sociodemographic variables of age, education, and sex, depressive symptoms, and migration background were analysed as potential predictors of performance. Our hypotheses were that (1) age, sex, education, and depression are predictors for cognition measured by the MoCA in both the German and the Turkish samples. Furthermore, we expect that "country of origin" is also a significant predictor for MoCA performance (2).

2. Materials and Methods

The present cross-sectional study was conducted with healthy native German (n = 419) and Turkish people living in North Rhine–Westphalia, Germany (n = 133) as part of the TRAKULA project, which aimed to develop a culture-sensitive nonverbal test battery. Recruitment took place in 2016 and 2017 after the study was approved by the ethical committee of the University Hospital Cologne (16-249). The sample can be described as old-aged according to the World Health Organization (WHO), since the majority of the subjects were over 55 years of age [24]. Participants were recruited at different institutions in North Rhine–Westphalia. During recruitment, attention was paid to a balanced and diverse selection of contact points for people of Turkish origin. Individuals were recruited mainly in Cologne and the surrounding area, where many people with a Turkish migration background live, for example, in cultural institutions, mosques, or places where the Turkish community is commonly found. Furthermore, advertisements were placed in doctors' offices and on our clinic website. Written informed consent was obtained from each participant. All participants completed questionnaires assessing sociodemographic details. Testing and recruiting were conducted in the participants' mother tongue by German–Turkish bilingual and bicultural psychologists (authors G.A. and Ü.S.S.) and trained bilingual students of medicine or psychology supervised by authors G.A. and Ü.S.S.

2.1. Inclusion and Exclusion Criteria

For all participants, inclusion criteria were a residency in Germany, being native German or being native Turkish and having a migration background, aged 50 years or older, normal, only slightly restricted, or corrected-to-normal vision and hearing, and the provision of written consent to participate in the study. Exclusion criteria were self-reported cognitive impairment or other past or current diagnosed neurological or psychiatric illnesses and cognitive disorders.

2.2. Assessment of Cognition and Depression

The MoCA [13] was administered using either the German or Turkish version [23,25]. Both versions have a maximum of 30 points and differ only in two subtests: memory (different words) and word fluency (letter "K" instead of "F"). The cultural adjustment that was made concerns the word "church" in the memory subtest, which was replaced by the word "mosque". The awarding of points in the individual subtests corresponds to the original version of the MoCA. Participants are classified as cognitively impaired when reaching less than 26 points in the German [25] and less than 21 points in the Turkish version [23]. The German version provides an education adjustment of one point for people who received less than 12 years of school education, whereas the Turkish version does not use an education adjustment, as proposed by Selekler et al. (2010) [23]. The cognitive domains tested in the MoCA are visuospatial/executive functions, naming, verbal short-term memory, attention, language, abstraction, and verbal long-term memory [13].

Furthermore, the participants received the German or Turkish version of the Geriatric Depression Scale Short Form (GDS-15; [26]) to assess depressive symptoms. The instrument consists of 15 yes/no items. Each answer counts as one point, and a maximum score of more than 5 points indicates possible, 6 to 10 points moderate, and 11 to 15 points severe depressive symptoms [26].

2.3. Statistics

Statistical analyses were performed using IBM SPSS Statistics 25 for Windows (IBM Corp, Armonk, NY, USA). Normal distribution was tested using the Kolmogorov–Smirnov test. Possible differences in demographic data (age, sex, and education) and depressive symptoms between the two groups (German and Turkish participants) and within groups were analysed using t-tests, or chi-square tests, where appropriate, each with a significance level of $\alpha = 0.05$. G*Power (http://www.gpower.hhu.de, accessed on June 2018) was used to estimate the achieved power with a post hoc analysis [27].

To analyse predictors of MoCA performance, linear regressions were conducted (one for the German and one for the Turkish sample), with the total MoCA score as a dependent variable and simultaneous entry of the predictors of age, sex, education, and depression.

To further analyse migration as a possible predictor of MoCA performance, an additional regression analysis was calculated for the combined German and Turkish sample, including age, sex, education, depression, and migration as possible predictors. Due to different sample sizes, the German and Turkish samples were matched. For this purpose, the first 133 German participants (mean age: 60.71, SD: 8.71; mean years of education: 11.99, SD: 1.90) were included in the total sample size and ordered according to sex to ensure a uniform distribution of women and men participants in the overall regression analysis. The education correction was not included in total MoCA scores used in the regression. The assumptions of multiple regression models were checked according to the suggestions of Field [28].

3. Results

3.1. Sample Characteristics

Demographic characteristics and results of the MoCA and GDS are shown in Table 1. Participants in the Turkish sample were significantly younger, less educated, showed significantly lower scores in the MoCA, and were significantly more depressed than participants in the German sample. German women were significantly less educated and more depressed than German men, whereas Turkish women were significantly younger, less educated, and showed lower performance in the MoCA than Turkish men did.

Table 1. Characteristics of the German and Turkish samples.

Sex	German Sample (*n* = 419)							Turkish Sample (*n* = 133)							*p*-Value
	♂ = 169 (40.33%)		♀ = 250 (59.67%)					♂ = 67 (50.38%)		♀ = 66 (49.62%)				0.059
	Men		Women		*p*-Value	Total German Sample		Men		Women		*p*-Value	Total Turkish Sample	
	M	SD	M	SD		M	SD	M	SD	M	SD		M	SD	
Age	62.10	8.84	61.43	9.04	0.451	61.70	8.96	61.29	8.98	58.22	7.06	**0.030**	59.74	8.19	**0.020**
Education (in school years)	12.15	1.81	11.43	2.02	**<0.001**	11.72	1.97	10.47	4.09	6.92	4.91	**<0.001**	8.68	4.84	**<0.001**
Cognitive Status	27.08	2.23	27.43	2.47	0.134	27.29	2.38	25.79	3.12	24.43	4.63	**0.050**	25.11	3.40	**<0.001**
Depression	1.41	1.64	1.79	2.23	**0.048**	1.64	2.02	3.77	2.99	3.91	3.09	0.794	3.84	3.03	**<0.001**
Age classes (N%)															
≤54 years	22.5%		25.2%			24.1%		25.8%		41.8%			33.8%		
55–59 years	23.7%		26%			25.1%		28.8%		25.4%			27.1%		
60–64 years	19.5%		20%			19.8%		16.7%		11.9%			14.3%		
65–69 years	14.2%		8.4%			10.7%		12.1%		10.4%			11.3%		
70–74 years	10.7%		10.4%			10.5%		4.5%		7.5%			6%		
75–79 years	4.1%		5.6%			5%		9.1%		3%			6%		
80–84 years	4.1%		1.6%			2.6%		0%		0%			0%		
≥85 years	1.2%		2.8%			2.1%		3.0%		0%			1.5%		
Education															
Primary school *	0.6%		0.4%			0.5%		21.2%		53.7%			37.6%		
Secondary Education	7.1%		15.2%			11.9%		39.4%		22.4%			30.8%		
Higher Education	91.7%		84%			87.1%		39.4%		23.9%			31.6%		

Note. Cognitive status was measured using the Montreal Cognitive Assessment Test (MoCA; Nasreddine et al., 2005 [13]) with a maximum of 30 points (German sample: 31 points, including correction for education). Depression was measured using the Geriatric Depression Scale (GDS; Yesavage et al., 1983 [26]) with a maximum of 15 points. * Primary school in Germany = 0–4 years; primary school in Turkey = 0–5 years.

3.2. Predictor Analysis

The regression model for the German sample is presented in Table 2. The entire regression model reached statistical significance ($F(4, 412) = 25.77; p < 0.001$). The explained variance of the model was 19.2% ($R^2 = 0.19; p < 0.001$). Predictors for MoCA performance were more years of education (B = 0.36), younger age (B = −0.05), being male (B = −0.46), and lower levels of depression scores (B = −0.21).

Table 2. Linear regression analysis for German sample on cognition measured with MoCA ($n = 417$).

	B	*SE B*	*ß*	*t*	*p*
Constant	26.52	1.27		20.87	<0.001
Education	0.36	0.06	0.29	5.86	<0.001
Age	−0.05	0.01	−0.19	−3.95	<0.001
Sex	−0.46	0.23	−0.09	−2.06	0.040
Depression	−0.21	0.05	−0.17	−3.91	<0.001
	$R^2 = 0.200$; adjusted $R^2 = 0.192$; $p < 0.001$				

Table 3 presents the regression model for the Turkish sample. This regression model was also significant ($F(4, 128) = 6.25; p < 0.001$), but in contrast with the German sample, the only significant predictor in the Turkish sample was more years of education (B = 0.33). The explained variance of this model was 13.7% ($R^2 = 0.137; p < 0.001$).

Table 3. Linear regression analysis for Turkish sample on cognition measured with MoCA ($n = 133$).

	B	*SE B*	*ß*	*t*	*p*
Constant	20.44	2.94		6.94	<0.001
Education	0.33	0.08	0.40	4.07	<0.001
Age	0.01	0.04	0.03	0.30	0.765
Sex	0.36	0.74	0.05	0.48	0.630
Depression	0.07	0.12	0.05	0.65	0.519
	$R^2 = 0.163$; corrected $R^2 = 0.137$; $p < 0.001$				

The regression model for the overall sample of German and Turkish participants, which included the predictors age, sex, education, and depression as well as migration, was significant ($F(5, 259) = 15.16; p < 0.001$), with having no migration background (B = 1.14) and higher education (B = 0.34) as significant predictors. The explained variance of this model was 21.1% ($R^2 = 0.211; p < 0.001$) (Table 4).

Table 4. Linear regression analysis for German and Turkish sample on Cognition measured with MoCA ($n = 264$).

	B	*SE B*	*ß*	*t*	*p*
Constant	22.04	1.68		13.14	<0.001
Ethnicity	1.14	0.46	0.17	2.49	0.013
Education	0.34	0.06	0.40	5.92	<0.001
Age	−0.01	0.02	−0.01	−0.22	0.827
Sex	−0.61	0.40	−0.09	−1.54	0.125
Depression	0.03	0.08	0.03	0.45	0.655
	$R^2 = 0.226$; adjusted $R^2 = 0.211$; $p < 0.001$				

4. Discussion

In this cross-sectional study with healthy older German adults and healthy older adults with Turkish migration background living in Germany, we investigated the impact of the sociodemographic variables age, sex, education, and migration background as well as depressive symptoms on cognitive performance, operationalized with the broadly used MoCA screening test. The main results are that MoCA performance (i) in the German sample was predicted by age, sex, and education, (ii) in the Turkish sample was only predicted by education, and (iii) in an overall sample was predicted by education and migration background.

First, results indicate that the German and Turkish samples differ significantly in sample characteristics; however, this finding reflects the current state of research: regarding sociodemographic variables, Turkish people are younger and less educated, which is in line with the results of a representative statistical survey on individuals with migration background in Germany [2]. Furthermore, they are more depressed, which is also in line with a previous representative study [29,30]. Finally, MoCA total scores were lower, which was expected according to other results on the cognitive state of migrants in different countries [31].

Our finding that MoCA performance in the German sample was predicted by more years of education, younger age, being male, and less depressive symptoms is in line with recent investigations in other population-based cohort studies that found significant predictors of cognition were also age, sex, and education [32–34]. The fact that younger age predicts MoCA performance has been described before [35] and can be well explained by the normal process of aging, which is accompanied by structural and functional brain changes leading to age-associated cognitive decline [36]. Further, the impact of education on cognition per se and the MoCA is also well known [37]. In contrast, sex as a possible predictor has less consistently been demonstrated and needs more consideration in future research [32,33,38–41]. Notably, as a typical pattern, German women were less educated than men in our sample, which can partly explain the results. Finally, our finding that lower depressive symptoms predict MoCA performance in the present study matches a whole body of evidence supporting the negative impact depressive symptoms have on cognition [42] and, more specifically, in elderly people [19]. In summary, in light of our data and other available data, the fact that the MoCA only provides a correction for education—at least for the German version—seems insufficient, and further versions that correct for the major influencing factors will improve precision of the screening process [34].

For the Turkish sample, the only significant predictor for MoCA performance was education, again demonstrating the consistency of this influence factor on cognition. In a study by Demir and Özcan (2015) [43], education was also an influencing factor on MoCA test performance in a Turkish population, but only in the language subtest. It is interesting to note that no differences could be detected in the other subtests. One reason for this could have been the small sample size of the study ($n = 50$). In another study on MoCA performance in Turkish migrants living in Germany [44], education and age were significant predictors. A possible reason for the lack of a significant age effect in our sample is the narrow age range, as 86.5% of the Turkish migrants were between 50 and 69 years and only 13.5% were 70 years or older (Table 1), whereas in the study by Krist et al. (2017) [44], individuals aged between 20 and 69 years were included. Sex was not a significant predictor in either study. Interestingly, depression, which was not assessed by Krist and colleagues, was not a significant predictor of MoCA performance in our study, although our Turkish sample was more depressed than our German sample. Thus, the impact of depression on cognitive performance in this population needs further investigation.

Our overall regression analysis, including both German and Turkish samples, again identifies education as having a substantial impact on MoCA performance. However, having or not having a migration background was an even stronger predictor of MoCA scores, showing that the migration process itself or life circumstances as a migrant predicts lower performance. Notably, Turkish migrants have worse working conditions, socioeconomic status, and health states in comparison with German people. However, it has been proposed that due to the harmful experience of migration, Turkish migrants are affected by faster aging processes, and this aspect was related to the fact that the prevalence of dementia is increasing more among this population in comparison with native Germans [7].

All of these aspects may explain why the migration background has such an influence on MoCA results even in healthy individuals. Importantly, our data indicate that normative data for the specific group of migrants will enhance the diagnostic accuracy of cognitive screenings. This aspect is also supported by further research. In a comparative study in the USA, different cut-off values for the MoCA were examined depending on ethnicity. In the detection of the optimal cut-off value in MCI, the authors found a point discrepancy of 1–2 points among non-Hispanic Whites, Hispanics, and non-Hispanic Blacks and a point discrepancy of 3 points in the detection of dementia [31]. Another systematic review that examined the cross-cultural application of MoCA also found a wide range of cut-offs, within and across different cultures [45].

Some limitations must be considered when interpreting the results of our study. First, information about health status (neurological, cognitive, and psychiatric disorders) was only assessed by self-reports. We cannot ensure that the study population did not include cognitively impaired participants, especially when considering MoCA test scores, as the MoCA cut-off score used for the Turkish sample was as low as 21 points, following the normative study by Selekler et al. (2010) [23]. When comparing our Turkish control group with the control group of Selekler et al. (2010) [23], there are differences of about two points. In the control group of Selekler et al. (2010), the average MoCA score was 23.50 (SD: 3.73), whereas our Turkish population averaged 25.11 points (SD: 3.40). Therefore, further studies should exclude cognitive impairment and dementia with a more elaborate clinical examination. In addition, an elaborated neuropsychological assessment is needed to replicate our findings to ensure generalizability.

Furthermore, a detailed status of migration (e.g., first- versus second-generation migration) was not assessed in the present study. Yet, this information could be used as a further predictor for cognitive performance, as second-generation immigrants may already be more integrated (e.g., reduction of language barriers). In the present study, we defined origin as "first language learned". However, various definitions of migration background exist and they are often not comparable [46,47]. As a result, our study may not be comparable to other investigations. Further studies should include a more specific operationalization of migration and its different variants. For example, in the study conducted by Krist et al. (2017) [44], Turkish migrants were able to select the test language (German or Turkish version of the MoCA), and the group that picked the German version showed significantly better MoCA scores than those who were tested in the Turkish version. This result indicated that the level of acculturation and integration differed (with those picking German being more integrated) and that this level was associated with cognitive performance. Another systematic review on bilingualism as an influencing factor in test performance in older adults indicates that migration status, acculturation level, and language of neuropsychological testing need to be considered when assessing bilingual older adults [48]. In line with this, the systematic review on migration and cognition conducted by Xu et al. (2017) [9] demonstrated a significant association between a higher level of acculturation and better performance in cognitive functioning among migrants. However, a recent study showed that more data are needed to better evaluate the utility of measuring acculturation level in neuropsychological assessment [49].

One main strength of the present study is the inclusion of a German sample as a control group to compare predictors of cognitive performance between migrants and non-migrants. Another strength is that our study included both sociodemographic variables and depressive symptoms as possible influence factors on cognition.

5. Conclusions

The present study investigated sociodemographic and affective predictors of MoCA performance in German individuals and Turkish migrants living in Germany. The most important finding is that migration background and education have an impact on cognitive performance: individuals with a Turkish migration background and individuals with lower education scored lower on the MoCA. The need for culturally sensitive instruments that take the educational situation into account has also been noted in other studies. The influence of education on the MMST test performance was shown to be similarly high in a Turkish population in Denmark [50]. In addition, a recent study on diagnosing dementia in patients with a migration background among German general practitioners showed that a language barrier could significantly complicate the diagnostic process [51]. Therefore, culture-sensitive and education-adjusted screening tools for the assessment of cognitive functioning in this population group are of high importance and these factors should be considered when developing new tools for the assessment of cognitive functioning or normative data for the MoCA for specific subpopulations. Notably, some work has been done in this respect in recent years. For example, Goudsmit et al. (2017) [52] developed the Cross-Cultural Dementia Screening (CCD) as a culture-fair test for addressing these problems in a Turkish population living in the Netherlands. Furthermore, our own working group has developed a culture-sensitive dementia test battery tool for Turkish migrants living in Germany that is based on culture-fair nonverbal materials [53,54].

More generally, as predictors of MoCA scores in separate samples of German individuals and Turkish migrants differed according to age, education, sex, and depression with only education having predictive value in both populations, our study further emphasizes that normative studies for specific populations are important. For clinical practice, the results of the MoCA should be interpreted with caution. The results can provide initial information, but new and culture-specific tests are still needed. For example, further studies are needed that apply a new weighting of the MoCA subtests with demographically adjusted standard scores. A better alternative would be a validated culture-fair test battery that considers group-specific characteristics and cut-offs as well.

Author Contributions: Conceptualization, G.A., Ü.S.S., J.K. and E.K.; methodology, G.A. and M.R.; formal analysis, G.A. and M.R.; investigation, G.A. and Ü.S.S.; resources, E.K., J.K. and O.A.O.; data curation, G.A. and Ü.S.S.; writing—original draft preparation, G.A.; writing—review and editing, M.R., Ü.S.S., H.L.-J., O.A.O., J.K. and E.K.; supervision, J.K. and E.K.; project administration, Ü.S.S., J.K., G.A. and E.K. All authors have read and agreed to the published version of the manuscript.

Funding: This research received no external funding.

Institutional Review Board Statement: The study was conducted according to the guidelines of the Declaration of Helsinki, and approved by the Ethics Committee Medical Faculty of the University of Cologne (16-249).

Informed Consent Statement: Informed consent was obtained from all subjects involved in the study.

Data Availability Statement: Data sharing not applicable.

Acknowledgments: The authors wish to thank all participants of the study. O.A.O. gratefully acknowledges the support from the Marga and Walter Boll Foundation.

Conflicts of Interest: The authors declare no conflict of interest.

References

1.   United Nations. *International Migration Report 2017*; United Nations: New York, NY, USA, 2017.

2.   Federal Statistical Office. *Bevölkerung und Erwerbstätigkeit Bevölkerung mit Migrationshintergrund—Ergebnisse des Mikrozensus 2016—Fachserie 1 Reihe 2.2*; Statistisches Bundesamt: Wiesbaden, Germany, 2017.

3.   Prince, M.; Wimo, A.; Maëlenn, G.; Gemma-Claire, A.; Yu-Tzu, W.; Matthew, P. *World Alzheimer Report 2015: The Global Impact of Dementia: An Analysis of Prevalence, Incidence, Cost and Trends*; Alzheimer's Disease International: London, UK, 2015.

4.   Bickel, H. *Die Häufigkeit von Demenzerkrankungen*; Dtsch Alzheimer Ges EV Informationsblatt: Dresden, Germany, 2020; pp. 1–7.

5.   Monsees, J.; Hoffmann, W.; Thyrian, J.R. Prevalence of dementia in people with a migration background in Germany. *Gerontol. Geriatr* **2018**. [CrossRef]

6.   Drewniok, A. Migration, Pflegebedürftigkeit und Demenz—ein Vergleich einer Standortbestimmung. *Pflegewissenschaft* **2014**, *16*, 452–466.

7.   Schimany, P.; Rühl, S.; Kohls, M. *Ältere Migrantinnen und Migranten: Entwicklungen, Lebenslagen, Perspektiven*; Bundesamt für Migration und Flüchtlinge: Nürnberg, Germany, 2012.

8.   Spallek, J.; Razum, O. Migration und Gesundheit. In *Soziologie von Gesundheit und Krankheit*; Richter, M., Hurrelmann, K., Eds.; Springer VS: Wiesbaden, Germany, 2016; pp. 153–166.

9. Xu, H.; Zhang, Y.; Wu, B. Association between migration and cognitive status among middle-aged and older adults: A systematic review. *BMC Geriatr.* **2017**, *17*, 184. [CrossRef]

10. Schenk, L. Migration und Gesundheit—Entwicklung eines Erklärungs- und Analysemodells für epidemiologische Studien. *Int. J. Public. Health* **2007**, *52*, 87–96. [CrossRef] [PubMed]

11. Brand, T.; Samkange-Zeeb, F.; Ellert, U.; Keil, T.; Krist, L.; Dragano, N.; Jöckel, K.-H.; Razum, O.; Reiss, K.; Greiser, K.H.; et al. Acculturation and health-related quality of life: Results from the German National Cohort migrant feasibility study. *Int. J. Public. Health* **2017**, *62*, 521–529. [CrossRef] [PubMed]

12. Krist, L.; Dornquast, C.; Reinhold, T.; Becher, H.; Jöckel, K.-H.; Schmidt, B.; Schramm, S.; Icke, K.; Danquah, I.; Willich, S.N.; et al. Association of Acculturation Status with Longitudinal Changes in Health-Related Quality of Life—Results from a Cohort Study of Adults with Turkish Origin in Germany. 2021. Available online: https://www.mdpi.com/1660-4601/18/6/2827 (accessed on 30 March 2021). [CrossRef]

13. Nasreddine, Z.S.; Phillips, N.A.; Bédirian, V.; Charbonneau, S.; Whitehead, V.; Collin, I.; Cummings, J.L.; Chertkow, H. The Montreal Cognitive Assessment, MoCA: A brief screening tool for mild cognitive impairment. *J. Am. Geriatr. Soc.* **2005**, *53*, 695–699. [CrossRef]

14. Daseking, M.; Petermann, F.; Waldmann, H.-C. Sex differences in cognitive abilities: Analyses for the German WAIS-IV. *Personal. Individ. Differ.* **2017**, *114*, 145–150. [CrossRef]

15. Li, R.; Singh, M. Sex differences in cognitive impairment and Alzheimer's disease. *Front. Neuroendocr.* **2014**, *35*, 385–403. [CrossRef]

16. Pinto, T.C.C.; Machado, L.; Bulgacov, T.M.; Rodrigues-Júnior, A.L.; Costa, M.L.G.; Ximenes, R.C.; Sougey, E.B. Influence of Age and Education on the Performance of Elderly in the Brazilian Version of the Montreal Cognitive Assessment Battery. *Dement. Geriatr. Cogn. Disord.* **2018**, *45*, 290–299. [CrossRef] [PubMed]

17. Tan, J.; Li, N.; Gao, J.; Wang, L.; Zhao, Y.; Yu, B.; Du, W.; Zhang, W.; Cui, L.; Wang, Q.; et al. Optimal cutoff scores for dementia and mild cognitive impairment of the Montreal Cognitive Assessment among elderly and oldest-old Chinese population. *J. Alzheimers Dis. JAD* **2015**, *43*, 1403–1412. [CrossRef] [PubMed]

18. Blair, M.; Coleman, K.; Jesso, S.; Desbeaumes Jodoin, V.; Smolewska, K.; Warriner, E.; Finger, E.; Pasternak, S.H. Depressive Symptoms Negatively Impact Montreal Cognitive Assessment Performance: A Memory Clinic Experience. *Can. J. Neurol. Sci.* **2016**, *43*, 513–517. [CrossRef]

19. Del Brutto, O.H.; Mera, R.M.; Del Brutto, V.J.; Maestre, G.E.; Gardener, H.; Zambrano, M.; Wright, C.B. Influence of depression, anxiety and stress on cognitive performance in community-dwelling older adults living in rural Ecuador: Results of the Atahualpa Project: Cognitive performance and psychological distress. *Geriatr. Gerontol. Int.* **2015**, *15*, 508–514. [CrossRef] [PubMed]

20. Greenberg, M.S.; Tanev, K.; Marin, M.-F.; Pitman, R.K. Stress, PTSD, and dementia. *Alzheimers Dement.* **2014**, *10*, S155–S165. [CrossRef] [PubMed]

21. Justice, N.J. The relationship between stress and Alzheimer's disease. *Neurobiol. Stress* **2018**, *8*, 127–133. [CrossRef] [PubMed]

22. Sotiropoulos, I.; Catania, C.; Pinto, L.G.; Silva, R.; Pollerberg, G.E.; Takashima, A.; Sousa, N.; Almeida, O.F.X. Stress Acts Cumulatively to Precipitate Alzheimer's Disease-Like Tau Pathology and Cognitive Deficits. *J. Neurosci.* **2011**, *31*, 7840–7847. [CrossRef]

23. Selekler, K.; Cangöz, B.; Uluc, S. Power of discrimination of Montreal Cognitive Assessment (MOCA) scale in Turkish patients with mild cognitive impairement and Alzheimer's disease. *Türk. Geriatri. Derg.* **2010**, *13*, 166–171.

24. Kowal, P.; Karen, P. Indicators for the Minimum Data Set Project on Ageing: A Critical Review in Sub-Saharan Africa. 2001. Available online: https://www.who.int/healthinfo/survey/ageing_mds_report_en_daressalaam.pdf (accessed on 26 May 2021).

25. Costa, A.S.; Fimm, B.; Friesen, P.; Soundjock, H.; Rottschy, C.; Gross, T.; Eitner, F.; Reich, A.; Schulz, J.B.; Nasreddine, Z.S.; et al. Alternate-form reliability of the Montreal cognitive assessment screening test in a clinical setting. *Dement. Geriatr. Cogn. Disord.* **2012**, *33*, 379–384. [CrossRef]

26. Yesavage, J.A.; Brink, T.L.; Rose, T.L.; Lum, O.; Huang, V.; Adey, M.; Leirer, V.O. Development and validation of a geriatric depression screening scale: A preliminary report. *J. Psychiatr. Res.* **1983**, *17*, 37–49. [CrossRef]

27. Erdfelder, E.; Faul, F.; Buchner, A. GPOWER: A general power analysis program. *Behav. Res. Methods Instrum. Comput.* **1996**, *28*, 1–11. [CrossRef]

28. Field, A.P. *Discovering Statistics Using SPSS: And Sex, Drugs and Rock "N" Roll*, 3rd ed.; SAGE Publications: Los Angeles, CA, USA, 2009.

29. Sahyazici, F.; Huxhold, O. Depressive Symptome bei Älteren Türkischen Migrantinnen und Migranten. In *Viele Welten Alterns*; Baykara-Krumme, H., Schimany, P., Motel-Klingebiel, A., Eds.; VS Verlag für Sozialwissenschaften: Wiesbaden, Germany, 2012; pp. 181–200.

30. Igde, E.; Heinz, A.; Schouler-Ocak, M.; Rössler, W. Depressive und somatoforme Störungen bei türkeistämmigen Personen in Deutschland. *Nervenarzt* **2019**, *90*, 25–34. [CrossRef]

31. Milani, S.A.; Marsiske, M.; Cottler, L.B.; Chen, X.; Striley, C.W. Optimal cutoffs for the Montreal Cognitive Assessment vary by race and ethnicity. *Alzheimers Dement. Amst. Neth.* **2018**, *10*, 773–781. [CrossRef] [PubMed]

32. Borland, E.; Nägga, K.; Nilsson, P.M.; Minthon, L.; Nilsson, E.D.; Palmqvist, S. The Montreal Cognitive Assessment: Normative Data from a Large Swedish Population-Based Cohort. *J. Alzheimers Dis.* **2017**, *59*, 893–901. [CrossRef] [PubMed]

33. Larouche, E.; Tremblay, M.-P.; Potvin, O.; Laforest, S.; Bergeron, D.; Laforce, R.; Monetta, L.; Boucher, L.; Tremblay, P.; Belleville, S.; et al. Normative Data for the Montreal Cognitive Assessment in Middle-Aged and Elderly Quebec-French People. *Arch. Clin. Neuropsychol.* **2016**. [CrossRef] [PubMed]

34. Thomann, A.E.; Goettel, N.; Monsch, R.J.; Berres, M.; Jahn, T.; Steiner, L.A.; Monsch, A.U. The Montreal Cognitive Assessment: Normative Data from a German-Speaking Cohort and Comparison with International Normative Samples. *J. Alzheimers Dis. JAD* **2018**, *64*, 643–655. [CrossRef] [PubMed]

35. Oren, N.; Yogev-Seligmann, G.; Ash, E.; Hendler, T.; Giladi, N.; Lerner, Y. The Montreal Cognitive Assessment in cognitively-intact elderly: A case for age-adjusted cutoffs. *J. Alzheimers Dis. JAD* **2015**, *43*, 19–22. [CrossRef]

36. Murman, D.L. The Impact of Age on Cognition. *Semin. Hear.* **2015**, *36*, 111–121. [CrossRef]

37. Mantri, S.; Nwadiogbu, C.; Fitts, W.; Dahodwala, N. Quality of education impacts late life cognition. *Int. J. Geriatr. Psychiatry* 2019. [CrossRef]

38. Hyde, J.S. Sex and cognition: Gender and cognitive functions. *Curr. Opin. Neurobiol.* **2016**, *38*, 53–56. [CrossRef]

39. Hyde, J.S. The gender similarities hypothesis. *Am. Psychol.* **2005**, *60*, 581–592. [CrossRef] [PubMed]

40. Upadhayay, N.; Guragain, S. Comparison of cognitive functions between male and female medical students: A pilot study. *J. Clin. Diagn. Res. JCDR* **2014**, *8*, BC12–BC15. [CrossRef] [PubMed]

41. Zell, E.; Krizan, Z.; Teeter, S.R. Evaluating gender similarities and differences using metasynthesis. *Am. Psychol.* **2015**, *70*, 10–20. [CrossRef] [PubMed]

42. Hammar, Å. Cognitive functioning in major depression—A summary. *Front. Hum. Neurosci.* **2009**, *3*, 26. [CrossRef] [PubMed]

43. Yancar Demir, E.; Özcan, T. Evaluating the relationship between education level and cognitive impairment with the Montreal Cognitive Assessment Test: Education, cognition, MoCA. *Psychogeriatrics* **2015**, *15*, 186–190. [CrossRef]

44. Krist, L.; Keller, T.; Sebald, L.; Yesil-Jürgens, R.; Ellert, U.; Reich, A.; Becher, H.; Heuschmann, P.U.; Willich, S.N.; Keil, T.; et al. The Montreal Cognitive Assessment (MoCA) in a population-based sample of Turkish migrants living in Germany. *Aging Ment. Health* **2017**, 1–8. [CrossRef] [PubMed]

45. O'Driscoll, C.; Shaikh, M. Cross-Cultural Applicability of the Montreal Cognitive Assessment (MoCA): A Systematic Review. *J Alzheimers Dis. JAD* **2017**, *58*, 789–801. [CrossRef]

46. Bijak, J. *Migration Forecasting: Beyond the Limits of Uncertainty (Data Briefing Series No. 6)*; Global Migration Data Analysis Centre (GMDAC): Berlin, Germany, 2016.

47. Kupiszewska, D.; Nowok, B. Comparability of Statistics on International Migration Flows in the European Union. In *International Migration in Europe: Data, Models and Estimates*; Raymer, J., Willekens, F., Eds.; Wiley: Chichester, UK; Hoboken, NJ, USA, 2008; pp. 41–72.

48. Celik, S.; Kokje, E.; Meyer, P.; Frölich, L.; Teichmann, B. Does bilingualism influence neuropsychological test performance in older adults? A systematic review. *Appl. Neuropsychol. Adult* **2020**, 1–19. [CrossRef]

49. Tan, Y.W.; Burgess, G.H.; Green, R.J. The effects of acculturation on neuropsychological test performance: A systematic literature review. *Clin. Neuropsychol.* **2021**, *35*, 541–571. [CrossRef]

50. Nielsen, T.R.; Vogel, A.; Gade, A.; Waldemar, G. Cognitive testing in non-demented Turkish immigrants—comparison of the RUDAS and the MMSE: Cognitive testing in Turkish immigrants. *Scand J. Psychol.* **2012**, *53*, 455–460. [CrossRef]

51. Tillmann, J.; Just, J.; Schnakenberg, R.; Weckbecker, K.; Weltermann, B.; Münster, E. Challenges in diagnosing dementia in patients with a migrant background—a cross-sectional study among German general practitioners. *BMC Fam. Pract.* **2019**, *20*, 34. [CrossRef]

52. Goudsmit, M.; Uysal-Bozkir, Ö.; Parlevliet, J.L.; van Campen, J.P.C.M.; de Rooij, S.E.; Schmand, B. The Cross-Cultural Dementia Screening (CCD): A new neuropsychological screening instrument for dementia in elderly immigrants. *J. Clin. Exp. Neuropsychol.* **2017**, *39*, 163–172. [CrossRef]

53. Seven, Ü.S.; Braun, I.V.; Kalbe, E.; Kessler, J. Demenzdiagnostik bei Menschen Türkischer Herkunft—TRAKULA. In *Lebenswelten von Menschen mit Migrationserfahrung und Demenz. 1. Aufl*; Dibelius, O., Feldhaus-Plumin, E., Piechotta-Henze, G., Eds.; Hogrefe: Bern, Switzerland, 2015; pp. 51–87.

54. Seven, Ü.S.; Onur, Ö.; Anapa, G.; Kalbe, E.; Kessler, J. TRAKULA—Transkulturelles Assessment Kognitiver Funktionen. Manuscript submitted.

International Journal of
*Environmental Research
and Public Health*

Article

Health Care Services Utilization of Persons with Direct, Indirect and without Migration Background in Germany: A Longitudinal Study Based on the German Socio-Economic Panel (SOEP)

Thomas Grochtdreis *, Hans-Helmut König and Judith Dams

Hamburg Center for Health Economics, Department of Health Economics and Health Services Research, University Medical Center Hamburg-Eppendorf, Martinistr. 52, 20246 Hamburg, Germany; h.koenig@uke.de (H.-H.K.); j.dams@uke.de (J.D.)
* Correspondence: t.grochtdreis@uke.de; Tel.: +49-(0)40-7410-52405

Abstract: There is ambiguous evidence with regard to the inequalities in health care services utilization (HCSU) among migrants and non-migrants in Germany. The aim of this study was to analyze the utilization of doctors and hospitalization of persons with direct and indirect migration background as well as those without in Germany. This study was based on data of the German Socio-Economic Panel using the adult sample of the years 2013 to 2019. HCSU was measured by self-reported utilization of doctors and hospitalization. Associations between HCSU and migration background were examined using multilevel mixed-effects logistic regression and zero-truncated multilevel mixed-effects generalized linear models. The odds ratios of utilization of doctors and hospitalization for persons with direct migration background compared with persons without migration background were 0.73 ($p < 0.001$) and 0.79 ($p = 0.002$), respectively. A direct migration background was associated with a 6% lower number of doctoral visits within three months compared with no migration background ($p = 0.023$). Persons with direct migration background still have a lower HCSU than persons without migration background in Germany. Access to health care needs to be ensured and health policy-makers are called upon to keep focus on the issue of inequalities in HCSU between migrants and non-migrants in Germany.

Keywords: surveys and questionnaires; health care; utilization; migrant; Germany

Citation: Grochtdreis, T.; König, H.-H.; Dams, J. Health Care Services Utilization of Persons with Direct, Indirect and without Migration Background in Germany: A Longitudinal Study Based on the German Socio-Economic Panel (SOEP). *Int. J. Environ. Res. Public Health* **2021**, *18*, 11640. https://doi.org/10.3390/ijerph182111640

Academic Editor: Paul B. Tchounwou

Received: 30 September 2021
Accepted: 4 November 2021
Published: 5 November 2021

Publisher's Note: MDPI stays neutral with regard to jurisdictional claims in published maps and institutional affiliations.

1. Introduction

In the year 2019, 26% of the total German population, about 21.2 million persons, had a migration background (based on a definition of migration background to people born without German nationality or if at least one parent was born without German nationality) [1]. Within the last ten years until 2019, this proportion of the population with migration background increased by 38%. Not only did the emergence of new migratory flows with the fall of the Iron Curtain and the expansion of the European Union in the last three decades lead to an increase in persons with direct migration background [2], but also the growing up of a new generation of persons, i.e., persons with indirect migration background, whose parents were work migrants, had bearing on this increase too [3,4].

A recent systematic literature review about inequalities in health care services utilization (HCSU) among migrants and non-migrants in Germany found a lower utilization among persons with migration background [5]. This lower utilization was shown, among others, for specialists, therapists, and medication, while the results for utilization of doctors and hospitalization were inconclusive. Furthermore, persons with direct migration background and females with migration background were identified as groups with a particular low HCSU. Another systematic literature review on HCSU of migrants in Europe; however, found that the probability of hospitalization of persons with migration background was higher compared with persons without migration background, whereas the results for utilization of doctors were also inconclusive [6].

73

According to the behavioral model by Andersen [7,8], HCSU is determined by predisposing, enabling, and need factors, where migration background might be seen as predisposing factor. Other predisposing factors are age, sex, and marital status; employment status is an enabling factor and health-related quality of life is a need factor that determines HCSU [7,8]. There is evidence that persons with direct migration background have a higher mental and lower physical health-related quality of life compared with persons without migration background in Germany [9–11], while no differences in health-related quality of life were observed for persons with indirect migration background [9–13]. Thus, differences in health-related quality of life might result in differences in HCSU of persons with and without migration background. It can, therefore, be assumed that persons with migration background have different needs and barriers with respect to health care than persons without migration background, and that migration background might not only be a predisposing factor, but also confounded with health and health-related quality of life, and other enabling and need factors [5,7,8]. Furthermore, it needs to be clarified whether the aforementioned assumptions are valid for both persons with direct migration background and for their descendants with indirect migration background [10,14,15].

In order to overcome the inconclusive results with respect to inequalities in HCSU and to follow the obstacles of differences in different needs and barriers with respect to health care among migrants and non-migrants in Germany, a study with a large nationally representative sample of persons with and without migration background is needed in order to be able to make a generally valid statement about the health care situation of migrants in Germany. However, earlier studies on the utilization of doctors and hospitalization of persons with and without migration background in Germany were based on regional samples [16–28], samples of children and adolescents or elderly [16,22,29–35], samples of women [18–21,24–27], or only samples on specific medical conditions [5,29,34,36,37]. Merely a study from 2011 by Glaesmer et al. [12] used a representative population survey of persons with direct and indirect migration background as well as those without to investigate differences in HCSU in Germany. The study, however, was not able to find any differences in the probability of utilization of doctors and hospitalization. Yet, a higher number of doctoral visits and nights in hospital of persons with direct migration background compared with persons without migration background was found.

The results of the aforementioned study should have called for a stronger health policy focus on access to health care services for persons with migration background. Because of those potentially unresolved issues, it is necessary to refocus research on HCSU of persons with migration background who have immigrated to Germany in the last three decades, but also on persons with indirect migration background. Based on the results of one earlier study [12], it is hypothesized that the migration background of more recently migrated persons and of those persons with indirect migration background is actually negatively associated with the probability of HCSU, and, if health care services were utilized, positively associated with the number of doctoral visits and nights in hospital. Therefore, the aim of this study was to analyze and compare the HCSU of persons with and without migration background in Germany. Thereby, the focus was on the utilization of doctors and hospitalization, as well as the number of doctoral visits and number of nights in hospital of persons with direct or indirect migration background and those without in a large representative sample.

2. Materials and Methods

2.1. Sample

The sample of this study was based on data of the German Socio-Economic Panel (SOEP) provided by the German Institute for Economic Research (DIW Berlin). The SOEP is a representative German household panel with over 20,000 participants surveyed annually since 1984, with 36 waves available up to 2021. As of the survey year 2013 (wave 29), two additional migrant samples (M1 and M2) were integrated into the SOEP to ensure the proportional representation of the previously underrepresented current generation of persons with migration background [38]. For the following analyses, the adult sample of the waves 29 to 36 (i.e., years 2013 to 2019) was used (n = 58,879; 251,930 observations). An analytical sample was generated by removing observations with missing information in the number of doctoral visits and number of nights in hospital (n = 44,403; 180,656 observations). Moreover, persons with missing information in sociodemographic characteristics were removed, resulting in a net sample of n = 43,921; 179,357 observations (75% of the original sample).

2.2. Measures

Persons without migration background and persons with direct/indirect migration background were distinguished based on a predefined variable of the SOEP. By combination of information on country of birth, citizenship, and of parental information, it was derived whether a person had an own migration experience or was born to at least one parent with direct migration background [39]. Concurrent with the definition of the European Migration Network [3,4], the DIW Berlin defined persons with direct migration background as persons with their own migration experience born without German citizenship, and persons with indirect migration background as persons without their own migration experience who were born to at least one parent with direct migration background [39]. Persons without migration background were persons born to parents without migration background.

In order to measure HCSU, participants of the SOEP were asked whether they had visited a doctor within the last three months and whether they had spent at least one night in hospital in the previous year. Furthermore, if they had visited a doctor within the last three months and if they had spent at least one night in hospital, they were asked how often they had visited a doctor within the last three months and how many nights in total they had spent in hospital within the last year, respectively. Regarding the utilization of doctors, no distinction was made in the SOEP between primary care physicians and specialists.

The sociodemographic characteristics age, sex (female and male), marital status (never married/single, married/in partnership, separated/divorced, and widowed), and employment status (employed fulltime, employed part-time, apprenticeship, marginally employed, other employment, and unemployed) were derived from the SOEP. For the purpose of illustration of the persons with direct/indirect migration background, nationality was also derived from the SOEP. Thereby, nationality was categorized into German, East European, South European, West and North European, African, Asian, and American/Oceanian countries of origin in accordance with the United Nations Standard Country or Area Codes for Statistical Use (M49) [40].

2.3. Statistical Analysis

Utilization of doctors within three months (yes/no) and hospitalization within the last year (yes/no) was dichotomized based on the questions on the utilization of a doctor within the last three months and on having spent at least one night in hospital within the last year. Furthermore, if persons utilized a doctor within the last three months, the number of doctoral visits within three months was used as a variable of HCSU. If persons spent at least one night in hospital within the last year, the number of nights in hospital within the last year was used.

Descriptive statistics of sociodemographic variables were calculated for persons without migration background and persons with direct and indirect migration background. Furthermore, differences in HCSU between persons without migration background and persons with direct or indirect migration background were calculated by sociodemographic characteristics (i.e., age, sex, marital status, employment status). The differences in HCSU by migration background were analyzed using Pearson's chi-squared test and Student's t-test. The descriptive statistics of sociodemographic variables and differences in HCSU were analyzed on the basis of cross-sectional data, using persons' data at first occurrence in the selected analytical sample.

Associations between utilization of doctors within three months and hospitalization within one year and migration background were examined using multilevel mixed-effects logistic regression with cluster robust standard errors [41]. The group structure for the random effects was identified by a central individual identifier, which was fixed over time. Furthermore, the sociodemographic factors comprising age, sex, marital status, employment status, and survey year (2013 to 2019) were used. Furthermore, interactions between migration background and sex, and migration background and survey year were added to the models as independent variables. The fixed-effects coefficients of the logistic regressions were reported as odds ratios (OR).

The associations between the number of doctoral visits within three months, the number of nights in hospital within the last year, and migration background were examined using zero-truncated multilevel mixed-effects generalized linear models (GLM) with a negative binomial family and log link function [41]. For the GLM, the same group structure for the random effects, independent variables, and interactions as for the logistic regressions was taken into account. GLM with a negative binomial family take into account the skewed distribution and overdispersion of HCSU data as dependent variables [42]. The results of the GLM were reported as exponentiated fixed-effects coefficients.

All analyses were performed using Stata/SE 16.1 (StataCorp, College Station, TX, USA). All applied statistics were two-sided. The level of significance was set at $\alpha = 0.05$.

3. Results

3.1. Sample Characteristics

The mean age of persons without migration background ($n = 32,535$) was 47 years. In comparison, persons with direct ($n = 8080$) and indirect migration background ($n = 3306$) were younger (42 and 30 years, both with $p < 0.001$). Of all persons without migration background and with indirect migration background, about half (52%) were female. Proportionally more persons with direct migration background were female (54%, $p = 0.004$). Persons with direct and indirect migration background differed in marital status and employment status (both with $p < 0.001$) compared with persons without migration background. Furthermore, the majority of persons with direct migration background had a German nationality (44%), followed by 18% and 16% with a nationality from a Southern European country and an Eastern European country, respectively. The vast majority of persons with indirect migration background had a German nationality (78%). The sociodemographic characteristics of the sample are shown in Table 1.

Of all persons without migration background, 72% ($n = 23,510$) utilized doctors within three months. Those who utilized doctors within three months had a mean number of doctoral visits of 2.38. Of all persons with direct and indirect migration background, 66% ($n = 5342$) and 67% ($n = 2200$) utilized doctors within three months, respectively. Those who utilized doctors within three months had mean numbers of doctoral visits of 2.05 and 2.07, respectively (Table S1 in the online Supplementary Materials).

Of all persons without migration background, 13% ($n = 4282$) were hospitalized within one year. Those who were hospitalized had a mean of 10.68 nights in hospital within one year. Of all persons with direct and indirect migration background, 10% were hospitalized within one year. Those who were hospitalized had a mean of 8.78 and 9.04 nights in hospital within one year, respectively (Table S2 in the online Supplementary Materials).

3.2. Doctoral Visits within Three Months

The logistic regression models showed that persons with direct migration background had lower odds of utilization of doctors within three months compared with persons without migration background (OR: 0.73, $p < 0.001$; Table 2). The odds ratio of the utilization of doctors within three months between persons with indirect migration background and persons without migration background was not statistically significant. Persons with a higher age (OR: 1.03; $p < 0.001$), females (OR: 1.60; $p < 0.001$), and persons not being employed fulltime (OR: 1.14–1.99; all with $p < 0.001$) had greater odds of hospitalization within one year in both models. Furthermore, the interaction of migration background and sex was statistically significant ($p < 0.001$) in the model comparing persons with direct and without migration background, indicating a modification of the association of direct migration background and the utilization of doctors by sex (Figure 1).

In the GLM, direct migration background was associated with a 6% reduction in the number of doctoral visits within three months compared with persons without migration background ($p = 0.023$), whereas no statistically significant association was found between indirect migration background and the number of doctoral visits within three months. The number of doctoral visits within three months was positively associated with a higher age ($p < 0.001$), female sex ($p < 0.001$), and not being employed fulltime or part-time (all with $p \leq 0.001$) in the model comparing persons with direct and without migration background (all with $p \leq 0.001$) and in the model comparing persons with indirect and without migration background (all with $p \leq 0.01$). In the model comparing persons with direct and without migration background, the interaction of migration background and sex was statistically not significant (Figure 2).

Table 1. Sociodemographic characteristics (years 2013 to 2019, $n = 43{,}921$).

Sociodemographic Characteristic	Persons without Migration Background ($n = 32{,}535$)	Persons with Direct Migration Background ($n = 8080$)	Persons with Indirect Migration Background ($n = 3306$)
Age: Mean (SD)	47.16 (17.98)	42.00 (14.16) ***	30.17 (12.07) ***
Grouped age: n (%)			
18–24	4641 (14.26)	773 (9.57) ***	1465 (44.31) ***
25–34	4038 (12.41)	1864 (23.07)	729 (22.05)
35–44	5573 (17.13)	2364 (29.26)	678 (20.51)
45–54	7123 (21.89)	1555 (19.25)	289 (8.74)
55–64	5002 (15.37)	905 (11.20)	113 (3.42)
$\geq$65	6158 (18.93)	619 (7.66)	32 (0.97)
Sex: n (%)			
Female	16,892 (51.92)	4338 (53.69) **	1713 (51.81)
Male	15,643 (48.08)	3742 (46.31)	1593 (48.19)
Marital status: n (%)			
Never married/single	9754 (29.98)	1661 (20.56) ***	1990 (60.19) ***
Married/in partnership	17,364 (53.37)	5386 (66.66)	1085 (32.82)
Separated/divorced	3779 (11.62)	838 (10.37)	211 (6.38)
Widowed	1638 (5.03)	195 (2.41)	20 (0.60)
Employment status: n (%)			
Employed fulltime	12,691 (39.01)	3171 (39.25) ***	1106 (33.45) ***
Employed part-time	4635 (14.25)	1075 (13.30)	346 (10.47)
Apprenticeship	1110 (3.41)	191 (2.36)	297 (8.98)
Marginally employed	1952 (6.00)	656 (8.12)	289 (8.74)
Other employment [1]	290 (0.89)	21 (0.26)	33 (1.00)
Unemployed	11,857 (36.44)	2966 (36.71)	1235 (37.36)
Nationality: n (%)			
German	32,535 (100.00)	3573 (44.22) ***	2573 (77.83) ***
East European	-	1419 (17.56)	10 (0.30)
South European	-	1260 (15.59)	376 (11.37)
West and North European [2]	-	382 (4.73)	43 (1.30)
African	-	178 (2.20)	9 (0.27)
Asian	-	1123 (13.90)	286 (8.65)
American/Oceanian	-	128 (1.58)	8 (0.24)
Stateless/ethnic minority	-	17 (0.21)	1 (0.03)

Comments: SD: Standard deviation; ** $p \leq 0.01$, *** $p \leq 0.001$; comparison of categorical characteristics of persons without migration background and with direct/indirect migration background was analyzed using Pearson's chi-squared test; comparison of mean age of persons without migration background and with direct/indirect migration background was analyzed using Student's t-test; [1] Near retirement with zero working hours, military service, community service, sheltered workshop; [2] Without German nationality.

Table 2. Multilevel mixed-effects logistic regressions of doctoral visits within three months (yes/no) and zero-truncated multilevel mixed-effects generalized linear models of number of doctoral visits within three months for persons without and with direct or indirect migration background (years 2013 to 2019, *n* = 43,921; 179,357 observations).

Independent Variables	Without Migration Background vs. with Direct Migration Background		Without Migration Background vs. with Indirect Migration Background	
	OR (SE) [†]	Exp(β) (SE) [‡]	OR (SE) [†]	Exp(β) (SE) [‡]
Migration background (Ref. without migration background)				
With direct migration background	0.73 (0.05) ***	0.94 (0.02) *		
With indirect migration background			0.99 (0.09)	0.99 (0.04)
Age	1.03 (0.00) ***	1.00 (0.00) ***	1.03 (0.00) ***	1.00 (0.00) ***
Sex (Ref. male)				
Female	1.60 (0.04) ***	1.08 (0.01) ***	1.60 (0.04) ***	1.08 (0.01) ***
Marital status (Ref. married/in partnership)				
Never married/single	1.16 (0.03) ***	0.98 (0.01)	1.18 (0.04) ***	0.97 (0.01) *
Separated/divorced	1.03 (0.03)	1.08 (0.01) ***	1.00 (0.04)	1.08 (0.01) ***
Widowed	1.02 (0.06)	1.00 (0.02)	1.04 (0.06)	0.99 (0.02)
Employment status (Ref. employed fulltime)				
Employed part-time	1.14 (0.03) ***	1.00 (0.01)	1.15 (0.03) ***	1.00 (0.01)
Apprenticeship	1.99 (0.11) ***	1.07 (0.02) ***	1.94 (0.10) ***	1.05 (0.02) **
Marginally employed	1.25 (0.04) ***	1.06 (0.01) ***	1.26 (0.05) ***	1.06 (0.01) ***
Other employment [1]	1.88 (0.20) ***	1.12 (0.04) ***	1.78 (0.19) ***	1.10 (0.04) **
Unemployed	1.70 (0.04) ***	1.24 (0.01) ***	1.65 (0.04) ***	1.22 (0.01) ***
Migration background * Sex (Ref. no migration background * male)				
Direct migration background * Female	1.21 (0.06) ***	1.01 (0.02)		
Indirect migration background * Female			1.32 (0.10) ***	1.08 (0.03) **
Survey year	Yes	Yes	Yes	Yes
Migration background * Survey year	Yes	Yes	Yes	Yes
Constant	0.48 (0.02) ***	2.24 (0.04) ***	0.49 (0.03) ***	2.27 (0.04) ***
Random effect: Person-ID Variance (Constant)	2.12 (0.04)	0.24 (0.00)	2.16 (0.04)	0.24 (0.00)

Comments: CI: confidence interval; SE: standard error; [1] near retirement with zero working hours, military service, community service, sheltered workshop; [†] Dependent variable: utilization of doctors within three months (yes/no); [‡] Dependent variable: number of doctoral visits (*n* = 31,052; 127,799 observations); * $p \leq 0.05$, ** $p \leq 0.01$, *** $p \leq 0.001$.

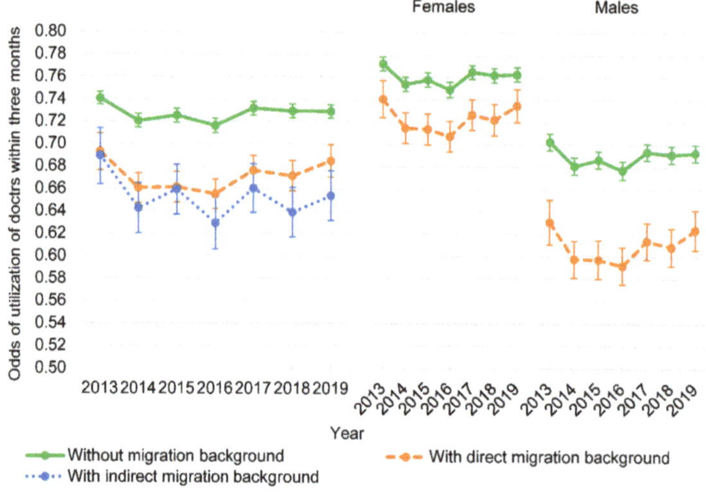

Figure 1. Adjusted odds of utilization of doctors within three months of persons without and with direct or indirect migration background, and for females and males without and with direct migration background (years 2013 to 2019, *n* = 43,921; 179,357 observations).

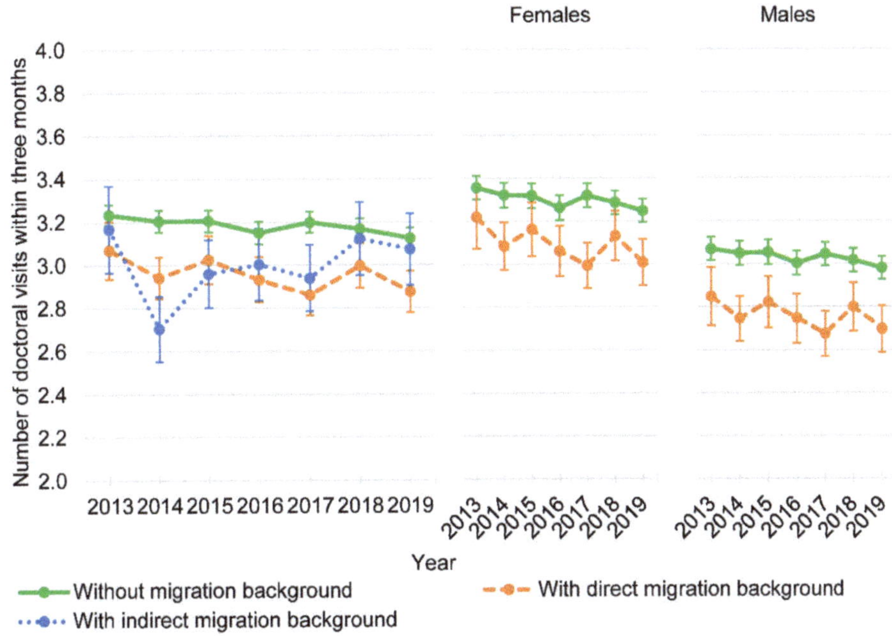

Figure 2. Adjusted zero-truncated number of doctoral visits within three months of persons without and with direct/indirect migration background, and of females and males without and with direct migration background (years 2013 to 2019, *n* = 31,052; 127,799 observations).

No time trend was observed between the years 2013 and 2019 with regard to the utilization of doctors, nor with regard to the number of doctoral visits for any of the groups of persons analyzed.

3.3. Nights in Hospital within the Last Year

Persons with direct migration background had lower odds of hospitalization within one year than persons without migration background (OR: 0.79, p = 0.002; Table 3) in the logistic regression models. The odds ratio of hospitalization within one year of persons with indirect migration background and persons without migration background was not statistically significant. Persons with a higher age (OR: 1.01; p < 0.001) and persons not being employed fulltime or part-time (OR: 1.33–2.73; all with p < 0.001) had greater odds of hospitalization within one year in both models. Greater odds of hospitalization within one year associated with female sex (OR: 1.05; p < 0.001) were only found in the model analyzing differences between persons with direct and without migration background. Compared with being married or in a partnership, having never been married or being single were both associated with lower odds of hospitalization within one year in both models (OR 0.79 and 0.77; both with p < 0.001), whereas being separated, divorced, or widowed were associated with higher odds of hospitalization within one year in both models (OR 1.16–1.23; all with p < 0.001). Furthermore, in the model comparing persons with direct and without migration background, the interaction of migration background and sex was statistically significant (p < 0.001), indicating a modification of the association of direct migration background and hospitalization by sex (Figure S1 in the online Supplementary Materials).

Table 3. Multilevel mixed-effects logistic regressions of hospitalization (yes/no) and zero-truncated multilevel mixed-effects generalized linear models of number of nights in hospital within the last year for persons without and with direct or indirect migration background (years 2013 to 2019, n = 43,921; 179,357 observations).

Independent Variables	Without Migration Background vs. with Direct Migration Background		Without Migration Background vs. with Indirect Migration Background	
	OR (SE) [†]	Exp(β) (SE) [‡]	OR (SE) [†]	Exp(β) (SE) [‡]
Migration background (Ref. without migration background)				
With direct migration background	0.79 (0.06) **	0.93 (0.06)		
With indirect migration background			0.82 (0.10)	1.16 (0.15)
Age	1.01 (0.00) ***	1.01 (0.00) ***	1.01 (0.00) ***	1.01 (0.00) ***
Sex (Ref. male)				
Female	1.05 (0.03) *	0.95 (0.02) *	1.05 (0.03)	0.95 (0.02) *
Marital status (Ref. married/in partnership)				
Never married/single	0.79 (0.03) ***	1.06 (0.03) *	0.77 (0.03) ***	1.05 (0.03)
Separated/divorced	1.23 (0.04) ***	1.17 (0.03) *	1.19 (0.04) ***	1.19 (0.03)
Widowed	1.17 (0.05) ***	1.09 (0.03) ***	1.16 (0.05) ***	1.06 (0.03) ***
Employment status (Ref. employed fulltime)				
Employed part-time	1.03 (0.04)	1.02 (0.03)	1.06 (0.04)	1.02 (0.03)
Apprenticeship	1.54 (0.11) ***	1.15 (0.09)	1.64 (0.11) ***	1.24 (0.09) **
Marginally employed	1.33 (0.06) ***	1.08 (0.04)	1.38 (0.07) ***	1.06 (0.05)
Other employment [1]	1.90 (0.22) ***	1.15 (0.12)	1.83 (0.22) ***	1.09 (0.11)
Unemployed	2.73 (0.07) ***	1.34 (0.03) ***	2.68 (0.08) ***	1.33 (0.03) ***
Migration background * Sex (Ref. no migration background * male)				
Direct migration background * Female	1.21 (0.07) ***	1.00 (0.05)	1.40 (0.13) ***	1.03 (0.09)
Survey year	Yes	Yes	Yes	Yes
Migration background * Survey year	Yes	Yes	Yes	Yes
Constant	0.04 (0.00) ***	3.53 (0.15) ***	0.04 (0.00) ***	3.58 (0.16) ***
Random effect: Person-ID Variance(Constant)	1.32 (0.04)	0.46 (0.01)	1.35 (0.04)	0.47 (0.01)

Comments: CI: confidence interval; OR: odds ratio; SE: standard error; [1] near retirement with zero working hours, military service, community service, sheltered workshop; [†] dependent variable: hospitalization within one year (yes/no); [‡] dependent variable: number of nights in hospital (n = 5464; 23,421 observations); * $p \leq 0.05$, ** $p \leq 0.01$, *** $p \leq 0.001$.

In the GLM, no statistically significant associations were found between direct or indirect migration background and the number of nights in hospital within the last year. The number of nights in hospital within the last year was positively associated with a higher age ($p < 0.001$) and negatively associated with female sex ($p = 0.012$ and 0.017) in both models (Figure S2 in the online Supplementary Materials).

No time trend was observed between the years 2013 and 2019 with regard to hospitalization, or with regard to the number of nights in hospital for any of the groups of persons analyzed.

4. Discussion

4.1. Main Findings

The aim of this study was to analyze the HCSU of persons with direct and indirect migration background compared with persons without migration background in Germany. Persons with direct migration background had lower odds of utilization of doctors within three months than persons without migration background. Lower odds of utilization of doctors within three months were particularly observed in men with direct migration background. Furthermore, for persons utilizing doctors within three months, the number of doctoral visits was lower for persons with direct migration background compared with persons without migration background. For persons with indirect migration background, no differences in the odds of utilization of doctors and in the number of doctoral visits were found. Hence, only a direct migration background can be seen as predisposing factor for determining a lower utilization of doctors as well as a lower number of doctoral visits [7,8]. Consequently, direct migration background might still be associated with fewer need factors determining utilization of doctors, as persons with direct migration background were, on average, healthier than the German population without migration background. However, this effect might not occur in persons with indirect migration background in connection with the utilization of doctors. Furthermore, male sex of persons with direct background can be seen as a predisposing factor for determining a lower utilization of doctors. Yet, it cannot be ruled out that other unobserved determinants of HCSU, such as health and health-related quality of life, and other enabling and need factors have had an influence on direct migration background as predisposing factor for determining a lower utilization of doctors.

With regard to hospitalization, persons with direct migration background had lower odds of hospitalization within one year than persons without migration background. However, no difference was found among hospitalized persons with direct migration background in the number of nights in hospital. For persons with indirect migration background, no differences were found in the odds of hospitalization, nor in the number of nights in hospital compared with persons without migration background. Hence, a direct migration background can also be seen as predisposing factor for determining a lower hospitalization, yet not for the number of nights in hospital [7,8]. Likewise, for determining hospitalization, direct but not indirect migration background might also still be associated with fewer need factors, as persons with direct migration background were, on average, healthier than the German population without migration background. For other predisposing and enabling factors determining HCSU, such as age, marital status, and employment status, it was controlled for in the logistic regression models and in the GLM. However, no inferences can be drawn on those potential determinants of HCSU with regard to migration background based on the current analyses, as they were not added as interactions to the logistic regression models and GLM as independent variables due to a lack of statistical significance. Furthermore, there are other predisposing, enabling, and need factors that determine HCSU, such as education, socio-economic status, or health status, which were not controlled for in the models [7,8]. However, it is known that the perceived need of health care is explainable by education and health beliefs [8], and health status is associated with HCSU. Not considering those determinants as independent variables in the models might have led to omitted variable bias.

4.2. Previous Research and Possible Explanations

One earlier study that also used a representative population survey of persons with direct or indirect migration background as well as those without in Germany could not confirm migration background as a predisposing factor for determining the utilization of doctors and hospitalization [12]. The odds of the utilization of general practitioners were not statistically significantly different, but the odds of the utilization of specialists were statistically significantly lower for persons with direct migration background compared with persons without migration background (OR 0.58), even lower than the odds of utilization of doctors found in the current study (OR 0.73). With regard to the number of doctoral visits and the number of nights in hospital, the study by Glasemer et al. [12] found a lower number for persons with direct migration background compared with persons without migration background.

A systematic literature review on the HCSU of persons with and without migration background in Germany found evidence for an overall lower utilization of specialists for persons with direct migration background [5]. With respect to hospitalization, the evidence found in the review was inconclusive. This is somewhat consistent with the results of the current study, as only differences in the odds of hospitalization were found for persons with direct migration background compared with persons without migration background, but not for persons with indirect migration background, nor in the number of nights in hospital within the last year. Together with the results of the study by Glaesmer et al. [12] and the systematic literature review [5], the current study can confirm continuing disparities in HSCU, especially for persons with direct migration background and connected with the utilization of doctors.

The mechanisms of a lower HCSU, in particular of persons with direct migration background, are not conclusively resolved. One possible reason for differences in HCSU might be inequalities in cultural preferences and health beliefs [5]. Other reasons can be inequalities in access to health care, for example due to a lack of information or communication barriers, or a lack of management of cultural diversity by health care workers [5,43,44]. Accordingly, inequalities are to be reduced by the health policy makers and health care workers making sure that cultural stereotypes are minimized, that health communication is target-specific, and that language barriers are removed [5,45].

In contrast to this, sex is widely known to be a predisposing factor determining HCSU [7,8]; also, in the German health care system, women utilize doctors more often than men [46,47]. Previous research has shown that sex is also associated with HCSU for persons with migration background [48]. However, in the current study, the negative association of utilization of doctors and male sex was even stronger among persons with direct migration background compared with persons without migration background. Possible explanations for this disproportionally low utilization of doctors by men with direct migration background may be a major hurdle with regard to health care services or merely the alleged absence of occasions of visiting a doctor, such as the unawareness of the availability of free preventive check-ups. Another explanation might be the greater proximity of women to the health care system, e.g., through regular gynecological preventive and pregnancy check-ups, as well as the occurrence of maternity health problems. Furthermore, in the current study, women with direct migration background had higher odds of hospitalization within one year than women without migration background. One possible explanation for this might be a higher birth rate among women with direct migration background [49–51]. The open questions and assumptions made with regard to the disproportionally high utilization of doctors and hospitalization by females with migration background still need to be confirmed on the basis of data other than those from the SOEP, which include reasons of utilization.

4.3. Generalizability

The proportion of persons with migration background in the sample of this study was 26%. This proportion corresponds to the proportion of the total German population with migration background. It has to be noted that the integration of the two additional migrant samples into the SOEP ensured this proportional representation of persons with migration background [38]. As the data of the SOEP used in the current study were representative of German households, it can be assumed that the results of this study can be generalized to a certain extent to all adult persons with and without migration background in Germany. However, it has to be acknowledged that 25% of the adult sample that was used for the analysis was removed due to missing information in the number of doctoral visits and number of nights in hospital. Thereby, a disproportionately large number of persons with direct migration background was excluded from the analyses. Furthermore, the persons of the sample that were removed from the analysis due to missing information were younger and more likely to be male. Thus, generalizability may be limited.

Furthermore, it is possible that migrants with better German language skills and with better integration and education were more likely to be included in the SOEP. However, the questionnaires of the SOEP are available in multiple languages and were further translated if necessary [38]. Nevertheless, generalizability of results may, therefore, be further limited.

4.4. Strengths and Limitations

To our knowledge, this was the first analysis of HCSU of persons with direct or indirect migration background as well as those without in Germany, based on a large longitudinal sample. The major strength of this study was the use of data from a representative German household panel that recently integrated additional migrant samples to ensure a proportional representation of persons with migration background. The use of multilevel mixed-effects logistic regressions and GLM with a negative binomial family can also be considered as strength of this study.

However, this study has some limitations that should be mentioned. First, HCSU might be biased due to seasonal effects of utilization of doctors, as the number of doctoral visits were inquired retrospectively only within a period of three months. Second, no migration-specific characteristics, i.e., years since arrival in Germany, language skills, or connection to Germany, were used as explanatory variables in the analyses. Since these variables are most likely to be correlated with migration background and since multicollinearity should be avoided in regression models, it can be assumed that these variables would have been excluded from the models anyway. Third, mixed-effects GLM with truncated zero values were used to analyze the associations between the number of doctoral visits within three months, the number of nights in hospital within the last year, and migration background. However, for zero-truncated data, a zero-truncated negative binomial model would have been more appropriate. Unfortunately, such model is not yet implemented for mixed-effects in Stata [42].

5. Conclusions

Persons with direct migration background still have lower odds of utilization of health care services than persons without migration background in Germany. Here, not only were odds of doctor utilization and hospitalization lower, but also the number of doctoral visits. Fortunately, no differences in the utilization of health care services were found for persons with indirect migration background. The call for a stronger health policy focus on access to health care for persons with direct migration background, especially for men, remains relevant with regard to the results of the present study. In addition, further research is needed to better understand the underlying causes and reasons for reduced HCSU of persons with direct migration background.

Supplementary Materials: The following are available online at https://www.mdpi.com/article/10.3390/ijerph182111640/s1; Table S1: Number of persons with utilization of doctors within three months and zero-truncated mean number of doctoral visits within three months by sociodemographic characteristics and migration background (waves 2013 to 2019, $n = 43,921$); Table S2: Number of persons with hospitalization and zero-truncated mean number of nights in hospital within the last year by sociodemographic characteristics and migration background (waves 2013 to 2019, $n = 43,921$); Figure S1: Adjusted odds of hospitalization within one year of persons without and with direct/indirect migration background, and of females and males without and with direct migration background (waves 2013 to 2019, $n = 43,921$; 179,357 observations); Figure S2: Adjusted zero-truncated number of nights in hospital within the last year of persons without and with direct/indirect migration background, and of females and males without and with direct migration background (waves 2013 to 2019, $n = 5464$; 23,421 observations).

Author Contributions: Conceptualization, T.G. and J.D.; visualization, T.G. and J.D.; review and editing of original draft, T.G., H.-H.K. and J.D.; data curation, T.G.; methodology, T.G.; formal analysis, T.G., H.-H.K. and J.D.; project administration, T.G.; writing of original draft, T.G.; supervision, J.D. and H.-H.K. All authors have read and agreed to the published version of the manuscript.

Funding: This research received no external funding.

Institutional Review Board Statement: This study was a secondary analysis of anonymized data, and therefore an ethics approval was not required. Detailed information on ethical clearance related to the German Socio-Economic Panel Study (SOEP) can be found on the website of the German Institute for Economic Research (DIW), Berlin (https://www.diw.de/soep, accessed on 2 November 2021).

Informed Consent Statement: Participants gave their informed consent prior to data collection. Detailed information on in-formed consent given by the participants related to the German Socio-Economic

Panel Study (SOEP) can be found on the website of the German Institute for Economic Research (DIW), Berlin (https://www.diw.de/soep, accessed on 2 November 2021).

Data Availability Statement: The code used during the current study is available from the corresponding author on reasonable request for all interested researchers. Interested parties may contact the Department of Health Economics and Health Services Research, University Medical Center Hamburg-Eppendorf (contact information: Thomas Grochtdreis, t.grochtdreis@uke.de, +49-40-7410–52405).

Acknowledgments: The data used in this study were made available by the German Socio-Economic Panel Study (SOEP) at the German Institute for Economic Research (DIW), Berlin. We thank the three anonymous reviewers for their valuable comments.

Conflicts of Interest: The authors declare no conflict of interest.

References

1. Statistisches Bundesamt (Destatis). Migration und Integration. Bevölkerung nach Migrationshintergrund und Geschlecht. Available online: https://www.destatis.de/DE/Themen/Gesellschaft-Umwelt/Bevoelkerung/Migration-Integration/Tabellen/liste-migrationshintergrund-geschlecht.html (accessed on 11 June 2021).
2. Salt, J. Trends in Europe's international migration. In *Migration and Health in the European Union*; Rechel, B., Mladovsky, P., Devillé, W., Rijks, B., Petrova-Benedit, R., Eds.; Open University Press: New York, NY, USA, 2011; pp. 17–36.
3. Razum, O.; Karrasch, L.; Spallek, J. Migration: A neglected dimension of inequalities in health? *Bundesgesundheitsblatt Gesundheitsforschung Gesundh.* **2016**, *59*, 259–265. [CrossRef] [PubMed]
4. Razum, O. Migration, Mortalität und der Healthy-migrant-Effekt. In *Gesundheitliche Ungleichheit: Grundlagen, Probleme, Perspektiven*; Richter, M., Hurrelmann, K., Eds.; VS Verlag für Sozialwissenschaften: Wiesbaden, Germany, 2009; pp. 267–282.
5. Klein, J.; von dem Knesebeck, O. Inequalities in health care utilization among migrants and non-migrants in Germany: A systematic review. *Int. J. Equity Health* **2018**, *17*, 160. [CrossRef] [PubMed]
6. Graetz, V.; Rechel, B.; Groot, W.; Norredam, M.; Pavlova, M. Utilization of health care services by migrants in Europe-a systematic literature review. *Br. Med. Bull.* **2017**, *121*, 5–18. [CrossRef] [PubMed]
7. Andersen, R.; Newman, J.F. Societal and individual determinants of medical care utilization in the United States. *Milbank Mem. Fund Q. Health Soc.* **1973**, *51*, 95–124. [CrossRef] [PubMed]
8. Andersen, R.; Davidson, P. Improving access to care in America: Individual and contextual indicators. In *Changing the US Health Care System: Key Issues in Health Services Policy and Management*; Kominski, G.F., Ed.; Jossey-Bass: San Fransisco, CA, USA, 2014.
9. Grochtdreis, T.; König, H.-H.; Dams, J. Health-related quality of life of persons with direct, indirect and no migration background in Germany: A cross-sectional study based on the German Socio-Economic Panel (SOEP). *Int. J. Environ. Res. Public Health* **2021**, *18*, 3665. [CrossRef]
10. Nesterko, Y.; Braehler, E.; Grande, G.; Glaesmer, H. Life satisfaction and health-related quality of life in immigrants and native-born Germans: The role of immigration-related factors. *Qual. Life Res.* **2013**, *22*, 1005–1013. [CrossRef]
11. Morawa, E.; Erim, Y. Health-related quality of life and sense of coherence among Polish immigrants in Germany and indigenous Poles. *Transcult Psychiatry* **2015**, *52*, 376–395. [CrossRef]
12. Glaesmer, H.; Wittig, U.; Braehler, E.; Martin, A.; Mewes, R.; Rief, W. Health care utilization among first and second generation immigrants and native-born Germans: A population-based study in Germany. *Int. J. Public Health* **2011**, *56*, 541–548. [CrossRef]
13. Nesterko, Y.; Turrion, C.M.; Friedrich, M.; Glaesmer, H. Trajectories of health-related quality of life in immigrants and non-immigrants in Germany: A population-based longitudinal study. *Int. J. Public Health* **2019**, *64*, 49–58. [CrossRef]
14. Igel, U.; Brähler, E.; Grande, G. The influence of perceived discrimination on health in migrants. *Psychiatr. Prax.* **2010**, *37*, 183–190. [CrossRef]
15. Razum, O.; Rohrmann, S. The healthy migrant effect: Role of selection and late entry bias. *Gesundheitswesen* **2002**, *64*, 82–88. [CrossRef] [PubMed]
16. Aarabi, G.; Reissmann, D.R.; Seedorf, U.; Becher, H.; Heydecke, G.; Kofahl, C. Oral health and access to dental care-a comparison of elderly migrants and non-migrants in Germany. *Ethn. Health* **2018**, *23*, 703–717. [CrossRef] [PubMed]
17. Aparicio, M.L.; Döring, A.; Mielck, A.; Holle, R. Differences between Eastern European immigrants of German origin and the rest of the German population in health status, health care use and health behaviour: A comparative study using data from the KORA-Survey 2000. *Soz. Prav.* **2005**, *50*, 107–118. [CrossRef] [PubMed]
18. Brenne, S.; David, M.; Borde, T.; Breckenkamp, J.; Razum, O. Are women with and without migration background reached equally well by health services? The example of antenatal care in Berlin. *Bundesgesundheitsblatt Gesundheitsforschung Gesundh.* **2015**, *58*, 569–576. [CrossRef]
19. David, M.; Borde, T.; Brenne, S.; Henrich, W.; Breckenkamp, J.; Razum, O. Caesarean Section Frequency among Immigrants, Second- and Third-Generation Women, and Non-Immigrants: Prospective Study in Berlin/Germany. *PLoS ONE* **2015**, *10*, e0127489. [CrossRef] [PubMed]

20. David, M.; Borde, T.; Brenne, S.; Ramsauer, B.; Henrich, W.; Breckenkamp, J.; Razum, O. Comparison of Perinatal Data of Immigrant Women of Turkish Origin and German Women-Results of a Prospective Study in Berlin. *Geburtshilfe Frauenheilkd* **2014**, *74*, 441–448. [CrossRef] [PubMed]

21. David, M.; Schwartau, I.; Anand Pant, H.; Borde, T. Emergency outpatient services in the city of Berlin: Factors for appropriate use and predictors for hospital admission. *Eur. J. Emerg. Med.* **2006**, *13*, 352–357. [CrossRef] [PubMed]

22. Geyer, S.; Peter, R.; Siegrist, J. Socioeconomic differences in children's and adolescents' hospital admissions in Germany: A report based on health insurance data on selected diagnostic categories. *J. Epidemiol. Community Health* **2002**, *56*, 109–114. [CrossRef] [PubMed]

23. Kavuk, I.; Weimar, C.; Kim, B.T.; Gueneyli, G.; Araz, M.; Klieser, E.; Katsarava, Z. One-year prevalence and socio-cultural aspects of chronic headache in Turkish immigrants and German natives. *Cephalalgia* **2006**, *26*, 1177–1181. [CrossRef]

24. Koller, D.; Lack, N.; Mielck, A. Social differences in the utilisation of prenatal screening, smoking during pregnancy and birth weight–empirical analysis of data from the Perinatal Study in Bavaria (Germany). *Gesundheitswesen* **2009**, *71*, 10–18. [CrossRef]

25. Reime, B.; Lindwedel, U.; Ertl, K.M.; Jacob, C.; Schücking, B.; Wenzlaff, P. Does underutilization of prenatal care explain the excess risk for stillbirth among women with migration background in Germany? *Acta Obstet. Gynecol. Scand.* **2009**, *88*, 1276–1283. [CrossRef]

26. Simoes, E.; Kunz, S.; Schmahl, F. Utilisation gradients in prenatal care prompt further development of the prevention concept. *Gesundheitswesen* **2009**, *71*, 385–390. [CrossRef] [PubMed]

27. Spallek, J.; Lehnhardt, J.; Reeske, A.; Razum, O.; David, M. Perinatal outcomes of immigrant women of Turkish, Middle Eastern and North African origin in Berlin, Germany: A comparison of two time periods. *Arch. Gynecol. Obstet.* **2014**, *289*, 505–512. [CrossRef] [PubMed]

28. Zeeb, H.; Baune, B.T.; Vollmer, W.; Cremer, D.; Krämer, A. Health situation of and health service provided for adult migrants—A survey conducted during school admittance examinations. *Gesundheitswesen* **2004**, *66*, 76–84. [CrossRef]

29. Bächle, C.; Haastert, B.; Holl, R.W.; Beyer, P.; Grabert, M.; Giani, G.; Icks, A. Inpatient and outpatient health care utilization of children and adolescents with type 1 diabetes before and after introduction of DRGs. *Exp. Clin. Endocrinol. Diabetes* **2010**, *118*, 644–648. [CrossRef] [PubMed]

30. Fassmer, A.M.; Luque Ramos, A.; Boiselle, C.; Dreger, S.; Helmer, S.; Zeeb, H. Tabakkonsum und Inanspruchnahme medizinischer Leistungen im Jugendalter–Eine Analyse der KIGGS Daten. *Gesundheitswesen* **2019**, *81*, 17–23. [CrossRef] [PubMed]

31. Gruber, S.; Kiesel, M. Inequality in health care utilization in Germany? Theoretical and empirical evidence for specialist consultation. *J. Public Health* **2010**, *18*, 351–365. [CrossRef]

32. Hirschfeld, G.; Wager, J.; Zernikow, B. Physician consultation in young children with recurrent pain-a population-based study. *PeerJ* **2015**, *3*, e916. [CrossRef]

33. Huber, J.; Lampert, T.; Mielck, A. Inequalities in health risks, morbidity and health care of children by health insurance of their parents (statutory vs. private health insurance): Results of the German KiGGS study. *Gesundheitswesen* **2012**, *74*, 627–638. [CrossRef] [PubMed]

34. Icks, A.; Rosenbauer, J.; Strassburger, K.; Grabert, M.; Giani, G.; Holl, R.W. Persistent social disparities in the risk of hospital admission of paediatric diabetic patients in Germany-prospective data from 1277 diabetic children and adolescents. *Diabet. Med.* **2007**, *24*, 440–442. [CrossRef]

35. Kamtsiuris, P.; Bergmann, E.; Rattay, P.; Schlaud, M. Use of medical services. Results of the German Health Interview and Examination Survey for Children and Adolescents (KiGGS). *Bundesgesundheitsblatt Gesundheitsforschung Gesundh.* **2007**, *50*, 836–850. [CrossRef]

36. Bermejo, I.; Frank, F.; Maier, I.; Hölzel, L.P. Health care utilisation of migrants with mental disorders compared with Germans. *Psychiatr. Prax.* **2012**, *39*, 64–70. [CrossRef]

37. Sundmacher, L.; Kopetsch, T. Waiting times in the ambulatory sector—The case of chronically ill patients. *Int. J. Equity Health* **2013**, *12*, 77. [CrossRef] [PubMed]

38. Brücker, H.; Kroh, M.; Bartsch, S.; Goebel, J.; Kühne, S.; Liebau, E.; Schupp, J. *The New IAB-SOEP Migration Sample: An Introduction into the Methodology and the Contents*. SOEP Survey Paper 216, Series C; DIW Berlin: Berlin/Nürnberg, Germany, 2014.

39. SOEP Group. *SOEP-Core v34–PPATHL: Person-Related Meta-Dataset. SOEP Survey Papers 762: Series D–Variable Descriptions and Coding*; DIW Berlin/SOEP: Berlin, Germany, 2019.

40. *United Nations Statistics Division Standard Country or Area Codes for Statistical Use (M49)*; United Nations: New York, NY, USA, 2020.

41. Rabe-Hesketh, S.; Skrondal, A. *Multilevel and Longitudinal Modeling Using Stata*; Stata Press Publication: College Station, TX, USA, 2012.

42. Hilbe, J.M. *Modeling Count Data*; Cambridge University Press: New York, NY, USA, 2014.

43. Bermejo, I.; Holzel, L.P.; Kriston, L.; Harter, M. Barriers in the attendance of health care interventions by immigrants. *Bundesgesundheitsblatt Gesundheitsforschung Gesundh.* **2012**, *55*, 944–953. [CrossRef] [PubMed]

44. Ciupitu, C.C.; Babitsch, B. Why is it not working? Identifying barriers to the therapy of paediatric obesity in an intercultural setting. *J. Child Health Care* **2011**, *15*, 140–150. [CrossRef] [PubMed]

45. Rechel, B.; Mladovsky, P.; Ingleby, D.; Mackenbach, J.P.; McKee, M. Migration and health in an increasingly diverse Europe. *Lancet* **2013**, *381*, 1235–1245. [CrossRef]

46. Babitsch, B.; Gohl, D.; von Lengerke, T. Re-revisiting Andersen's Behavioral Model of Health Services Use: A systematic review of studies from 1998-2011. *Psychosoc. Med.* **2012**, *9*, Doc11. [CrossRef]
47. Thode, N.; Bergmann, E.; Kamtsiuris, P.; Kurth, B.M. Predictors for ambulatory medical care utilization in Germany. *Bundesgesundheitsblatt Gesundheitsforschung Gesundh.* **2005**, *48*, 296–306. [CrossRef]
48. Alemi, Q.; Stempel, C.; Koga, P.M.; Smith, V.; Danis, D.; Baek, K.; Montgomery, S. Determinants of Health Care Services Utilization among First Generation Afghan Migrants in Istanbul. *Int. J. Environ. Res. Public Health* **2017**, *14*, 201. [CrossRef]
49. Baykara-Krumme, H.; Milewski, N. Fertility Patterns Among Turkish Women in Turkey and Abroad: The Effects of International Mobility, Migrant Generation, and Family Background. *Eur. J. Popul.* **2017**, *33*, 409–436. [CrossRef]
50. Pötsch, O. Fertility in Germany before and after the 2011 census: Still no trend reversal in sight. *Comp. Popul. Stud.* **2016**, *41*, 87–118. [CrossRef]
51. Kaus, W.; Mundil-Schwarz, R. Die Ermittlung der Einwohnerzahlen und der demografischen Strukturen nach dem Zensus 2011. *WISTA Wirtsch. Stat.* **2015**, *4*, 18–38.

International Journal of
***Environmental Research
and Public Health***

MDPI

Article

Health-Related Quality of Life of Persons with Direct, Indirect and No Migration Background in Germany: A Cross-Sectional Study Based on the German Socio-Economic Panel (SOEP)

Thomas Grochtdreis *, Hans-Helmut König and Judith Dams

Department of Health Economics and Health Services Research, Hamburg Center for Health Economics, University Medical Center Hamburg-Eppendorf, Martinistr. 52, 20246 Hamburg, Germany; h.koenig@uke.de (H.-H.K.); j.dams@uke.de (J.D.)
* Correspondence: t.grochtdreis@uke.de; Tel.: +49-40-7410-52405; Fax: +49-40-7410-40546

Abstract: Global migration towards and within Europe remains high, shaping the structure of populations. Approximately 24% of the total German population had a migration background in 2017. The aim of the study was to analyze the association between migration background and health-related quality of life (HrQoL) in Germany. The analyses were based on 2014 and 2016 data of the German Socio-Economic Panel. Differences in sociodemographic characteristics between migrant and non-migrant samples were equal by employment of the entropy balancing weights. HrQoL was measured using the physical (PCS) and mental (MCS) component summary scores of the SF-12v2. Associations between PCS and MCS scores and migration background were examined using Student's *t*-test. The mean PCS and MCS scores of persons with migration background (n = 8533) were 51.5 and 50.9, respectively. Persons with direct migration background had a lower PCS score (-0.55, $p < 0.001$) and a higher MCS score ($+1.08$, $p < 0.001$) than persons without migration background. Persons with direct migration background differed with respect to both physical and mental HrQoL from persons without migration background in the German population. Differences in HrQoL for persons with indirect migration background had $p = 0.305$ and $p = 0.072$, respectively. Causalities behind the association between direct migration background and HrQoL are to be determined.

Keywords: SF-12; surveys and questionnaires; health; quality of life; migrant

Citation: Grochtdreis, T.; König, H.-H.; Dams, J. Health-Related Quality of Life of Persons with Direct, Indirect and No Migration Background in Germany: A Cross-Sectional Study Based on the German Socio-Economic Panel (SOEP). *Int. J. Environ. Res. Public Health* **2021**, *18*, 3665. https://doi.org/10.3390/ijerph18073665

Academic Editors: Heiko Becher, Volker Winkler and Hajo Zeeb

Received: 25 February 2021
Accepted: 29 March 2021
Published: 1 April 2021

Publisher's Note: MDPI stays neutral with regard to jurisdictional claims in published maps and institutional affiliations.

1. Introduction

In Germany, about 19.3 million persons had a migration background in the year 2017, corresponding to a proportion of about 23.6% of the total German population [1]. By definition of the German Institute for Economic Research (DIW Berlin), persons with a direct migration background are persons with their own migration experience born without German citizenship, and persons with an indirect migration background are persons without their own migration experience who were born to at least one parent with direct migration background [2]. This definition is concurrent with the definition of migrants and second-generation migrants by the European Migration Network [3,4]. In general, the spectrum of migration to Germany and the histories of persons with migration background have changed in the past decades, inter alia, through the increasing globalization and through the generations of persons growing up who were born to parents who became sedentary in Germany after passing through the status of guest worker [5,6]. In the 1990s and 2000s, with the fall of the Iron Curtain, the development of new economic flows between European regions, the emergence of new areas of origin, and the entry into the European Union of 12 countries, new migratory flows have emerged in Europe [7].

As the proportion of persons with migration background is increasing throughout Europe, there is also a need for monitoring and extending knowledge on migrants' health and quality of life, not least for the provision of adequate and accessible health care services [8,9]. Reduced health and quality of life might mutually affect engagement in education, work, and social activities and thus have an influence on the integration of persons with migration background as a whole [8]. Reasons for a reduced health and quality of life might be informal barriers to accessing health care, such as a complex interaction between language, communication, and sociocultural factors, but also an interlink with migration and ethno-cultural diversity [8,10,11]. However, the influence of migration, especially for the new generation of persons with migration background and for persons with indirect migration background, is not yet fully resolved.

Earlier studies on health and quality of life of persons with migration background in Germany mainly focused on the healthy-migrant effect and the health and quality of life of migrant workers [12–15], whereas one recent study focused on the trajectories of health-related quality of life (HrQoL) in persons with and without migration background in Germany [16]. In further studies that analyzed the association of HrQoL with migration background in Germany, no association with physical HrQoL has been found, whereas mental HrQoL was negatively associated with migration background [14,16,17]. Indeed, persons with migration background are commonly known to be comparatively healthy, but yet this might not be true for recently migrated persons and persons with indirect migration background, as the sociodemographic diversity of persons with migration background and particular health challenges might be not the same [18]. Also, persons with indirect migration background might have greater health challenges than persons with direct migration background [18–20].

However, not much is known about potential differences in HrQoL of those persons who migrated during the new migratory flows that have emerged in Europe and those persons who are descendants of parents with direct migration background who became sedentary in Germany compared to those persons without migration background. Knowledge about differences in HrQoL between different person groups is particularly of interest for research and policy-makers in order to be able to focus on possible target-specific healthcare services and clinical implications. Therefore, it is necessary to refocus research on health and quality of life of persons with migration background, with a special focus on those persons with direct migration background that had migrated in the 1990s and 2000s as well as on those persons with indirect migration background. Our hypothesis is that, in contrast to the positive effects of migration on HrQoL that have been found for the generation of migrant workers [12–15], the migration background of the more recently migrated persons and those persons with indirect migration background is negatively associated with HrQoL. The aim of this study, therefore, was to analyze the associations between direct, indirect, and no migration background and the physical and mental HrQoL of persons in Germany.

2. Materials and Methods

2.1. Sample

The sample of this study was based on cross-sectional data of the German Socio-Economic Panel (SOEP), provided by the German Institute for Economic Research (DIW Berlin). The SOEP is a representative German household panel with over 20,000 participants annually since 1984, with 35 waves available by 2019. In order to ensure that the previously underrepresented current generation of persons with migration background was represented in the SOEP proportionally to their share of the German population, two additional migrant samples (M1 and M2) were integrated into the SOEP [21].

For the samples M1 and M2, households of persons who had immigrated to Germany during the years 1994 to 2009 and 2010 to 2013 were selected, respectively. By 2019, five waves of the M1 sample and three waves of the M2 sample were available (waves 30 to 34). However, as HrQoL was surveyed only in even years since the year 2002, data sets from the waves 31 and 33 (2014 and 2016) were used (n = 108,903; 100%). Out of these waves, a sample was generated by removing persons with missing information in HrQoL and by using only the initial measurement of HrQoL in the respective wave in order to avoid interdependence of repeated measurements (n = 30,174; 28%). Furthermore, persons with missing information in sociodemographic characteristics were removed, resulting in a net sample of n = 29,642 (27%). A flow chart of the selection process is presented in Figure 1.

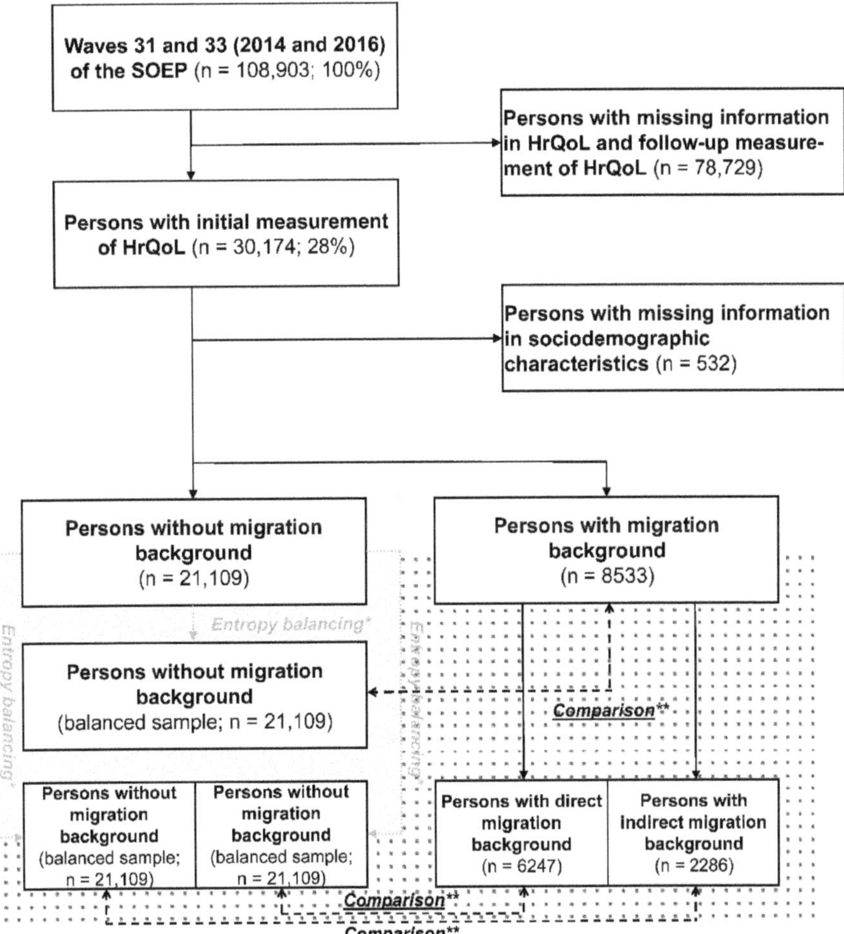

Figure 1. Flow chart of the selection and reweighting process. SOEP: German Socio-Economic Panel, HrQoL: health-related quality of life; * For the sample of persons without migration background, three different weights were derived using entropy balancing, thus differences in means and standard errors of sociodemographic characteristics between persons with and without migration background, persons with direct and without migration background, and persons with indirect and without migration background were equal by employment of the entropy balancing weights in the explanatory models. ** Differences in HrQoL with respect to migration background, direct migration background, and indirect migration background were evaluated by comparison with the respectively reweighted sample of persons without migration background.

2.2. Measures

In order to distinguish between persons with direct and indirect migration background and persons without migration background, a predefined SOEP variable was used. In consistence with the definition of direct and indirect migration background of the DIW Berlin, this variable combined information on country of birth, citizenship, and parental information to derive whether a person had an own migration experience or was born to at least one parent with direct migration background [2]. Persons born in another country than Germany were assigned a direct migration background, and persons born in Germany whose father and/or mother had a migration background were assigned an indirect migration background. Persons born in Germany without parents with migration background were assigned no migration background. Furthermore, it was distinguished between persons with and without migration background, whereby persons with migration background consisted of persons with direct and indirect migration background.

In order to measure HrQoL, a modified version of the standardized questionnaire SF-12v2 was used in the SOEP [22,23]. The SF-12 consists of 12 items with 8 subscales: physical functioning, physical role limitations, bodily pain, general health, vitality, social functioning, emotional role limitations, and mental health [24]. In the SOEP, the SF-12 question about 'work interference due to pain' was removed, whereas one additional SF-36 question about 'severe physical pain' was included. Furthermore, the SF-12 questionnaire in the SOEP deviates to some extent in the layout and in the form and order of the questions [23]. The eight subscales of the SF-12 were Z-transformed by norm-based scoring using mean values and standard deviations of a German normative sample [22,25]. According to Ware et al. [24], a physical component summary (PCS) score was calculated by combining the items of the dimensions physical functioning, role limitations, social functioning, and pain. Furthermore, by combining the items of the dimensions social functioning, emotional role limitations, and mental health, a mental component summary (MCS) score was calculated. Thus, the PCS and the MCS scores represent physical and mental HrQoL on scales ranging between 0 and 100 (with higher scores representing better HrQoL), respectively.

The sociodemographic characteristics age (18–29, 30–39, 40–49, and $\geq$50), sex (female and male), marital status (never married/single, married/in partnership, separated/divorced, and widowed), and employment status (employed fulltime, employed part-time, apprenticeship, marginally employed, and unemployed) were derived from the SOEP. Furthermore, nationality was categorized into German, East European, South European, West and North European, African, Asian, and American/Oceanian countries of origin according to the geographic regions of the Standard Country or Area Codes for Statistical Use (M49) of the United Nations [26].

2.3. Statistical Analysis

In order to evaluate the differences in HrQoL with respect to migration background, direct and indirect migration background, and no migration background, the non-migrant sample was preprocessed in such a way that all samples were balanced with respect to the sociodemographic characteristics. Therefore, three different weights were derived for the non-migrant sample using entropy balancing with the predictors age, sex, marital status, and employment status on the basis of the three migrant (sub-)samples. In the subsequent explanatory models, differences in means and standard errors of those sociodemographic characteristics between persons with and without migration background, persons with direct and without migration background, and persons with indirect and without migration background were equal by employment of the entropy balancing weights [27]. The respective data of persons with migration background, direct and indirect migration background, were used as reference and remained unchanged. A flow chart of the reweighting process is presented in Figure 1.

Descriptive statistics of sociodemographic variables were calculated for the unbalanced samples. PCS and MCS scores were calculated by sociodemographic characteristics for the migrant samples and the balanced non-migrant samples. Furthermore, differences in PCS and MCS scores between persons without migration background and persons with direct and indirect migration background were calculated by sociodemographic characteristics. Differences in PCS and MCS scores by migration background were analyzed using Student's *t*-test. Weights derived by entropy balancing were included for adjustment of differences in sociodemographic characteristics between persons without migration background and persons with direct and indirect migration background.

All analyses were performed using Stata/SE 16.1 (StataCorp, College Station, TX, USA). Entropy balancing was performed using the Stata command 'ebalance' [28]. All applied statistics were two-sided. In total, 13 tests for statistical significance of group differences mean PCS/MCS scores were conducted per sample. Therefore, the level of significance was set at $\alpha = 0.004$ (0.05/13) to correct for multiple significance tests to avoid a type I error [29].

3. Results

3.1. Sample Characteristics

Before entropy balancing, persons with migration background (n = 8533) differed in age, marital status, employment status (all with $p < 0.001$) compared with persons without migration background (n = 21,109), whereas no difference in sex was observed (both 54%; $p = 0.844$). With a mean age of 39 years (42 years/29 years), persons with (direct/indirect) migration background were younger than persons without migration background, who were on average 50 years old ($p < 0.001$). Furthermore, the majority of persons with migration background had a German nationality (53.4%), followed by 15.6% and 13.4% with a nationality from a Southern European country and an Eastern European country, respectively. The sociodemographic characteristics of the sample pre-balancing are shown in Table 1.

Table 1. Sociodemographic characteristics of the sample, pre-balancing (survey years 2014 and 2016).

Sociodemographic Characteristic	Persons without Migration Background Pre-Balancing (n = 21,109)	Persons with Migration Background (n = 8533)	Persons with Direct Migration Background (n = 6247)	Persons with Indirect Migration Background (n = 2286)
Age: Mean (SE)	49.55 (0.12) **	38.73 (0.16) **	42.21 (0.18) **	29.25 (0.22) **
Sex: N (%)				
Female	11,370 (53.86)	4580 (53.67)	3383 (54.15)	1197 (52.36)
Male	9739 (46.14)	3953 (46.33)	2864 (45.85)	1089 (47.64)
Grouped age: N (%)				
18–29	3200 (15.16) **	2438 (28.57) **	1129 (18.07) **	1309 (57.26) **
30–39	3066 (14.52)	2454 (28.76)	1856 (29.71)	598 (26.16)
40–49	4642 (21.99)	1885 (22.09)	1620 (25.93)	265 (11.59)
≥50	10,201 (48.33)	1756 (20.58)	1642 (26.28)	114 (4.99)
Marital status: N (%)				
Never married/single	5210 (24.68) **	2638 (30.92) **	1210 (19.37) **	1428 (62.47) **
Married/in partnership	12,009 (56.89)	5038 (59.04)	4315 (69.07)	723 (31.63)
Separated/divorced	2611 (12.37)	692 (8.11)	568 (9.09)	124 (5.42)
Widowed	1279 (6.06)	165 (1.93)	154 (2.47)	11 (0.48)
Employment status: N (%)				
Employed fulltime	7908 (37.46) **	3207 (37.58) **	2444 (39.12) **	763 (33.38) **
Employed part-time	3098 (14.68)	1070 (12.54)	838 (13.41)	232 (10.15)
Apprenticeship	573 (2.71)	382 (4.48)	151 (2.42)	231 (10.10)
Marginally employed	1348 (6.39)	747 (8.75)	526 (8.42)	221 (9.67)
Unemployed	8182 (38.76)	3127 (36.65)	2288 (36.63)	839 (36.70)
Nationality [1]: N (%)				
German	21,109 (100.00) **	4554 (53.37) **	2798 (44.79) **	1756 (76.82) **
East European	-	1139 (13.35)	1131 (18.10)	8 (0.35)
South European	-	1311 (15.36)	1015 (16.25)	296 (12.95)
West and North European [2]	-	284 (3.33)	260 (4.16)	24 (1.05)
African	-	125 (1.46)	121 (1.94)	4 (0.17)
Asian	-	1004 (11.77)	814 (13.03)	190 (8.31)
American/Oceanian	-	98 (1.12)	91 (1.46)	7 (0.31)
Stateless	-	18 (0.21)	17 (0.27)	1 (0.04)

Comments: SE: Standard error; comparison of mean age of persons with and without migration background was analyzed using Student's *t*-test; comparison of categorical characteristics of persons with and without migration background was analyzed using Pearson's chi^2 test; comparison of mean age of persons with direct and indirect migration background was analyzed using Student's *t*-test; comparison of categorical characteristics of persons with direct and indirect migration background was analyzed using Pearson's chi^2 test; [1] Nationality was not considered for balancing; [2] Without German nationality; ** $p \leq 0.001$.

After balancing, the migrant samples were similar to the non-migrant samples with respect to sociodemographic characteristics. The majority of the total sample was female (53.7%) and was either employed fulltime (37.6%) or unemployed (36.7%). Furthermore,

59.0% were married or in a partnership, 30.9% had never been married or were single, and 8.1% were separated or divorced.

3.2. Differences in PCS Scores between Persons with and without Migration Background

The difference in PCS scores between persons with and without migration background had $p = 0.009$ (51.5 vs. 51.9; Table 2). However, women with migration background had lower PCS scores than women without migration background (51.0 vs. 51.6, $p < 0.001$). Persons with migration background aged 50 years and older had a lower PCS score than persons without migration background of the same age (44.2 vs. 46.0, $p < 0.001$). Furthermore, persons being married or in a partnership with migration background had a higher PCS score than persons without migration background with the same marital status (50.4 vs. 51.1, $p < 0.001$).

Table 2. Mean PCS and MCS scores by sociodemographic characteristics and migration background (survey years 2014 and 2016).

Sociodemographic Characteristic	Mean PCS (SE)		Mean MCS (SE)	
	Persons without Migration Background (Balanced Sample; n = 21,109)	Persons with Migration Background (n = 8533)	Persons without Migration Background (Balanced Sample; n = 21,109)	Persons with Migration Background (n = 8533)
Total sample	51.87 (0.08)	51.52 (0.11)	49.96 (0.09) **	50.87 (0.10) **
Sex				
Female	51.57 (0.12) **	50.95 (0.15) **	48.86 (0.13) **	49.76 (0.14) **
Male	52.22 (0.12)	52.19 (0.15)	51.24 (0.13) **	52.15 (0.15) **
Grouped age				
18–29	55.28 (0.15)	55.35 (0.14)	49.60 (0.20) **	50.99 (0.19) **
30–39	53.19 (0.18)	53.51 (0.17)	49.51 (0.20) **	50.71 (0.19) **
40–49	51.20 (0.15)	50.79 (0.22)	49.92 (0.16)	50.56 (0.23)
≥50	46.03 (0.12) **	44.22 (0.26) **	51.18 (0.12)	51.24 (0.25)
Marital status				
Never married/single	54.58 (0.13)	55.09 (0.15)	49.60 (0.16) **	50.73 (0.18) **
Married/in partnership	51.12 (0.11) **	50.40 (0.14) **	50.46 (0.12) **	51.32 (0.13) **
Separated/divorced	48.91 (0.27)	48.18 (0.41)	47.61 (0.30)	48.63 (0.43)
Widowed	44.00 (0.14)	42.73 (0.91)	50.42 (0.48)	48.77 (0.91)
Employment status				
Employed fulltime	53.02 (0.11)	53.16 (0.14)	50.73 (0.12) **	52.14 (0.15) **
Employed part-time	52.45 (0.17)	51.67 (0.28)	49.94 (0.19)	50.34 (0.29)
Apprenticeship	54.94 (0.29)	55.04 (0.36)	50.90 (0.41)	50.65 (0.47)
Marginally employed	52.56 (0.29)	52.16 (0.33)	48.82 (0.35)	49.99 (0.33)
Unemployed	49.96 (0.17)	49.21 (0.20)	49.34 (0.18)	49.98 (0.19)
Nationality				
German	51.87 (0.08)	51.60 (0.14)	49.96 (0.09)	50.42 (0.14)
East European	-	52.39 (0.26)	-	52.63 (0.26)
South European	-	51.39 (0.27)	-	51.13 (0.26)
West and North European [1]	-	51.47 (0.57)	-	51.07 (0.60)
African	-	52.20 (0.84)	-	50.82 (0.76)
Asian	-	50.16 (0.33)	-	50.62 (0.31)
American/Oceanian	-	53.41 (0.93)	-	50.52 (0.93)
Stateless	-	47.34 (2.52)	-	48.12 (2.39)

PCS: Physical Component Summary; MCS: Mental Component Summary; SE: standard error; comparison of mean PCS and MCS scores by migration background were analyzed using Student's *t*-test; [1] without German nationality; ** $p \leq 0.001$.

3.3. Differences in MCS Scores between Persons with and without Migration Background

The mean MCS score of persons without migration background was 50.0 (Table 2). The mean MCS score of persons with migration background was higher (50.9, $p < 0.001$). Both women and men with migration background had higher MCS scores than women and men without migration background (49.8 vs. 48.9 and 52.3 vs. 51.2, both with $p < 0.001$). Furthermore, the MCS scores of persons aged less than 50 years was higher for persons with migration background compared with persons without migration background. Thereby, the MCS scores of persons with migration background decreased with higher age (51.0 to 50.6), whereas for persons without migration background, MCS scores increased with higher age (49.6 to 49.9, all with $p < 0.001$). Persons being married or in a partnership as well as never married or single with migration background had a higher MCS score than persons without migration background with the same marital status (50.7 vs. 49.6 and 51.3 vs. 50.7, all with $p < 0.001$).

3.4. Differences in PCS and MCS Scores between Persons with Direct/Indirect and without Migration Background

Persons with direct migration background (n = 6247) had a lower PCS score compared with persons without migration background (-0.52, $p = 0.001$; Table 3). The MCS score was higher ($+1.11$, $p < 0.001$). The differences in PCS and MCS scores of persons with indirect migration background compared with persons without migration background had $p = 0.305$ and $p = 0.072$, respectively ($+0.19$ and $+0.41$, respectively). Mean PCS and MCS scores by sociodemographic characteristics by migration background are shown in Tables S1 and S2 in the online Supplementary Materials.

Table 3. Differences in PCS and MCS scores by sociodemographic characteristics between persons with direct/indirect and without migration background (survey years 2014 and 2016).

Sociodemographic Characteristic	Mean Diff. [1] in PCS (SE)		Mean Diff. [1] in MCS (SE)	
	Persons with Direct Migration Background (n = 6247)	Persons with Indirect Migration Background (n = 2286)	Persons with Direct Migration Background (n = 6247)	Persons with Indirect Migration Background (n = 2286)
Total sample	−0.52 (0.16) **	0.19 (0.19)	1.11 (0.16) **	0.41 (0.23)
Sex				
Female	−0.84 (0.22) **	−0.05 (0.28)	1.12 (0.22) **	0.37 (0.32)
Male	−0.15 (0.22)	0.45 (0.25)	1.10 (0.22) **	0.47 (0.32)
Grouped age				
18–29	0.06 (0.28)	−0.01 (0.23)	2.05 (0.35) **	0.91 (0.32) *
30–39	0.33 (0.27)	0.39 (0.36)	1.75 (0.31) **	−0.48 (0.43)
40–49	−0.58 (0.30)	1.08 (0.56)	0.71 (0.30)	0.05 (0.63)
≥50	−1.82 (0.30) **	−0.70 (1.04)	0.11 (0.29)	0.30 (0.90)
Marital status				
Never married/single	0.40 (0.29)	0.40 (0.21)	1.61 (0.32) **	0.79 (0.30)
Married/in partnership	−0.76 (0.19) **	0.16 (0.35)	1.05 (0.19) **	−0.29 (0.37)
Separated/divorced	−0.47 (0.54)	−1.79 (1.04)	1.30 (0.57)	0.10 (1.09)
Widowed	−1.18 (1.06)	−2.45 (3.52)	−1.97 (1.08)	1.97 (2.70)
Employment status				
Employed fulltime	0.06 (0.20)	0.28 (0.29)	1.67 (0.22) **	0.61 (0.36)
Employed part-time	−0.90 (0.36)	−0.39 (0.63)	0.87 (0.37)	−1.29 (0.69)
Apprenticeship	−0.33 (0.67)	0.31 (0.53)	0.29 (0.88)	−0.50 (0.71)
Marginally employed	−0.55 (0.52)	−0.17 (0.57)	1.19 (0.54)	0.86 (0.72)
Unemployed	−1.02 (0.31) **	0.33 (0.34)	0.64 (0.30)	0.84 (0.42)
Nationality				
German	−1.26 (0.21) **	0.41 (0.21)	0.82 (0.21) **	0.04 (0.26)

PCS: Physical Component Summary; MCS: Mental Component Summary; SE: standard error; comparison of mean PCS scores by migration background were analyzed using Student's *t*-test; [1] mean difference between persons with direct/indirect migration background and persons without migration background (balanced samples); * $p \leq 0.004$, ** $p \leq 0.001$.

4. Discussion

4.1. Main Findings

The aim of this study was to analyze the associations between migration background and physical and mental HrQoL. Mental HrQoL was higher, and physical HrQoL was lower among persons with direct migration background compared with persons without migration background in Germany. No differences in HrQoL were observed between persons with indirect and without migration background.

Direct comparison of HrQoL of persons with migration background and persons without migration background based on samples from the German general population might be biased, because persons with migration background are younger, less likely to be female, more likely to be single, and more often unemployed [30,31]. In order to reduce the imbalance in sociodemographic characteristics in samples from general populations, it is possible to either control for imbalances by using multiple regression models or reweight parts of the samples using propensity score matching methods or entropy balancing [27]. To date, sample-reweighting techniques have been applied only rarely in HrQoL studies based on population surveys. In the last decade, European studies based on National Health and Wellness Surveys used propensity score matching to compare persons with a certain disease with controls without this disease in order to estimate HrQoL differences [32–34]. Previous studies that analyzed the association of HrQoL and migration background in Germany used regression analysis to control for imbalances in sociodemographic characteristics [14,16,17]. In the current study, the reweighting of non-migrant comparison samples by entropy balancing resulted in a balance in sociodemographic characteristics. Pre-balancing, the persons of the migrant samples were younger compared with those of the non-migrant sample. Also, martial and employment statuses were differently distributed over migrant and non-migrant samples. Thus, reweighting the comparison samples for imbalances in sociodemographic characteristics was necessary to avoid bias in HrQoL differences, as older age, unemployment, and having never been married, separated, or divorced were found to be associated with lower HrQoL [35]. By employment of the entropy balancing weights in the explanatory models, differences in physical and mental HrQoL between persons with and without migration background can be regarded as unbiased, at least with respect to specific sociodemographic characteristics. Thereby, the weights also take into account differences in the variances of physical and mental HrQoL between the migrant and non-migrant groups. Furthermore, compared to other preprocessing techniques, valuable information is retained from the data, as no information has been discarded by non-matching [27].

Compared with an unbalanced German representative normative sample that has been used to compute SF-12 summary scores, persons with migration background had a higher physical HrQoL (51.5 vs. 50.0) [23]. The mental HrQoL of persons with migration background was also higher compared with that of the German normative sample (50.9 vs. 50.0). As the current sample of persons with migration background and the normative sample were not balanced with respect to sociodemographic characteristics, the difference in physical HrQoL might be explained by the younger age of persons with migration background compared with the persons of the normative sample (39 vs. 48 years) and the negative association between older age and physical HrQoL.

4.2. Previous Research and Possible Explanations

Earlier studies from Germany that were based on cross-sectional and longitudinal data found inconclusive results concerning HrQoL of migrant populations. Concerning physical HrQoL, in one sample that was based on longitudinal data of the SOEP, baseline physical HrQoL was higher for persons with direct migration background compared with persons without migration background [16]. The current study and two other studies, however, found lower physical HrQoL for persons with direct migration background compared with persons without migration background [14,36]. One representative population-based study found no difference in physical HrQoL between persons with direct migration background and persons without migration background [17]. No study, included the current study, found any association between physical HrQoL and indirect migration background [14,16,17,36]. Earlier statements concerning different physical HrQoL of migrant and non-migrant samples contained adverse employment situations of migrants and non-migrants as a possible explanation for the difference [16,37]. However, in the current sample, the difference in physical HrQoL persisted after balancing out differences in employment status between non-migrant and migrant samples, indicating another (unobserved) reason for this difference in physical HrQoL.

Concerning mental HrQoL, one study that used a chain sampling technique found lower mental HrQoL for persons with Polish migration background compared with persons without migration background, whereas the current study found higher mental HrQoL for persons with direct migration background compared with persons without migration background [36]. Other studies did not find any associations between mental HrQoL and direct migration background [14,16,17]. In the sample that was based on longitudinal data of the SOEP, a higher baseline mental HrQoL was found for persons with indirect migration background compared with persons without migration background [16], whereas no difference was found in the current study or in any of the other studies [14,17,36].

All this amounts to the fact that the current generation of persons with migration background who had immigrated to Germany during the years 1994 to 2013 still have a lower physical HrQoL compared with persons without migration background. In addition, those persons with migration background seem to be better off with respect to mental HrQoL compared with persons without migration background and probably also compared with older migrant generations. A possible explanation for the difference in physical and mental HrQoL might be that there is a certain probability of interpreting the meaning and answering items of the SF-12 in a different way by persons with different nationalities [38,39]. Furthermore, the lower physical HrQoL might be affected by unknown migration-specific characteristics or an inadequate access to healthcare services for persons with migration background [40]. Finally, there is a distinct need for investigation of the reasons of the better mental HrQoL of persons with migration background and subsequently their reinforcement.

4.3. Generalizability

The data of the SOEP used in the current study were representative of German households. The additional two migrant samples M1 and M2 that were integrated into the SOEP, however, were selected to represent the countries of origin of migrants which recently became increasingly important; thus, certain migrant groups were overrepresented [21]. Groups of countries from new Eastern European Union and southern European Union member states and Arab and Islamic countries were overrepresented in the sample, as immigration from those countries increased significantly in the last decade. Furthermore, households of the so-called guest workers were overrepresented in order to represent their descendants better in the SOEP. In 2017, about 23.6% of the total population had a migration background in Germany [1]. This proportion was lower than the proportion of persons with migration background in the sample of the current study (28.8%).

As this study can be considered exploratory, statistical significance of group differences and associations with respect to HrQoL found using statistical tests should be interpreted with caution. According to Wasserstein et al. [41,42], conclusions should not be drawn only on statistical significance in conjunction with arbitrary levels of significance, and differences and associations are neither present nor absent just because of statistical (in)significance. Indeed, the differences and associations with respect to HrQoL should be benchmarked against their real-life relevance. Such a benchmark could be the minimal (clinically) important difference of the PCS and MCS, which were commonly defined to be 3.5 to 5.0 (e.g., [43,44]). With respect to this benchmark, the associations with regard to HrQoL found in the current study should be interpreted restrainedly.

4.4. Strengths and Limitations

To our knowledge, this was the first study of HrQoL between persons with migration background and persons without migration background in Germany that used data from the migrant samples from the SOEP. One major strength of this study is the use of a large migrant and non-migrant sample that was based on a representative German household panel.

However, this study also has some limitations. First, the analyses of this study were based on cross-sectional data, and therefore information on the temporal ordering of causes and effects was not evaluated. Furthermore, only data sets from the years 2014 and 2016 were available at the time of the analysis. However, as the aim of the study was to explore associations between migration background and the physical and mental HrQoL, the recency of data was not a truly decisive factor. Future confirmatory studies should nevertheless include also forthcoming SOEP data sets and preferably analyze the data longitudinally. Second, better integrated and more highly educated persons with migration background, i.e., with better German language skills, might have been included in the migrant samples of the SOEP more probably. In order to reduce this bias, the questionnaires of the SOEP were translated into English, Russian, Turkish, Polish and Romanian, and the option of taking an interpreter was given for the interviews [21]. Furthermore, by using entropy balancing as a sample-reweighting technique, the balanced sample of persons without migration background was similar to the samples of persons with migration background with respect to specific sociodemographic characteristics. Nevertheless, the results of this study might not be generalizable to all persons with migration background in Germany. Third, as this study primarily aimed to analyze associations between migration background and HrQoL, no migration-specific characteristics, such as years since arrival in Germany, age at arrival, or citizenship, were considered.

5. Conclusions

After the reduction of imbalance in sociodemographic characteristics between the migrant and non-migrant samples, persons with direct migration background had a lower physical HrQoL and a higher mental HrQoL than persons without migration background. It has to be highlighted that persons who are descendants of parents with direct migration background who became sedentary in Germany did not differ with respect to physical and mental HrQoL compared to persons without migration background. Appropriate measures with respect to target-specific healthcare services and clinical implications should be taken by researchers and policy-makers in order to address the reduced physical HrQoL of persons with direct migration background who migrated during the new migratory flows that have emerged in Europe. However, with respect of the exploratory character of this study and the doubtful benchmark of those associations, such advice should be adopted with caution. Notwithstanding, further research is needed in order to determine the causalities behind the lower physical HrQoL and the higher mental HrQoL of persons with direct migration in Germany.

Supplementary Materials: The following are available online at https://www.mdpi.com/article/10.3390/ijerph18073665/s1; Table S1: Mean PCS scores by sociodemographic characteristics by migration background (survey years 2014 and 2016); Table S2: Mean MCS scores by sociodemographic characteristics by migration background (survey years 2014 and 2016).

Author Contributions: Conceptualization, T.G. and J.D.; visualization, T.G. and J.D.; review and editing of original draft, T.G., H.-H.K., and J.D.; data curation, T.G.; methodology, T.G.; formal analysis, T.G., H.-H.K., and J.D.; project administration, T.G.; writing of original draft, T.G.; supervision, J.D. and H.-H.K. All authors have read and agreed to the published version of the manuscript.

Funding: This research received no external funding.

Institutional Review Board Statement: This study was a secondary analysis of anonymized data, and therefore an ethics approval was not required. Detailed information on ethical clearance related to the German Socio-Economic Panel Study (SOEP) can be found on the website of the German Institute for Economic Research (DIW), Berlin (https://www.diw.de/soep).

Informed Consent Statement: Participants gave their informed consent prior to data collection. Detailed information on informed consent given by the participants related to the German Socio-Economic Panel Study (SOEP) can be found on the website of the German Institute for Economic Research (DIW), Berlin (https://www.diw.de/soep).

Data Availability Statement: The code used during the current study is available from the corresponding author on reasonable request for all interested researchers. Interested parties may contact the Department of Health Economics and Health Services Research, University Medical Center Hamburg-Eppendorf (contact information: Thomas Grochtdreis, t.grochtdreis@uke.de, +49-40-7410-52405).

Acknowledgments: The data used in this study were made available by the German Socio-Economic Panel Study (SOEP) at the German Institute for Economic Research (DIW), Berlin. We thank the three anonymous reviewers for their valuable comments.

Conflicts of Interest: The authors declare no conflict of interest.

References

1. Statistisches Bundesamt. *Statistisches Jahrbuch 2018. Deutschland und Internationales*; Statistisches Bundesamt: Wiesbaden, Germany, 2018.
2. SOEP Group. *SOEP-Core v34—PPATHL: Person-Related Meta-Dataset. SOEP Survey Papers 762: Series D—Variable Descriptions and Coding*; DIW; SOEP: Berlin, Germany, 2019.
3. European Migration Network. Second-Generation Migrant. Available online: https://ec.europa.eu/home-affairs/what-we-do/networks/european_migration_network/glossary_search/second-generation-migrant_en (accessed on 12 March 2021).
4. European Migration Network. Migrant. 2021. Available online: https://ec.europa.eu/home-affairs/what-we-do/networks/european_migration_network/glossary_search/migrant_en (accessed on 12 March 2021).
5. Razum, O.; Karrasch, L.; Spallek, J.; Razum, M.O.; Spallek, J.P.D.J. Migration. *Bundesgesundheitsblatt Gesundh. Gesundh.* **2016**, *59*, 259–265. [CrossRef]
6. Razum, O. Migration, Mortalität und der Healthy-migrant-Effekt. In *Gesundheitliche Ungleichheit: Grundlagen, Probleme, Perspektiven*; Richter, M., Hurrelmann, K., Eds.; VS Verlag für Sozialwissenschaften: Wiesbaden, Germany, 2009; pp. 267–282.
7. Salt, J. Trends in Europe's international migration. In *Migration and Health in the European Union*; Rechel, B., Mladovsky, P., Devillé, W., Eds.; Open University Press: New York, NY, USA, 2011; pp. 17–36.
8. Nørredam, M.; Krasnik, A. Migrants' access to health services. In *Migration and Health in the European Union*; Rechel, B., Mladovsky, P., Devillé, W., Eds.; Open University Press: New York, NY, USA, 2011; pp. 67–78.
9. Rechel, B.; Mladovsky, P.; Devillé, W. Monitoring the health of migrants. In *Migration and Health in the European Union*; Rechel, B., Mladovsky, P., Devillé, W., Eds.; Open University Press: New York, NY, USA, 2011; pp. 81–98.
10. Durieux-Paillard, S. Differences in language, religious beliefs and culture: The need for culturally responsive health services. In *Migration and Health in the European Union*; Rechel, B., Mladovsky, P., Devillé, W., Eds.; Open University Press: New York, NY, USA, 2011; pp. 203–212.
11. Führer, A.; Tiller, D.; Brzoska, P.; Korn, M.; Gröger, C.; Wienke, A. Health-Related Disparities among Migrant Children at School Entry in Germany. How does the Definition of Migration Status Matter? *Int. J. Environ. Res. Public Health* **2019**, *17*, 212. [CrossRef]
12. Andersen, R.; Newman, J.F. Societal and Individual Determinants of Medical Care Utilization in the United States. *Milbank Mem. Fund Q. Health Soc.* **1973**, *51*, 95. [CrossRef] [PubMed]
13. Igel, U.; Brähler, E.; Grande, G. Der Einfluss von Diskriminierungserfahrungen auf die Gesundheit von MigrantInnen. *Psychiatr. Prax.* **2010**, *37*, 183–190. [CrossRef]
14. Nesterko, Y.; Braehler, E.; Grande, G.; Glaesmer, H. Life satisfaction and health-related quality of life in immigrants and native-born Germans: The role of immigration-related factors. *Qual. Life Res.* **2012**, *22*, 1005–1013. [CrossRef]
15. Razum, O.; Rohrmann, S. The healthy migrant effect: Role of selection and late entry bias. *Das Gesundh.* **2002**, *64*, 82–88. [CrossRef]
16. Nesterko, Y.; Turrión, C.M.; Friedrich, M.; Glaesmer, H. Trajectories of health-related quality of life in immigrants and non-immigrants in Germany: A population-based longitudinal study. *Int. J. Public Health* **2018**, *64*, 49–58. [CrossRef]
17. Glaesmer, H.; Wittig, U.; Braehler, E.; Martin, A.; Mewes, R.; Rief, W. Health care utilization among first and second generation immigrants and native-born Germans: A population-based study in Germany. *Int. J. Public Health* **2011**, *56*, 541–548. [CrossRef]
18. Rechel, B.; Mladovsky, P.; Devillé, W.; Rijks, B.; Petrova-Benedict, R.; McKee, M. Migration and health in the European Union: An introduction. In *Migration and Health in the European Union*; Rechel, B., Mladovsky, P., Devillé, W., Eds.; Open University Press: New York, NY, USA, 2011; pp. 3–13.
19. Ingleby, D. *European Research on Migration and Health. Background Paper Developed within the Framework of the IOM Project "Assisting Migrants and Communities (AMAC): Analysis of Social Determinants of Health and Health Inequalities"*; International Organization for Migration: Geneva, Switzerland, 2009.
20. Gushulak, B. Monitoring Migrants' Health. In *Health of Migrants—The Way Forward*; Report of a Global Consultation, Madrid, Spain, 3–5 March 2010; World Health Organization: Geneva, Switzerland, 2010; pp. 28–42.
21. Brücker, H.; Kroh, M.; Bartsch, S.; Goebel, J.; Kühne, S.; Liebau, E.; Trübswetter, P.; Tucci, I.; Schupp, J. *The New IAB-SOEP Migration Sample: An Introduction into the Methodology and the Contents*; SOEP Survey Paper 216, Series, C; DIW: Berlin, Germany; Nürnberg, Germany, 2014.
22. Andersen, H.H.; Mühlbacher, A.; Nübling, M. *Die SOEP-Version des SF 12 als Instrument Gesundheitsökonomischer Analysen*; German Institute for Economic Research: Berlin, Germany, 2007.
23. Schupp, J.; Wagner, G.; Nübling, M.; Andersen, H.H.; Mühlbacher, A. Computation of Standard Values for Physical and Mental Health Scale Scores Using the SOEP Version of SF12v2. *Schmollers Jahrb.* **2007**, *127*, 171–182.
24. Ware, J.J.; Kosinski, M.; Keller, S.D. A 12-Item Short-Form Health Survey: Construction of scales and preliminary tests of reliability and validity. *Med. Care* **1996**, *34*, 220–233. [CrossRef]
25. Nübling, M.; Andersen, H.H.; Mühlbacher, A. *Entwicklung eines Verfahrens zur Berechnung der Körperlichen und Psychischen Summenskalen auf Basis der SOEP-Version des SF 12 (Algorithmus)*; German Institute for Economic Research: Berlin, Germany, 2006.

26. United Nations Statistics Division. *Standard Country or Area Codes for Statistical Use (M49)*; United Nations: New York, NY, USA, 2006.

27. Hainmueller, J. Entropy Balancing for Causal Effects: A Multivariate Reweighting Method to Produce Balanced Samples in Observational Studies. *Political Anal.* **2012**, *20*, 25–46. [CrossRef]

28. Hainmüller, J.; Xu, Y. Ebalance: A Stata Package for Entropy Balancing. *J. Stat. Softw.* **2013**, *54*, 1–18. [CrossRef]

29. Dickhaus, T. Classes of multiple test procedures. In *Simultaneous Statistical Inference: With Applications in the Life Sciences*; Springer: Berlin/Heidelberg, Germany, 2014; pp. 29–45.

30. Statistisches Bundesamt. *Bevölkerung und Erwerbstätigkeit. Schutzsuchende Ergebnisse des Ausländerzentralregisters 2018*; Statistisches Bundesamt: Wiesbaden, Germany, 2018.

31. Statistisches Bundesamt. *Statistisches Jahrbuch 2019. Deutschland und Internationales*; Statistisches Bundesamt: Wiesbaden, Germany, 2019.

32. daCosta DiBonaventura, M.; Yuan, Y.; Wagner, J.-S.; L'Italien, G.J.; Lescrauwaet, B.; Langley, P. The burden of viral hepatitis C in Europe: A propensity analysis of patient outcomes. *Eur. J. Gastroenterol. Hepatol.* **2012**, *24*, 869–877. [CrossRef] [PubMed]

33. DiBonaventura, M.; Richard, L.; Kumar, M.; Forsythe, A.; Flores, N.M.; Moline, M. The Association between Insomnia and Insomnia Treatment Side Effects on Health Status, Work Productivity, and Healthcare Resource Use. *PLoS ONE* **2015**, *10*, e0137117. [CrossRef] [PubMed]

34. Vo, P.; Fang, J.; Bilitou, A.; Laflamme, A.K.; Gupta, S. Patients' perspective on the burden of migraine in Europe: A cross-sectional analysis of survey data in France, Germany, Italy, Spain, and the United Kingdom. *J. Headache Pain* **2018**, *19*, 1–11. [CrossRef] [PubMed]

35. Grochtdreis, T.; König, H.-H.; Riedel-Heller, S.G.; Dams, J. Health-Related Quality of Life of Asylum Seekers and Refugees in Germany: A Cross-Sectional Study with Data from the German Socio-Economic Panel. *Appl. Res. Qual. Life* **2020**, 1–19. [CrossRef]

36. Morawa, E.; Erim, Y. Health-related quality of life and sense of coherence among Polish immigrants in Germany and indigenous Poles. *Transcult. Psychiatry* **2014**, *52*, 376–395. [CrossRef]

37. Seifert, W. Occupational and Economic Mobility and Social Integration of Mediterranean Migrants in Germany. *Eur. J. Popul.* **1997**, *13*, 1–16. [CrossRef]

38. Gibbons, C.J.; WHOQOL Group; Skevington, S.M. Adjusting for cross-cultural differences in computer-adaptive tests of quality of life. *Qual. Life Res.* **2018**, *27*, 1027–1039. [CrossRef] [PubMed]

39. Benítez-Borrego, S.; Mancho-Fora, N.; Farràs-Permanyer, L.; Urzúa-Morales, A.; Guàrdia-Olmos, J. Differential Item Functioning of WHOQOL-BREF in nine Iberoamerican countries. *Rev. Iberoam. Psicol. Salud* **2016**, *7*, 51–59. [CrossRef]

40. Razum, O.; Saß, A.-C. Migration und Gesundheit: Interkulturelle Öffnung bleibt eine Herausforderung. *Bundesgesundheitsblatt Gesundheitsforschung Gesundheitsschutz* **2015**, *58*, 513–514. [CrossRef] [PubMed]

41. Amrhein, V.; Greenland, S.; McShane, B. Scientists rise up against statistical significance. *Nat. Cell Biol.* **2019**, *567*, 305–307. [CrossRef]

42. Wasserstein, R.L.; Schirm, A.L.; Lazar, N.A. Moving to a World Beyond "$p < 0.05$". *Am. Stat.* **2019**, *73*, 1–19. [CrossRef]

43. Fontana, M.A.; Lyman, S.; Sarker, G.K.; Padgett, D.E.; MacLean, C.H. Can Machine Learning Algorithms Predict Which Patients Will Achieve Minimally Clinically Important Differences from Total Joint Arthroplasty? *Clin. Orthop. Relat. Res.* **2019**, *477*, 1267–1279. [CrossRef] [PubMed]

44. Busse, J.W.; Bhandari, M.; A Einhorn, T.; Schemitsch, E.; Heckman, J.D.; Iii, P.T.; Leung, K.-S.; Heels-Ansdell, D.; Makosso-Kallyth, S.; Della Rocca, G.J.; et al. Re-evaluation of low intensity pulsed ultrasound in treatment of tibial fractures (TRUST): Randomized clinical trial. *BMJ* **2016**, *355*, i5351. [CrossRef]

International Journal of
*Environmental Research
and Public Health*

MDPI

Article

Feasibility of a Culturally Adapted Dietary Weight-Loss Intervention among Ghanaian Migrants in Berlin, Germany: The ADAPT Study

Stephen Amoah [1,*], Ruth Ennin [2], Karen Sagoe [2], Astrid Steinbrecher [3], Tobias Pischon [3,4,5], Frank P. Mockenhaupt [2] and Ina Danquah [1,6,7]

[1] Institute for Social Medicine, Epidemiology and Health Economics, Charité–Universitaetsmedizin Berlin, Corporate Member of Freie Universitaet Berlin, Humboldt-Universitaet zu Berlin, and Berlin Institute of Health, 10117 Berlin, Germany; ina.danquah@uni-heidelberg.de

[2] Institute of Tropical Medicine and International Health, Charité–Universitaetsmedizin Berlin, Corporate Member of Freie Universitaet Berlin, Humboldt-Universitaet zu Berlin, and Berlin Institute of Health, 13353 Berlin, Germany; enninruth@yahoo.com (R.E.); kaysagoe@yahoo.co.uk (K.S.); frank.mockenhaupt@charite.de (F.P.M.)

[3] Molecular Epidemiology Research Group, Max Delbrück Center for Molecular Medicine in the Helmholtz Association (MDC), 13125 Berlin, Germany; asteinbrech@web.de (A.S.); tobias.pischon@mdc-berlin.de (T.P.)

[4] Charité–Universitaetsmedizin Berlin, Corporate Member of Freie Universitaet Berlin, Humboldt-Universitaet zu Berlin, and Berlin Institute of Health, 10117 Berlin, Germany

[5] DZHK (German Centre for Cardiovascular Research), Partner Site Berlin, 10785 Berlin, Germany

[6] Department of Molecular Epidemiology, German Institute of Human Nutrition Potsdam-Rehbruecke (DIfE), 14558 Nuthetal, Germany

[7] Heidelberg Institute of Global Health, University Heidelberg, 69120 Heidelberg, Germany

* Correspondence: stephen.amoah@charite.de

Citation: Amoah, S.; Ennin, R.; Sagoe, K.; Steinbrecher, A.; Pischon, T.; Mockenhaupt, F.P.; Danquah, I. Feasibility of a Culturally Adapted Dietary Weight-Loss Intervention among Ghanaian Migrants in Berlin, Germany: The ADAPT Study. *Int. J. Environ. Res. Public Health* **2021**, *18*, 510. https://doi.org/10.3390/ijerph18020510

Received: 20 December 2020
Accepted: 4 January 2021
Published: 9 January 2021

Publisher's Note: MDPI stays neutral with regard to jurisdictional claims in published maps and institutional affiliations.

Abstract: Background: Dietary weight-loss interventions often fail among migrant populations. We investigated the practicability and acceptability of a culturally adapted dietary weight-loss intervention among Ghanaian migrants in Berlin. Methods: The national guidelines for the treatment of adiposity were adapted to the cultural characteristics of the target population, aiming at weight-loss of $\geq$2.5 kg in 3 months using food-based dietary recommendations. We invited 93 individuals of Ghanaian descent with overweight or obesity to participate in a 12-weeks intervention. The culturally adapted intervention included a Ghanaian dietician and research team, one session of dietary counselling, three home-based cooking sessions with focus on traditional Ghanaian foods, weekly smart-phone reminders, and monthly monitoring of diet and physical activity. We applied a 7-domains acceptability questionnaire and determined changes in anthropometric measures during clinic-based examinations at baseline and after the intervention. Results: Of the 93 invitees, five participants and four family volunteers completed the study. Reasons for non-participation were changed residence (13%), lack of time to attend examinations (10%), and no interest (9%); 64% did not want to give any reason. The intervention was highly accepted among the participants (mean range: 5.3–6.0 of a 6-points Likert scale). Over the 12 weeks, median weight-loss reached −0.6 kg (range: +0.5, −3.6 kg); the diet was rich in meats but low in convenience foods. The median contribution of fat to daily energy intake was 24% (range: 16–40%). Conclusions: Acceptance of our invitation to the intervention was poor but, once initiated, compliance was good. Assessment centers in the participants' vicinity and early stakeholder involvement might facilitate improved acceptance of the invitation. A randomized controlled trial is required to determine the actual effects of the intervention.

Keywords: obesity; weight loss; diet; lifestyle; African migrants; Germany

Int. J. Environ. Res. Public Health **2021**, *18*, 510. https://doi.org/10.3390/ijerph18020510   https://www.mdpi.com/journal/ijerph

1. Introduction

Adiposity is a growing public health problem, already affecting more than 2 billion adults worldwide. Of these, over 650 million individuals have obesity (body mass index (BMI) $\geq$ 30.0 kg/m^2) [1]. Current projections indicate that by the year 2030, about 58% of the world's adult population will have overweight or obesity (BMI $\geq$ 25.0 kg/m^2) [2]. This will fuel the development of diabetes mellitus, cardiovascular disease, and cancers [3,4]. Among the growing group of sub-Saharan African migrant populations in Europe, overweight and obesity occur more frequently than in the European host populations [5,6]. Already, more than half a million people of African origin live in Germany [7] and their numbers are anticipated to increase rapidly [8]. Ghanaians form one of the largest groups of sub-Saharan African migrants in Europe [9,10]. In fact, around 46,000 Ghanaian migrants live in Germany [10,11] of whom 20% reside in Berlin [10]. General obesity (BMI $\geq$ 30.0 kg/m^2) is prevalent in 14% of Ghanaian men and in 39% of Ghanaian women living in Berlin. For abdominal obesity (waist circumference > 102 cm for men and >88 cm for women), these figures are 15% in men and 71% in women [12].

Lifestyle modification constitutes the first-line treatment for obesity because it is safe and usually effective [13], but dietary interventions often fail among migrant populations because specific cultural needs are neglected [14]. Evidence from African Americans and Asian migrants in Europe emphasize the importance of cultural adaptations for weight-loss programs to produce better outcomes than generalized interventions [15–17]. These strategies may not be transferrable to West-African migrants in Europe because of their linguistic, educational and migration-related characteristics.

Given the obesity-related health problems and no previous interventions targeting migrants from sub-Saharan Africa in Germany, this feasibility study aimed at evaluating the practicability and the acceptability of a culturally adapted dietary weight-loss intervention among a group of well-characterized Ghanaian adults in Berlin. As a secondary objective, we aimed at exploring the weight-loss effect on changes in cardio-metabolic risk factors.

2. Materials and Methods

2.1. Study Population and Design

For the present ADAPT study (Feasibility of a Culturally Adapted Dietary Weight-Loss Intervention Among Ghanaian Migrants in Berlin), participants of the multi-center, cross-sectional Research on Obesity and Diabetes among African Migrants (RODAM) study [18] were re-invited in Berlin via telephone in September 2017. Owing to the well-established difficulties of enrolling migrant groups in population-based studies [12], we used documented contact details of previous RODAM participants in Berlin (n = 547) who were eligible (n = 93). Shopping vouchers (10 €) were offered for each completed examination visit as incentives to participate. In brief, the RODAM study was implemented between 2012 and 2015 and comprised Ghanaians aged 25–70 years living in rural and urban Ghana as well as in Amsterdam, Berlin and London. The study used standardized instruments for data collection at all the study sites, comprising questionnaire-based interviews, physical examination, and biological sample collection.

The inclusion criteria were BMI $\geq$ 25.0 kg/m^2 or waist circumference >94 cm for men or >80 cm for women, Ghanaian migrant status (defined as being born in Ghana or having two parents born in Ghana), age $\geq$ 25 years, and the cooperation of the family cook or volunteer supporting the participant's behavioral change. The exclusion criteria were known diabetes, receiving long-term oral corticosteroids or weight-loss medication, and current pregnancy.

2.2. Ethics Statement

The study protocol was reviewed and approved by the Ethics Committee of Charité-Universitaetsmedizin Berlin (EA1/151/17). The study was registered retrospectively at the German Registry for Clinical Trials (DRKS00013767). Prospective registration was not performed, because the study has not been planned as a randomized controlled trial.

Therefore, the study was registered after the participants were recruited. The authors confirm that all ongoing and related trials for this intervention are registered. In order to finish the program before Christmas, the baseline assessments started on 2 October 2017, and the last visit was performed on 18 December 2017. All participants gave informed written consent prior to their enrolment.

2.3. Intervention Program

We adapted the guidelines for the treatment of adiposity by the German Society of Adiposity [19] (Table 1). To achieve behavioral changes at the individual level, we applied goal setting, behavioral contracting, and tailored health communication. These strategies were drawn from the Social Cognitive Theory of behavioral changes [20] and the stages of change construct of the Transtheoretical Model [21]. The adaptations of the treatment guidelines were based on the concept of Resnicow et al. [22], entailing an appropriate structure, process and strategy in adaptation.

Table 1. Adaptations of the guidelines for the treatment of adiposity by the German Society of Adiposity (DAG).

Variable	German Society of Adiposity (DAG) Guidelines	ADAPT Intervention	Reasons for Adaptation
Participants	Individuals with adiposity	Ghanaian adult migrants (defined as born in Ghana or both parents born in Ghana) with either general overweight/obesity or abdominal overweight/obesity and one adult family volunteer. Main cook agrees to co-operate.	Recruit Ghanaian migrants with high prevalence rates of adiposity; encourage support from family members, particularly from those who are responsible for the family meals; encourage healthier lifestyle in the entire family
Inclusion criteria	Body mass index (BMI) $\geq$ 30.0 kg/m^2 or waist circumference $\geq$ 88 cm for women and $\geq$102 cm for men, if BMI 25.0 < 30.0 kg/m^2	Body mass index (BMI) $\geq$ 30.0 kg/m^2 or waist circumference $\geq$ 88 cm for women and $\geq$102 cm for men, if BMI 25.0 < 30.0 kg/m^2	Potential recruits may have central obesity, but have a low BMI; acknowledge the important role of central body fat accumulation
Setting	General practitioner	Community for recruitment, ethnically matched practitioner for examination, home setting for intervention	Encourage community and family involvement; increase compliance; reduce attrition
Duration of the intervention	3 months	3 months intensive intervention period with 1 group contact, 3 family-based contacts and weekly mobile phone reminders	Facilitate motivation, compliance, self-efficacy, family involvement, and sustainability
Weight loss goal	$\geq$5% of initial body weight, if BMI 25.0 < 30.0 kg/m^2; $\geq$10% of initial body weight, if BMI $\geq$ 30.0 kg/m^2	$\geq$2.5 kg in the intervention group	Realistic for Ghanaian migrants and still relevant to improve the cardio-metabolic profile
Physical activity (PA)	>30 min/day ($\approx$ 1200–1800 kcal/week); mainly endurance sports; for individuals with BMI $\geq$ 30.0 kg/m^2, increase PA in daily routine (e.g., walking, taking stairs); PA counselling: health-beneficial effects of physical activity beyond weight loss and PA goal setting	>30 min/day ($\approx$ 1200–1800 kcal/week); increase PA in daily routine (e.g., brisk walking, taking stairs); Group counselling and lifestyle poster: health-beneficial effects of physical activity beyond weight loss; PA goal setting: pedometer; PA self-contracting: weekly mobile phone text messages	Most relevant; achievable recommendations, accounting for work load and family time; encouragement of self-chosen outdoor or gym activity in a group or alone; incorporates goal setting, behavioral contracting, and tailored health communication
Dietary intervention and targets	Dietary advice by general practitioner: daily energy deficit of 500 kcal; reduction of total fat and/or reduction of carbohydrates	Group counselling, lifestyle poster: reduced energy intake, not specific in nutrients; consultation with a dietician in the language of choice (German, English, local Ghanaian); reducing the intakes of frequently consumed foods that are rich in fats and carbohydrates; 3 home-based cooking sessions focusing on cooking methods, portion sizes, food choices, and fat amount for cooking; diet goal setting: 24-h dietary recall protocols; diet self-contracting: weekly mobile phone text messages	Bilingual dietary counselling available; Achievable and comprehensible approach, given the low level of formal education and health literacy in the study population; Engage the available family in a domestic setting especially those who prepare the family meals; Incorporates goal setting, behavioral contracting, and tailored health communication

Linguistic, constituent-involving and socio-cultural adaptations seem to be the most successful for weight-loss and dietary changes [15,16,23]. Therefore, we focused on the socio-cultural context and the languages of Ghanaians living in Berlin (Table 1). In this regard, trained personnel conducted questionnaire-based interviews in the participant's preferred language, either English or a local Ghanaian language. The goal of this culturally adapted dietary intervention was to achieve weight-loss of at least 2.5 kg [24] which is based on international guidelines of a minimum 5% weight-loss of body weight during the 3-month period. The individual schedule for the study participants is shown in Table 2.

In brief, the participants and their family volunteers received group counselling by an ethnically matched dietician. The counselling focused on the reduction of energy-dense foods, fewer eating occasions, and smaller portion sizes. Participants were encouraged to increase the consumption of fruits and vegetables. In addition, they received an information poster about a healthy Ghanaian diet and regular physical activity. The latter followed the recommendations of moderate to vigorous physical activity for at least 30 min per day. We organized monthly home-based cooking sessions with a dietician, and sent weekly smartphone reminders covering the participants' dietary and activity goals. Lifestyle was monitored by culture-sensitive dietary assessment methods and by subjective and objective measurements of physical activity, respectively.

Table 2. Individual intervention and examination schedule of the ADAPT study.

Week	1	2	3	4	5	6	7	8	9	10	11	12
Examinations												
Anthropometry	X											X
Oral glucose tolerance test	X											X
Blood pressure	X											X
Laboratory analyses												
Blood glucose (0, 30, 120 min)	X											X
C-Peptide, Insulin (0, 30, 120 min)	X											X
HbA1c	X											X
Fasting blood lipids	X											X
Intervention												
Group counselling	X											
Info poster	X											
Cooking session			X				X				X	
Smartphone reminder			X	X	X	X	X	X	X	X	X	X
ActivPAL set			X				X				X	
ActivPAL collect				X				X				X
24 h dietary recall								X				X
WHO STEPS activity questionnaire								X				X
Acceptability questionnaire							X					X

X represents the week in which an activity was undertaken.

2.4. Recruitment

We used documented contact details of previous RODAM participants in Berlin ($n = 547$) [12] who fulfilled the inclusion criteria ($n = 93$). We invited them by phone. Reasons for non-participation were documented. Upon agreement, an appointment at the study center was scheduled for the baseline examination. The individual schedule for interviews and physical examinations is presented in Table 2. Figure 1 provides the CONSORT flow chart of the recruitment success. Of the 93 invited eligible participants, 16 were scheduled for the baseline examination. Finally, 6 individuals and 4 family volunteers were enrolled in the intervention study, translating into a participation rate of 6.5%.

2.5. Assessments of Demographics, Acceptability, and Lifestyle

Trained personnel conducted questionnaire-based interviews with the active study participants but not the family volunteers. These were performed in the participant's preferred language, either English or a local Ghanaian language. Demographic characteristics included age and sex.

2.5.1. Acceptability

The acceptability questionnaire was administered in weeks 7 and 12. We used a questionnaire that was based on the theoretical framework of acceptability, comprising seven component constructs [25]:

- Affective attitude: I enjoyed the diet and sports program;
- Burden: I easily integrated the diet and sports program in my daily life;
- Ethicality: The diet and sports program was important for me;

- Intervention coherence: I easily understood the diet and sports program;
- Opportunity costs: I am convinced by the diet and sports program;
- Perceived effectiveness: The diet and sports program will improve my health;
- Self-efficacy: The diet and sports program will help me to change my lifestyle.

There were six response categories to avoid the possibility of neutral answering, ranging from "strongly disagree" to "strongly agree".

CONSORT 2010 Flow Diagram

***modified for non-randomized trial design**

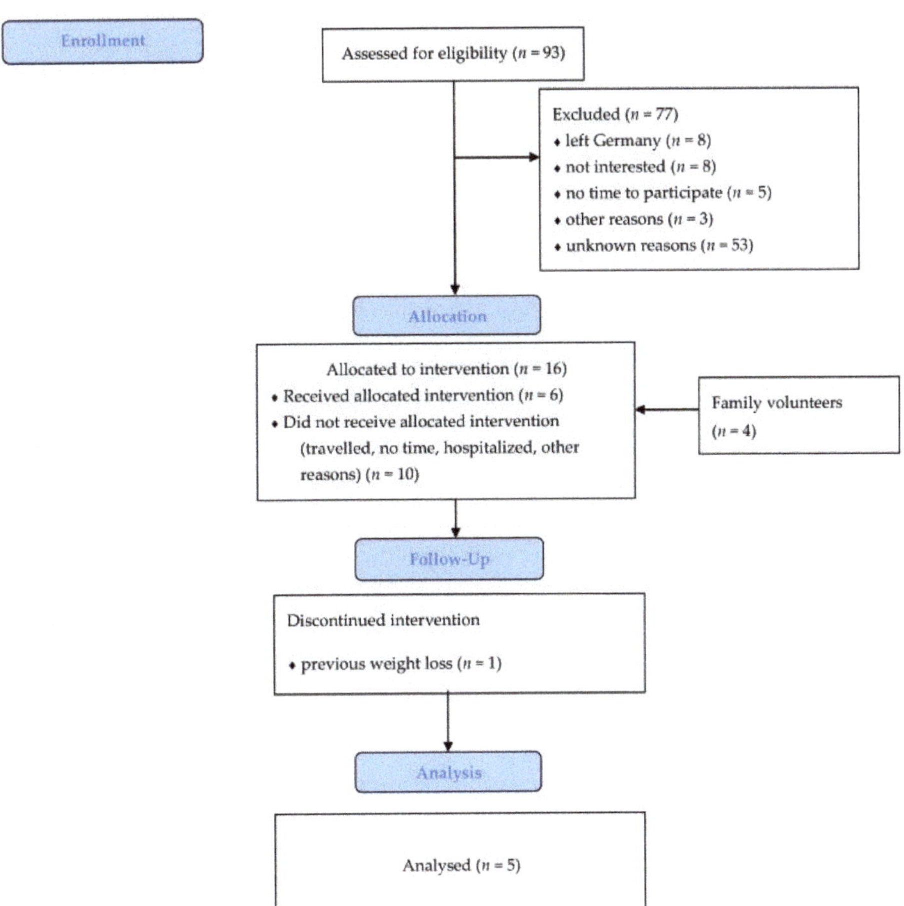

Figure 1. CONSORT Flow chart.

2.5.2. Dietary Behavior

For information about the baseline habitual diet, we used data of the previous RO-DAM study, that had been collected with the semi-quantitative Ghana-Food Propensity Questionnaire (Ghana-FPQ) [26]. The Ghana-FPQ queries about the usual intake frequencies of food groups in predefined portion sizes during the preceding 12 months. It covers 134 items reflecting both indigenous Ghanaian and typical German foods. The German Nutrient Database (BLS) and the West African Food Composition Table were used to calculate the intakes of energy (kcal/d) and macronutrients (% of daily energy intake). During the course of the ADAPT study, 24-h dietary recalls (24HDRs) were conducted according to the 5-Steps Multiple Pass Method [27] at two points in time. Participants provided information on eating times, types of foods and beverages consumed in the past 24 h, and portion sizes. Specific information about brands and recipes were also recorded. Common Ghanaian household utensils were used to estimate the portion sizes.

2.5.3. Physical Activity

Again, we used self-reported physical activity data of the previous RODAM study as our baseline information. Physical activity had been assessed by means of the WHO STEPwise approach to chronic disease risk factor Surveillance (STEPS) questionnaire [28]. The same tool was applied in the present ADAPT study in weeks 7 and 12. The STEPS questionnaire gathers information on physical activity in three settings (at work, travel to and from places and recreational) and sedentary behavior. Metabolic equivalents of task (MET)-hours were calculated. In addition to the self-reported physical activity data, we carried out objective measurements using a monitoring device (ActivPAL activity monitor; PAL Technologies Ltd., Glasgow, UK). This tool collects information about static and dynamic acceleration. The measurements were done for a week's period on three different occasions. The lightweight device was worn discretely on the participants' thigh for up to one week to quantify sedentary, upright and ambulatory activities as well as total MET-hours/day.

2.6. Physical Examinations

Similar to the questionnaire-based interviews, physical examinations were conducted only among the active study participants but not their family volunteers. A trained nutrition scientist conducted the physical examinations among participants in light clothes and without shoes. The measurements comprised body weight (kg; SECA 877), height (cm; SECA 217), waist circumference (cm) and hip circumference (cm) using a measuring tape. Systolic and diastolic blood pressures (mmHg) were measured (Boso Medicus Control; Bosch + Sohn GmbH, Jungingen, Germany) in triplicates after an appropriate resting time. The mean of the last two measurements was used for analysis.

2.7. Statistical Analysis

Baseline characteristics are presented as median and range for continuous variables and as percentage for categorical data. We calculated differences between baseline and follow-up data for secondary outcomes, i.e., weight-loss and lifestyle factors. All analyses were performed using Microsoft Excel 2016 (Microsoft Cooperation, Washington, DC, USA).

3. Results

3.1. Study Population

Two men and four women attended the ADAPT baseline examination. Table 3 presents the characteristics of the participants of the year 2014 (RODAM Study) and September 2017 (ADAPT baseline examination). The median age at baseline was 51 years (range: 25–62 years). The median BMI was 29.9 kg/m^2 (range: 23.3–35.1 kg/m^2) and the median waist circumference was 98.3 cm (range: 86.0–100.0 cm).

Table 3. Baseline characteristics of the participants of the ADAPT feasibility study.

Characteristics	2014		2017	
	Median/Percentage	Range/Number	Median/Percentage	Range/Number
n	100%	6	100%	6
Age (years)	47.5	22.0–58.0	50.6	25.0–61.5
Sex (male)	33.3%	2	33.3%	2
Weight (kg)	75.5	64.0–83.9	77.4	62.8–87.6
Body mass index (kg/m^2)	29.7	25.7–31.3	29.9	23.3–35.1
Waist circumference (cm)	92.2	83.1–105.1	98.3	86.0–100.0
Physical activity (MET-h/week)	195	0.0–392		
Energy intake (kcal/d) *	2384	922–3361		

* Energy intake was calculated based on the Ghana Food Propensity Questionnaire (Ghana). MET, metabolic equivalents of task.

3.2. Practicability and Acceptability

We contacted 93 eligible individuals by phone. As depicted in Figure 1, the main reasons for non-participation were change of residence (13%), lack of time to attend clinic-based examinations (as opposed to their nearest Ghanaian practitioner; 10%), or no interest (9%); 64% of the non-participants did not want to give reasons for their decision. After the baseline examination, one individual actively withdrew from the study, because the person reported previous weight-loss and did not want to lose more weight. Thus, the analytical sample for all follow-up assessments comprised 5 individuals.

Figure 2 shows the acceptability of the intervention programme according to the 7-items acceptability questionnaire in week 7 (Figure 2A) and week 12 (Figure 2B), using a 6-points Likert scale. In week 7, intervention coherence, opportunity costs, and self-efficacy reached the maximum score points, followed by perceived effectiveness (5.0), affective attitude (4.0), burden (4.0), and ethicality (4.0). These figures further improved until week 12 (Figure 2B).

3.3. Weight-Loss and Lifestyle Characteristics

The changes in anthropometric measures, lifestyle characteristics and clinical variables between baseline examination and follow-up are shown in Table 4.

Table 4. Differences in anthropometric and lifestyle characteristics between the ADAPT baseline and follow-up.

Characteristics.	Median/Percentage	Range/Number
Anthropometry		
Δ weight (kg)	−0.6	0.5, −3.6
Δ body mass index (kg/m^2)	−0.3	0.2, −1.2
Δ waist circumference (cm)	−1.3	4.1, −4.5
Lifestyle characteristics		
Δ physical activity (MET-h/week)	65	−24, 249
Δ energy intake (kcal/d) *	−1480	−3300, −127

* Energy intake at baseline was measured by the Ghana Food Propensity Questionnaire and at follow-up by 24-h dietary recall. Δ-Change in values.

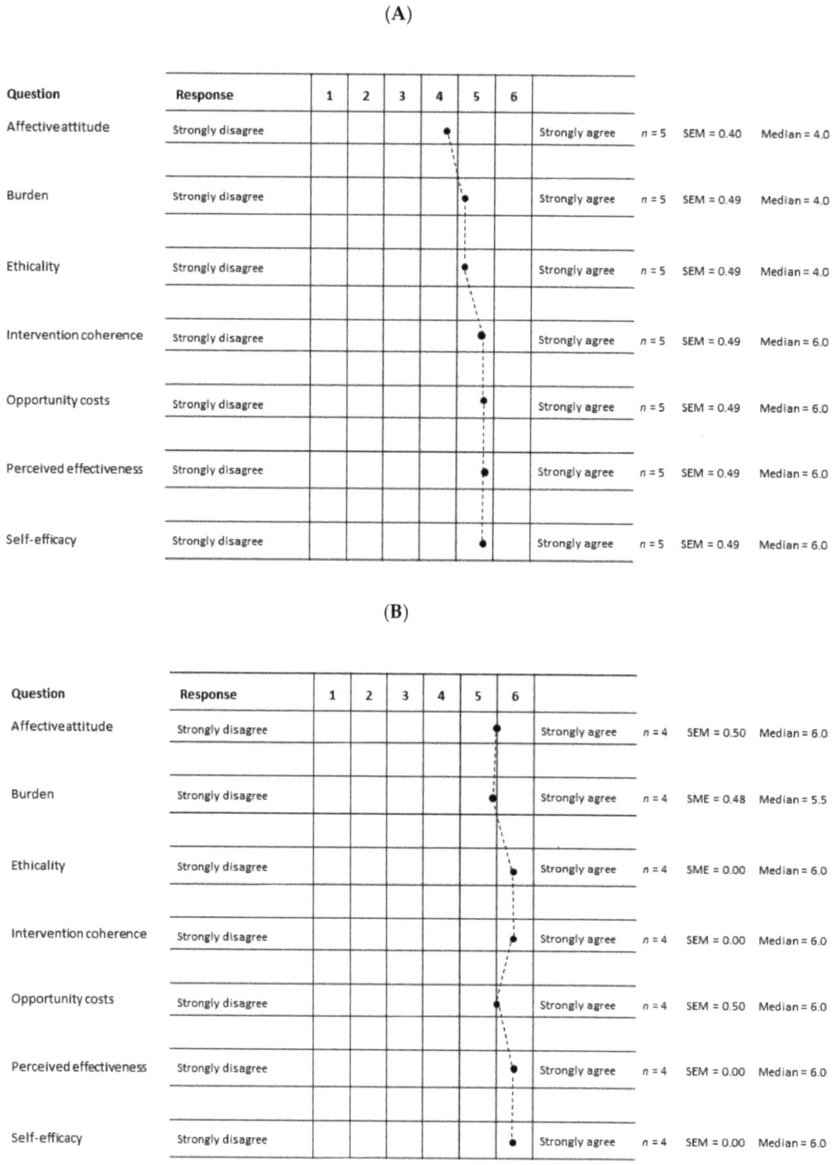

(A)

Question	Response	1	2	3	4	5	6				
Affective attitude	Strongly disagree							Strongly agree	n = 5	SEM = 0.40	Median = 4.0
Burden	Strongly disagree							Strongly agree	n = 5	SEM = 0.49	Median = 4.0
Ethicality	Strongly disagree							Strongly agree	n = 5	SEM = 0.49	Median = 4.0
Intervention coherence	Strongly disagree							Strongly agree	n = 5	SEM = 0.49	Median = 6.0
Opportunity costs	Strongly disagree							Strongly agree	n = 5	SEM = 0.49	Median = 6.0
Perceived effectiveness	Strongly disagree							Strongly agree	n = 5	SEM = 0.49	Median = 6.0
Self-efficacy	Strongly disagree							Strongly agree	n = 5	SEM = 0.49	Median = 6.0

(B)

Question	Response	1	2	3	4	5	6				
Affective attitude	Strongly disagree							Strongly agree	n = 4	SEM = 0.50	Median = 6.0
Burden	Strongly disagree							Strongly agree	n = 4	SME = 0.48	Median = 5.5
Ethicality	Strongly disagree							Strongly agree	n = 4	SME = 0.00	Median = 6.0
Intervention coherence	Strongly disagree							Strongly agree	n = 4	SEM = 0.00	Median = 6.0
Opportunity costs	Strongly disagree							Strongly agree	n = 4	SEM = 0.50	Median = 6.0
Perceived effectiveness	Strongly disagree							Strongly agree	n = 4	SEM = 0.00	Median = 6.0
Self-efficacy	Strongly disagree							Strongly agree	n = 4	SEM = 0.00	Median = 6.0

Figure 2. Acceptability of the culturally adapted dietary weight-loss intervention in week 7 (**A**) and in week 12 (**B**). Black dots indicate means. SEM, standard error of the mean.

After 12 weeks, the median weight-loss was −0.6 kg (range: +0.5; −3.6 kg), and median BMI and median waist circumference tended to be lower (Table 4). For lifestyle characteristics, the RODAM data served as the baseline information: median energy intake had been 2384 kcal/d (range: 992, 3361 kcal/d), and median energy expenditure had been 195 MET-h/week (range: 0–392 MET-h/week) (Table 3). The median difference in energy intake between these baseline data and the ADAPT follow-up information was −1480 kcal/d (range: −3330, −127 kcal/d). For physical activity, the median difference was 65 MET-h/week (range: −24, 249 MET-h/week) (Table 4).

Figure 3 presents the food group consumption and macronutrient intakes during study conduct. In the 3-months intervention period, the dominating food groups were carbohydrate-rich items (bread, cereals, potatoes, rice and pasta), vegetables, meat and fish. The participants rarely consumed fruits and convenience foods, and never consumed energy-containing beverages (Figure 3A). Carbohydrates, fat and protein contributed each 39%, 24% and 18% to daily energy intake (Figure 3B).

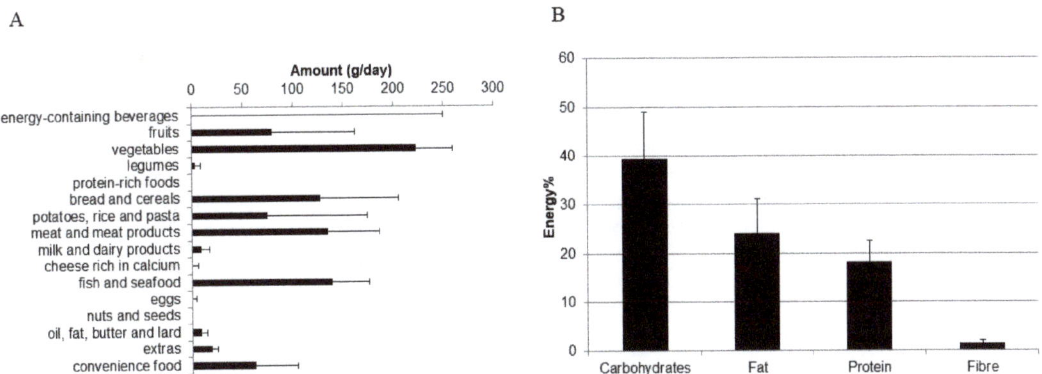

Figure 3. Median intakes of food groups (g/d) (**A**) and macronutrients (energy %) (**B**), based on the means of two 24-h dietary recalls. Error bars represent standard errors of the mean (SEM).

Figure 4 presents the individual physical activity of the participants during the course of the ADAPT study, based on self-report and by objective measurements. The median self-reported energy expenditure was 144 MET-h/week (range: 20, 478 MET-h/week), while the median energy expenditure by ActivPAL was 249 MET-h/week (range: 238, 270 MET-h/week).

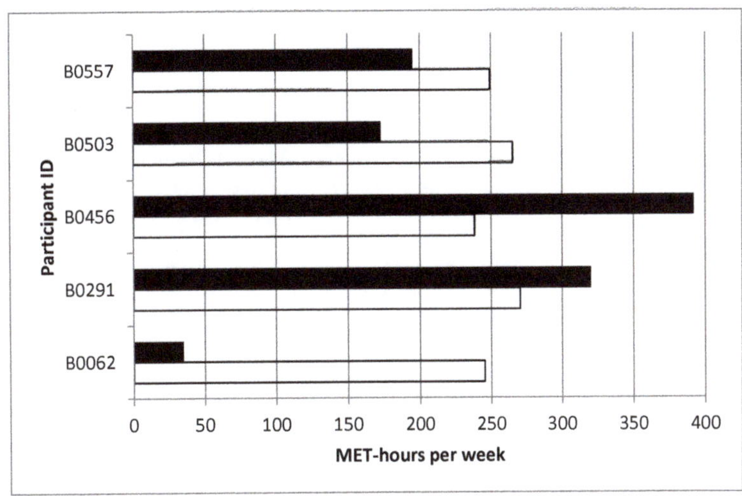

Figure 4. Median physical activity of weeks 7 and 12 for each ADAPT Study participant (participants codes, *y*-axis) by questionnaire-based self-report (black bars) and by objective measurement (white bars) (MET-hours per week).

4. Discussion

This feasibility study examined the practicability and the acceptability of a 12-weeks culturally adapted, dietary weight-loss intervention among Ghanaian migrants with overweight in Berlin. Our study was the first attempt in Germany to provide a culturally adapted lifestyle intervention for West-African migrants. Innovative adaptations comprised linguistic, constituent-involving and socio-cultural components: a Ghanaian research team and an ethnically matched dietician to facilitate culturally appropriate communication and culinary knowledge to deliver nutrition education, respectively. Nudging, through regular text and image messaging, enhanced self-empowering and frequent assessments of dietary behavior and physical activity during the intensive intervention period facilitated self-contracting to achieve the individual weight-loss goals.

4.1. Practicability and Acceptability

For ethnic minorities in the United States, Nierkens et al. have reported response rates ranging from 31% to 97% for culturally adapted interventions aiming at smoking cessation, diet, and physical activity [16]. The lower participation rate of 6.5% in the present study was similar to the one seen in the RODAM baseline recruitment in 2015 [12], and may be attributed to mistrust and competing demands, as indicated in other African American population groups, too [29]. Indeed, in the previous RODAM Study, we experienced initial response rates of less than 5% following written invitation. In the ADAPT study, 19% of the contacted individuals claimed to have no time or no interest for such a program. This was also seen among Asian migrant populations in the UK, where community-orientated personal approaches for recruitment were most successful (83% response rate) [30]. Also, early sensitization and involvement of community leaders have contributed to enhanced enrolment of African migrants into health interventions in the Netherlands and in the UK [31,32]. Therefore, the present study has built on the documented contacts from former RODAM participants in Berlin.

Still, this target population appears to be highly mobile as indicated by the proportions of individuals who moved (8/93) or traveled (3/93). The poor intervention uptake may also stem from unawareness in the target population for adiposity as a risk factor of chronic diseases. In fact, in a qualitative study with Ghanaian migrants in Amsterdam (n = 46), few respondents associated hypertension with adiposity, even though many had overweight [33]. Moreover, Ghanaian adults perceive a certain degree of overweight as a sign of wealth, fertility, and beauty, particularly for and among women [34]. From this perspective, there seems to be no evidential need to engage in weight-loss activities. Lastly, food choices are not only influenced by individual factors, such as biological, demographic, psychosocial and situational aspects. Rather, interpersonal, environmental and political determinants take a growing role in the decision for food [35]. This system's pressure might have generated reservations in the Ghanaian community about participating in the offered dietary intervention.

With regard to mistrust, this may be manifest regarding the German health system and its actors. Study participants rather preferred Ghanaian practitioners as study physicians. Therefore, future intervention studies aiming at dietary weight-loss among Ghanaian migrants should focus on early stakeholder involvement and evidential communication to create trust. In addition, we need to offer low-threshold interventions that minimize the time, the costs, and the potential of mistrust in the participants by employing ethnically matched practitioners, nurses and dieticians.

Notably, the intervention program was rated as highly convenient by those individuals who completed the program. While this could indicate selection of the most motivated people, it may also signal the cultural acceptability of our adapted program.

4.2. Weight-Loss and Lifestyle

The present feasibility study indicates that the culturally adapted dietary intervention may reduce body weight, BMI and waist circumference in this population over a period of 12 weeks, although the study lacks a comparison group. This corroborates previous findings that culturally tailored and facilitated interventions produce better outcomes than generalized interventions [15–17,36]. For instance, a tailored study among African Americans showed that body weight was reduced in the intervention group (mean difference: -2 ± 3.2 kg), but not in the standard care group (mean difference: 0.20 ± 2.9 kg, $p = 0.02$) over a period of 6 months [15]. To establish the actual health effects of the present intervention, a larger study with a comparison group is definitely required.

Regarding the dietary behavior among Ghanaian migrants, core food items of the traditional diet are maintained even several years after migration, including starchy roots and tubers, bread, rice, and leafy vegetables [26]. These typical dietary habits are still seen during the present intervention, and also cover low intakes of health-beneficial fruits and dairy products. This indicates that intensified efforts are required that go beyond the implemented dietary modifications, aiming at meals to support the integration of new, healthy food groups, paralleling the reduction of portion sizes for a negative energy balance. Also, there were hardly any changes to physical activity, indicating that additional promotion of regular physical activity is required.

4.3. Limitations

Since the study did not aim to test an effect of the intervention, no sample size calculations were performed. Owing to the small sample size, which impairs external validity of our findings, we refrained from concluding any effects of the intervention regarding biomedical data. The lack of a comparison group limits the interpretation of our acceptability results, because the poor participation rate could either result from the intervention program per se or from the reported reasons. Selection bias might have occurred, if study participants differ from non-participants in unobserved characteristics. The acceptability questionnaire used in this study may involve subjectivity bias in which the participants might have different interpretations of items resulting in different endorsements of ratings. For the assessment of dietary intake, detailed information about eating times, portion sizes, recipes and their preparations were gathered by culture-specific assessment tools [26]. However, two 24HDRs may not have captured day-to-day and weekend-to-weekday variations of food consumption. Different instruments were used during the RODAM study and the ADAPT project, which also complicated the comparison of energy intakes and macronutrients consumption.

5. Conclusions

This feasibility study examined the practicability and the acceptability of a 12-weeks culturally adapted, dietary weight-loss intervention among Ghanaian migrants with overweight in Berlin. The study showed a low participation rate, but 5 out of 6 enrolled participants completed the intervention and rated the program as highly convenient. The dietary behavior during the intervention period still relied on starchy foods and animal-based products and was low in fruit consumption. Yet, the proportions of consumed vegetables and the contributions of macronutrients to energy intake adhered to international dietary guidelines.

The present culturally adapted dietary weight-loss intervention for Ghanaian migrants in Germany has the potential to reduce adiposity and, thus, to prevent cardio-metabolic conditions in this vulnerable population group. In future studies, early stakeholder involvement, advocacy by community leaders, and sensitization of the Ghanaian community prior to the implementation are key factors for the success of this program. For the establishment of the actual weight-loss and its cardio-metabolic effects, a randomized, controlled trial is required.

Author Contributions: Conceptualization, I.D.; Formal analysis, S.A., R.E., K.S., A.S., T.P., F.P.M. and I.D.; Funding acquisition, I.D.; Investigation, S.A., R.E., K.S. and I.D.; Methodology, F.P.M. and I.D.; Project administration, S.A., R.E. and K.S.; Resources, A.S. and T.P.; Supervision, F.P.M. and I.D.; Writing—original draft, S.A.; Writing—review & editing, R.E., K.S., A.S., T.P., F.P.M. and I.D. All authors have read and agreed to the published version of the manuscript.

Funding: This research was funded by the Diabetes-Stiftung, grant number 392/12/16.

Institutional Review Board Statement: The study was conducted according to the guidelines of the Declaration of Helsinki, and approved by the Ethics Committee of Charité-Universitaetsmedizin Berlin (EA1/151/17, 10 August 2017).

Informed Consent Statement: Informed consent was obtained from all subjects involved in the study.

Data Availability Statement: The data presented in this study are available on request from the corresponding author. The data are not publicly available due to information that can potentially reveal the participants' identity.

Acknowledgments: The authors are very grateful to the dietician and Ghanaian volunteers for participating in this project.

Conflicts of Interest: The authors declare no conflict of interest.

References

1. Ng, M.; Fleming, T.; Robinson, M.; Thomson, B.; Graetz, N.; Margono, C.; Mullany, E.C.; Biryukov, S.; Abbafati, C.; Abera, S.F.; et al. Global, regional, and national prevalence of overweight and obesity in children and adults during 1980–2013: A systematic analysis for the Global Burden of Disease Study 2013. *Lancet* **2014**, *384*, 766–781. [CrossRef]
2. Kelly, T.; Yang, W.; Chen, C.S.; Reynolds, K.; He, J. Global burden of obesity in 2005 and projections to 2030. *Int. J. Obes.* **2008**, *32*, 1431–1437. [CrossRef] [PubMed]
3. Guh, D.P.; Zhang, W.; Bansback, N.; Amarsi, Z.; Birmingham, C.L.; Anis, A.H. The incidence of co-morbidities related to obesity and overweight: A systematic review and meta-analysis. *BMC Public Health* **2009**, *9*, 88. [CrossRef] [PubMed]
4. Must, A.; Spadano, J.; Coakley, E.H.; Field, A.E.; Colditz, G.; Dietz, W.H. The disease burden associated with overweight and obesity. *JAMA* **1999**, *282*, 1523–1529. [CrossRef]
5. Misra, A.; Ganda, O.P. Migration and its impact on adiposity and type 2 diabetes. *Nutrition* **2007**, *23*, 696–708. [CrossRef]
6. Agyemang, C.; Owusu-Dabo, E.; De Jonge, A.; Martins, D.; Ogedegbe, G.; Stronks, K. Overweight and obesity among Ghanaian residents in The Netherlands: How do they weigh against their urban and rural counterparts in Ghana? *Public Health Nutr.* **2009**, *12*, 909–916. [CrossRef]
7. Organization for Economic Cooperation and Development (OECD). Key Indicators on International Migration. 2014. Available online: http://www.oecd.org/els/mig/keyindicatorsoninternationalmigration.htm (accessed on 24 October 2017).
8. Federal Office for Migration and Refugees (BAMF). Migration Report Germany 2013. 2014. Available online: https://www.bamf.de/SharedDocs/Anlagen/DE/Publikationen/Migrationsberichte/migrationsbericht-2013.html (accessed on 24 October 2017).
9. Kohnert, D. *African Migration to Europe: Obscured Responsibilities and Common Misconceptions*; 2007; Available online: https://doi.org/10.2139/ssrn.989960 (accessed on 24 October 2017).
10. German Technical Cooperation (GIZ). *Die ghanaische Diaspora in Deutschland—Ihr Beitrag zur Entwicklung*; GTZ: Eschborn, Germany, 2009.
11. Tonah, S. Ghanaians Abroad and Their Ties Home. Cultural and Religious Dimensions of Transnational Migration. In Proceedings of the Transnationalisation and Development(s): Towards a North-South Perspective, Bielefeld, Germany, 31 May–1 June 2007.
12. Agyemang, C.; Meeks, K.; Beune, E.; Owusu-Dabo, E.; Mockenhaupt, F.P.; Addo, J.; de Graft Aikins, A.; Bahendeka, S.; Danquah, I.; Schulze, M.B.; et al. Obesity and type 2 diabetes in sub-Saharan Africans—Is the burden in today's Africa similar to African migrants in Europe? The RODAM study. *BMC Med.* **2016**, *14*, 166. [CrossRef]
13. Jensen, M.D.; Ryan, D.H.; Apovian, C.M.; Ard, J.D.; Comuzzie, A.G.; Donato, K.A.; Hu, F.B.; Hubbard, V.S.; Jakicic, J.M.; Kushner, R.F.; et al. 2013 AHA/ACC/TOS guideline for the management of overweight and obesity in adults: A report of the American College of Cardiology/American Heart Association Task Force on Practice Guidelines and The Obesity Society. *J. Am. Coll. Cardiol.* **2014**, *63 Pt B*, 2985–3023. [CrossRef]
14. Scott, P.; Rajan, L. Eating habits and reactions to dietary advice among two generations of Caribbean people: A South London study, part 1. *Pract. Diabetes* **2000**, *17*, 183–186. [CrossRef]
15. Kong, A.; Tussing-Humphreys, L.M.; Odoms-Young, A.M.; Stolley, M.R.; Fitzgibbon, M.L. Systematic review of behavioural interventions with culturally adapted strategies to improve diet and weight outcomes in African American women. *Obes. Rev.* **2014**, *15* (Suppl. 4), 62–92. [CrossRef]

16. Nierkens, V.; Hartman, M.A.; Nicolaou, M.; Vissenberg, C.; Beune, E.J.; Hosper, K.; van Valkengoed, I.G.; Stronks, K. Effectiveness of cultural adaptations of interventions aimed at smoking cessation, diet, and/or physical activity in ethnic minorities. A systematic review. *PLoS ONE* **2013**, *8*, e73373.

17. Tovar, A.; Renzaho, A.M.; Guerrero, A.D.; Mena, N.; Ayala, G.X. A Systematic Review of Obesity Prevention Intervention Studies among Immigrant Populations in the US. *Curr. Obes. Rep.* **2014**, *3*, 206–222. [CrossRef] [PubMed]

18. Agyemang, C.; Beune, E.; Meeks, K.; Owusu-Dabo, E.; Agyei-Baffour, P.; Aikins, A.D.G.; Dodoo, F.; Smeeth, L.; Addo, J.; Mockenhaupt, F.P.; et al. Rationale and cross-sectional study design of the Research on Obesity and type 2 Diabetes among African Migrants: The RODAM study. *BMJ Open* **2015**, *4*, e004877. [CrossRef] [PubMed]

19. Hauner, H.; Moss, A.; Berg, A.; Bischoff, S.C.; Colombo-Benkmann, M.; Ellrott, T.; Heintze, C.; Kanthak, U.; Kunze, D.; Stefan, N.; et al. Interdisziplinäre Leitlinie der Qualität S3 zur "Prävention und Therapie der Adipositas". *Adipositas-Ursachen Folgeerkrankungen Ther.* **2014**, *8*, 179–221. [CrossRef]

20. Bandura, A. *Social Foundations of thought and Action: A Social Cognitive Theory*; Prentice-Hall, Inc.: Englewood Cliffs, NJ, USA, 1986.

21. Prochaska, J.O.; DiClemente, C.C. Transtheoretical therapy: Toward a more integrative model of change. *Psychother. Theory Res. Pract.* **1982**, *19*, 276. [CrossRef]

22. Ahluwalia, J.; Baranowski, T.; Braithwaite, R.; Resnicow, K. Cultural sensitivity in public health: Defined and demystified. *Ethn. Dis.* **1999**, *9*, 10–21.

23. Renzaho, A.M.; Mellor, D.; Boulton, K.; Swinburn, B. Effectiveness of prevention programmes for obesity and chronic diseases among immigrants to developed countries—A systematic review. *Public Health Nutr.* **2010**, *13*, 438–450. [CrossRef]

24. Bhopal, R.S.; Douglas, A.; Wallia, S.; Forbes, J.F.; Lean, M.E.; Gill, J.M.; McKnight, J.A.; Sattar, N.; Sheikh, A.; Wild, S.H.; et al. Effect of a lifestyle intervention on weight change in south Asian individuals in the UK at high risk of type 2 diabetes: A family-cluster randomised controlled trial. *Lancet Diabetes Endocrinol.* **2014**, *2*, 218–227. [CrossRef]

25. Sekhon, M.; Cartwright, M.; Francis, J.J. Acceptability of healthcare interventions: An overview of reviews and development of a theoretical framework. *BMC Health Serv. Res.* **2017**, *17*, 88. [CrossRef]

26. Galbete, C.; Nicolaou, M.; Meeks, K.A.; Aikins, A.D.-G.; Addo, J.; Amoah, S.K.; Smeeth, L.; Owusu-Dabo, E.; Klipstein-Grobusch, K.; Bahendeka, S.K.; et al. Food consumption, nutrient intake, and dietary patterns in Ghanaian migrants in Europe and their compatriots in Ghana. *Food Nutr. Res.* **2017**, *61*, 1341809. [CrossRef]

27. Conway, J.M.; Ingwersen, L.A.; Vinyard, B.T.; Moshfegh, A.J. Effectiveness of the US Department of Agriculture 5-step multiple-pass method in assessing food intake in obese and nonobese women. *Am. J. Clin. Nutr.* **2003**, *77*, 1171–1178. [CrossRef] [PubMed]

28. World Health Organization. *The WHO STEPwise Approach to Noncommunicable Disease Risk Factor Surveillance (STEPS)*; World Health Organization: Geneva, Switzerland, 2003.

29. George, S.; Duran, N.; Norris, K. A systematic review of barriers and facilitators to minority research participation among African Americans, Latinos, Asian Americans, and Pacific Islanders. *Am. J. Public Health* **2014**, *104*, e16–e31. [CrossRef] [PubMed]

30. Wallia, S.; Bhopal, R.S.; Douglas, A.; Sharma, A.; Hutchison, A.; Murray, G.; Gill, J.; Sattar, N.; Lawton, J.; Tuomilehto, J.; et al. Culturally adapting the prevention of diabetes and obesity in South Asians (PODOSA) trial. *Health Promot. Int.* **2013**, *29*, 768–779. [CrossRef] [PubMed]

31. Agyemang, C.; Nicolaou, M.; Boateng, L.; Dijkshoorn, H.; Van De Born, B.-J.; Stronks, K. Prevalence, awareness, treatment, and control of hypertension among Ghanaian population in Amsterdam, the Netherlands: The GHAIA study. *Eur. J. Prev. Cardiol.* **2013**, *20*, 938–946. [CrossRef] [PubMed]

32. Elam, G.; McMunn, A.; Nazroo, J. *Feasibility Study for Health Surveys among Black African People Living in England: Final Report—Implications for the Health Surveys for England*; Department of Health: London, UK, 2001.

33. Beune, E.J.A.J.; Haafkens, J.A.; Schuster, J.S.; Bindels, P.J.E. 'Under pressure': How Ghanaian, African-Surinamese and Dutch patients explain hypertension. *J. Hum. Hypertens.* **2006**, *20*, 946–955. [CrossRef] [PubMed]

34. Tuoyire, D.A.; Kumi-Kyereme, A.; Doku, D.T.; Amo-Adjei, J. Perceived ideal body size of Ghanaian women: "Not too skinny, but not too fat". *Women Health* **2018**, *58*, 583–597.

35. Symmank, C.; Mai, R.; Hoffmann, S.; Stok, F.M.; Renner, B.; Lien, N.; Rohm, H. Predictors of food decision making: A systematic interdisciplinary mapping (SIM) review. *Appetite* **2017**, *110*, 25–35. [CrossRef]

36. Walker, R.J.; Smalls, B.L.; Bonilha, H.S.; Campbell, J.A.; Egede, L.E. Behavioral interventions to improve glycemic control in African Americans with type 2 diabetes: A systematic review. *Ethn. Dis.* **2013**, *23*, 401–408.

International Journal of
Environmental Research and Public Health

MDPI

Article

Eritrean Refugees' and Asylum-Seekers' Attitude towards and Access to Oral Healthcare in Heidelberg, Germany: A Qualitative Study

Yonas Semere Kidane [1], Sandra Ziegler [2], Verena Keck [1], Janine Benson-Martin [1,3], Albrecht Jahn [1,*], Temesghen Gebresilassie [1] and Claudia Beiersmann [1]

1   Heidelberg Institute of Global Health, Heidelberg University Hospital, 69120 Heidelberg, Germany; yonas.kidane@alumni.uni-heidelberg.de (Y.S.K.); verena.keck@t-online.de (V.K.); janinebensonmartin@gmail.com (J.B.-M.); temesgenbk@gmail.com (T.G.); beiersmann@uni-heidelberg.de (C.B.)
2   Department of General Practice and Health Services Research, Section Health Equity Studies & Migration, Heidelberg University Hospital, 69120 Heidelberg, Germany; Sandra.Ziegler@med.uni-heidelberg.de (S.Z.)
3   Gesundheitsamt Enzkreis, The Public Health Office Enzkreis, 75177 Pforzheim, Germany
*   Correspondence: albrecht.jahn@uni-heidelberg.de; Tel.: +49-(0)-6221-56-4886

Citation: Kidane, Y.S.; Ziegler, S.; Keck, V.; Benson-Martin, J.; Jahn, A.; Gebresilassie, T.; Beiersmann, C. Eritrean Refugees' and Asylum-Seekers' Attitude towards and Access to Oral Healthcare in Heidelberg, Germany: A Qualitative Study. *Int. J. Environ. Res. Public Health* **2021**, *18*, 11559. https://doi.org/10.3390/ijerph182111559

Academic Editor: Takaaki Tomofuji

Received: 27 September 2021
Accepted: 1 November 2021
Published: 3 November 2021

Publisher's Note: MDPI stays neutral with regard to jurisdictional claims in published maps and institutional affiliations.

Abstract: Oral health concerns in Eritrean refugees have been an overlooked subject. This qualitative study explored the access of Eritrean refugees and asylum-seekers (ERNRAS) to oral health care services in Heidelberg, Germany, as well as their perceptions and attitudes towards oral health care. It involved 25 participants. We employed online semi-structured interviews ($n = 15$) and focus group discussions ($n = 2$). The data was recorded, transcribed, and analysed, using thematic analysis. The study found out that most of the participants have a relatively realistic perception and understanding of oral health. However, they have poor dental care practices, whilst a few have certain misconceptions of the conventional oral hygiene tools. Along with the majority's concerns regarding psychosocial attributes of poor oral health, some participants are routinely consuming Berbere (a traditional spice-blended pepper) to prevent bad breath. Structural or supply-side barriers to oral healthcare services included: communication hurdles; difficulty in identifying and navigating the German health system; gaps in transculturally, professionally, and communicationally competent oral health professionals; cost of dental treatment; entitlement issues (asylum-seekers); and appointment mechanisms. Individual or demand-side barriers comprised: lack of self-sufficiency; issue related to dental care beliefs, trust, and expectation from dentists; negligence and lack of adherence to dental treatment follow-up; and fear or apprehension of dental treatment. To address the oral health burdens of ERNRAS, it is advised to consider oral health education, language-specific, inclusive, and culturally and professionally appropriate healthcare services.

Keywords: oral health care; dental; access; attitude; Eritrea; refugees; asylum-seekers; qualitative

1. Introduction

To date, literature regarding the magnitude of oral health burdens of the widely dispersed Eritrean refugees and asylum seekers is scarce. The world is experiencing a surging number of forcefully displaced persons with 70.8 million in 2019. They were either refugee (25.9 million), internally-displaced persons (41.3 million), or asylum-seekers (3.6 million) [1]. The UN High Commissioner for Refugees (UNHCR) defined a refugee as "someone who is unable or unwilling to return to their country of origin owing to a well-founded fear of being persecuted for reasons of race, religion, nationality, membership of a particular social group, or political opinion [2]". An asylum-seeker, though, is "someone whose claim has not yet been finally decided on by the country in which the claim is submitted [3]". By the end of 2018, 10% of the world's refugees resided in Europe [4] with Germany hosting the largest number [5]. Eritrea, despite its small population of

Int. J. Environ. Res. Public Health **2021**, *18*, 11559. https://doi.org/10.3390/ijerph182111559   https://www.mdpi.com/journal/ijerph

an estimated 3.5 million in 2018 [6], is ranked ninth as a country of origin of refugees with almost 15% of its population living in the diaspora [5]. By the end of 2018, there were 55,300 Eritrean refugees in Germany [1] and Eritreans were ninth by nationality of all applicants seeking protection in Germany [7].

Even as many refugees were able to escape from threats of persecution; many often failed to avoid the risks associated with poor health conditions. There is evidence to suggest that the general health of refugees is inferior in comparison to that of the host population [8–10]. Correspondingly, the oral health of refugees and asylum-seekers was poor in comparison to that of the general host country's population [11–14]. Though data is scarce, it was found that refugees and asylum seekers have a higher prevalence of oral disease and lower oral health status than their counterpart native Germans [15,16].

According to the World Health Organization (WHO), oral disease burdens are one of the leading health problems that refugees experience [17]. Among refugees, there is a high prevalence of major oral diseases such as dental caries, periodontal disease, malocclusion, missing and fractured teeth, orofacial trauma, and orofacial malignancies [11,13,18–21]. A study among newly-arrived refugees in Massachusetts indicated that oral diseases were the most common complaint in children and the second most common in adults [22]. Another study in Brussels, Belgium, also showed that dental conditions were the second most frequent diagnosis following respiratory tract infections [23]. This suggests that dental care is suggested as a pressing healthcare need of many refugees and asylum seekers [14,24,25].

Among the principal factors contributing to poor oral health are: ill-equipped or inaccessible dental healthcare in their Country of Origin (COO); lack of dental care in migration transit or refugee camps; and poor personal or cultural dental care practice [24,26–28]. While many appear to have reasonable perceptions and understanding regarding the significance of good oral health as holistic health [28], their overall oral health knowledge, attitudes, and good practice remain unsatisfactory [29–31]. Literature indicates, however, that the principal source of poor oral health status among refugees is, more often than not, actually related to limited access to dental care in the host country [24,26,28]. The German legislation is restrictive. For example, in Germany, access to dental care services according to '*§4 and § 6 Asylum-Seekers' Benefits Act (AsylbLG)* is limited during the first 18 months after arrival [32]. In essence, the legislation severely limits access to general and oral health care [32]. Other barriers described as limitations to care in the host country include language and communication issues; fear of dental treatment; anxiety and trust issues; high treatment costs; low income; distance to a dental clinic; quality of care; restricted treatment choice; long waiting lists and time; low oral health literacy; and other cultural and psychological barriers [12,28,30,33–35].

According to the WHO, oral health is a crucial indicator of overall health, wellbeing, and quality of life [36]. Many of the oral conditions, comprising periodontitis, tooth loss, dental caries, and oropharyngeal infections, share modifiable risk factors (high-sugar diets, poor oral hygiene and care, and excessive alcohol and tobacco use) with the leading non-communicable diseases including diabetes, cancer, and cardiovascular and respiratory diseases [37–40]. When the connection of oral health to quality of life was measured [36], the Oral Health-Related Quality of Life (OHRQoL) of refugees was found to be very low [41,42]. Despite this knowledge, oral health care is frequently neglected and undervalued as a vital healthcare service to refugees [43]. The UNHCR has yet to assign it within the significant health framework for the refugee population [44]. To date, Canada remains the only country ever to develop specific guidelines for oral health screening of refugees and immigrants [45]. While refugees' oral health care remains a pressing issue, studies are scarce [24,30]. In addition, almost all of the studies concluded that oral health is less understood and above all less accessible to these disadvantaged populations [15,24,28,29].

Thus far, no qualitative research has been published in Germany looking at the refugees' perspectives, understanding, experience, and the main difficulties associated with oral healthcare access and utilization. In particular, no study has focused on these aspects of oral health among Eritrean refugees in Europe or in Germany.

Therefore, this study aims to close this research gap. It explores the access of Eritrean refugees and asylum-seekers (ERNRAS) to oral health care services in Heidelberg, Germany, as well as their perceptions and attitudes towards oral health care.

2. Materials and Methods

This study was conducted using the qualitative research method. As this study seeks to understand an individual's experiences regardless of any preconceived ideas [46], as such, when describing and interpreting the data, participants' own perceptions, understandings, and perspectives were taken into account.

2.1. Study Setting

The study took place in the city of Heidelberg, Baden-Württemberg, Germany. Due to the influx of refugees in 2015 and Baden-Württemberg being the second largest state, it received a high number of asylum applications for three consecutive years (2016–2018) [47,48]. On the outskirts of the city of Heidelberg, the Patrick Henry Village (PHV), a former US army housing area, is used as a refugee arrival and registration centre [49]. Heidelberg city itself is the home to more than 450 refugees and asylum seekers, including many ERNRAS [50].

2.2. Sampling Procedures

Participants were selected through exponential non-discriminative snowball sampling, in which the first participant recruited provides multiple referrals. Each new reference offers further information for referral until sufficient participants are enlisted [51]. As the principal researcher (Y.S.K.) is Eritrean, he was able to socialise and exchange addresses with potential participants before the COVID-19 pandemic. He visited places where Eritreans usually gather such as Eritrean restaurants, and Eritrean church services in Heidelberg. After a series of snowballing and referrals, the first author had invited 31 ERNRAS to participate in the study of which 25 finally participated.

Of these, 84%, were refugees (individuals granted a refugee status), and 16% were asylum seekers. On average, participants had lived in Germany for 4.4 years (range: 2–7 years), and the majority, 76%, have lived there for more than three years. The study sample consisted mostly of men, 76%; the mean age of participants was 29 years (Range: 19–52); with more than half having a secondary level education. The majority of participants (60%) were employed at the time of the interview. Most of them, 72%, are unmarried. The sample characteristics are shown in Table 1.

2.3. Data Collection Instrument and Procedures

Data collection took place in April and May 2020. The principal researcher (Y.S.K.) carried out fifteen individual in-depth interviews (IDI) and two focus group discussions (FGDs) (with 6 participants each) in Tigrinya (Eritrean official language). The data collection tools consisted of a semi-structured in-depth interview (IDI), as well as a focus group discussion (FGD) guide. These guides were developed by the principal researcher with assistance from the co-authors (C.B., J.B.M.). A pilot test was performed among three Eritreans (friends and colleagues of Y.K.S.) before final adoption. The guides covered a wide range of topics within the following subject areas: perception of oral healthcare, understanding of oral health determinants, dental care behaviour, and barriers of access to oral healthcare services (see Appendix A, Table A1).

Data collection was originally planned face-to-face, however was conducted online via video conference software (Skype or WhatsApp) because of the COVID-19 pandemic restrictions. Each interview lasted an average of 40 min, with the FGDs lasting one hour and fifteen minutes. The FGDs were protocolled by an assistant (T.G.). The conversations were audio-recorded in Tigrinya, transcribed in Tigrinya, and translated into English for analysis by the first author (Y.S.K.).

Table 1. Socio-demographic characteristics of participants.

		n	
Age range (Years)			
	18–25	11	44%
	26–35	9	36%
	36–45	5	16%
	46–55	1	4%
Gender			
	Male	19	76%
	Female	6	24%
Educational Level (Years attending school)			
	Primary (1–5)	2	8%
	Middle (6–8)	4	16%
	Secondary (9–12)	14	56%
	Higher (13+)	5	20%
Marital Status			
	Married	7	28%
	Unmarried	18	72%
Employment Status			
	Employed	15	60%
	Unemployed	10	40%
Stay in Germany (Years)			
	≤3	7	28%
	>3	19	72%
Place of Residence			
	Heidelberg	15	60%
	Eppelheim	3	12%
	Plankstadt	2	8%
	Dossennheim	3	12%
	Bammental	2	8%
Refugee Status			
	Refugee	21	84%
	Asylum seeker	4	16%

2.4. Data Analysis

In order to organize, process, and manage the data, NVivo 12 (QSR International, Melbourne, Australia), a qualitative data analysis programme, was used. Data were analysed using thematic content analysis. This method supports identifying, analysing, and interpreting patterns of meanings (themes) within qualitative data [52]. A framework developed by Levesque and colleagues in 2013 [53], comprehensively addresses access to health care. The framework was used to code deductively for predefined themes in the data. According to the framework, healthcare accessibility involves five supply-side dimensions (provider-side): Approachability, Availability and Accommodation, Acceptability, Affordability, and Appropriateness. Correspondingly, there are also five dimensions (conceptualized as abilities) paralleling on the demand-side (user-side): the Ability to perceive, the Ability to seek, the Ability to reach, the Ability to pay, and the Ability to engage [53]. In addition to this deductive coding, the researcher read, explored, and coded the dataset for patterns and themes that emerged inductively.

Finally, eight major themes could be identified—three themes, mainly on the perception of oral healthcare, understanding of oral health determinants, and dental care behaviour of ERNRAS. The five subsequent themes depicted are in line with the five principal dimensions of access to healthcare and their equivalent users' abilities adopted from the aforementioned conceptual framework developed by Lévesque et al. (see Figure 1).

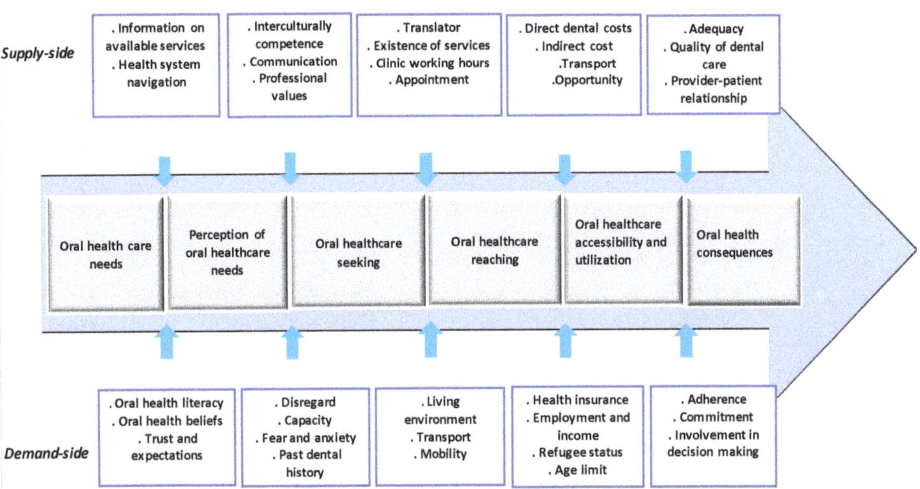

Figure 1. Conceptualization of oral healthcare access among ERNRAS in Heidelberg, Germany, adopted from the access framework of Levesque et al. (2013) [53].

2.5. Ethical Consideration

This study was conducted in accordance with the Declaration of Helsinki, and the protocol was approved by the Ethical Commission of the Medical Faculty of Heidelberg, Germany (protocol number: S-207/2020). Study participation was voluntary, and all had given their written informed consent. Confidentiality was protected by the use of pseudonyms when storing, analysing, and reporting the data. No incentives or compensation was given to the participants.

3. Results

The findings of this study are presented according to eight emerging themes and sub-themes. For a detailed overview please see Appendix B, Table A2.

3.1. Perception of Oral Health Care

In exploring the perception and perspective of Eritrean refugees and asylum seekers, good oral health was described as having white teeth, a pleasing smile without any decayed, broken or crooked teeth, and also no gum bleeding or bad breath: *"I would say, we shouldn't have a dental cavity, bleeding gums, bad mouth odour, or neither broken nor crooked tooth" (FGD-1).* They further commented on the vital significance of oral hygiene as part of personal hygiene and the valuable benefit of regular oral healthcare in boosting self-esteem and social approval. One of the respondents remarked: *"Earlier, I lost one of my front teeth from a fall injury; it was so horrible to see myself in front of a mirror (laughter) [. . .]. I used to feel so embarrassed in public and I used to cover my mouth all the time. It was so awful to see your tooth missing" (FGD-1).* Commenting on social uneasiness and suffering from bad breath, some participants also cited and believed that eating a traditionally prepared spice-blended pepper (Berbere) neutralises bad breath. Berbere is a traditional Eritrean spice blended pepper powder, mainly containing chilies, garlic, fenugreek—it is an ingredient in most Eritrean dishes:

"I believe Berbere protects you from bad mouth odour! As for me, I am getting Berbere from home [Eritrea] solely prepared by my mother, and I am consuming it every day. You know what [. . .]? as Berbere is my routine food, I do not have any terribly smelling mouth like others do" (IDI-13).

3.2. Understanding of oral Health Determinants

The study sought participants' views on the determinants of oral healthcare. The majority noted the close association between oral health and general health and remarked critically on the seriousness of oral diseases. One member of the discussion said: *"As molar tooth pain can go the head […], it makes it so dangerous and risky to your life. And from what I heard and also experienced […], there is no other pain cause more anguish than dental pain" (FGD-2)*. Most of the participants believed that the risk factor for their poor oral health was related to their lifestyle: poor oral hygiene routine, sweet and starchy food consumption, trauma, and tobacco smoking. The majority, however, have stressed that the dietary transition from a fibrous and low-sugar traditional diet back in Eritrea to a high-sugar or processed food in Germany as the major risk factor:

> *"Life in Europe is somehow different from our country. Most of us here [in Germany] we tend to change our lifestyle. We start to eat differently, like sweet and packaged food that are not common in our country; starting from me, smoking isn't also uncommon. I believe that those things are the reason for my poor oral health" (IDI-4).*

3.3. Dental Care Behaviour

As we enquired about personal and professional dental care practices, most of the respondents acknowledged the fundamental function of routine oral cleanness in preventing and reducing dental diseases. Almost all mentioned exercised some form of oral hygiene routines that varied from once, 6 (24%), or twice, 16 (64%), a day to an irregular basis, 3 (12%), using toothbrushes and toothpaste. In addition, eight (32%), also spoke about their habit of mouth washing in addition to toothbrushing or separately. However, the majority are either not using, or unaware of, dental flossing as a complementary oral hygiene method: *"I have no comments on this method of cleaning teeth. […] honestly, I know nothing and have also never used it" (FGD-2)*.

Few participants, five (20%), have been using tooth twigs (Mewets), a traditional Eritrean teeth cleaning tool, similar to Miswak (Asia, Africa, and the Middle East) [54], which is prepared mainly from two tree branches, those of the Olive tree (Olea Europea subspecies. Africana) and of the Sand olive (Dononaea Angustifolia). The dimensions are 6–10 cm long and 4–10 mm thick. The stick is applied to the teeth to scrub the surface in a horizontal or vertical motion until the twig split thereby allowing one to clean between the teeth as well as massaging and cleaning the gums [55]. When asked about their perception of using Mewets in Germany, their responses were mixed, as most of them acknowledged their habit of applying Mewets as the only tooth cleaning tool in Eritrea but had now changed to other methods: *"I have never used the twig in this country [Germany]; I couldn't find the right tree. I don't have any choice but to use the toothbrush" (IDI-6).*

Regarding the frequency of dental attendance, only, two (8%), participants cited visiting the dentist regularly and diligently on a bi-annual basis while two (8%) admitted that they have never attended a dental clinic in their lifetime. The majority's main reason for a dental visit was as a result of dental emergencies: *"The only time I went to my dentist was [..], the day that I experienced very serious dental pain" (FGD-2).*

Some participants raised doubts over their current regular oral hygiene tools and materials. Two (8%), participants of the FGD commented negatively regarding the regular use of toothbrush and paste: *"If we use the toothbrush frequently, with toothpaste after every meal, I believe that it may damage our tooth" (FGD-1)*. One female participant also assumed that the regular utilization of a dental toothbrush widened the gaps between her teeth. As well, irregular, or intermittent use of a toothbrush was considered as a risk factor for bad breath by another participant: *"If we habitually brush our teeth and stop, we may expose ourselves to bad mouth odour" (IDI-13)*. Most of the participants also voiced their concerns regarding dental flossing: *"I know about the thread […], I believe, if you keep on doing it, you can harm your gums now and then" (FGD-1).*

3.4. Approachability and Ability to Perceive

This theme refers to the capacity of refugees or asylum-seekers to discover dental care services and the availability of adequate oral health information sources that influence an individual's judgment of access to dental care facilities [53]. Obtaining clear information, locating dental services, and navigating the German health system, was found to be a complex and inconsistent endeavour for most of the ERNRAS. Many of the newly arrived ERNRAS remarked on the challenges associated with finding reliable information on health services, or a person to guide them through the health system. A recently-arrived mother of three, asylum-seeker, commented: *"No one would show or take you to a dental clinic. You have to find it on your own; and it was so difficult to understand and to find out where the dental clinics are" (IDI-2)*. The majority of the respondents also reported on their difficulties of navigating the health system in Germany in general: *"You have no idea […]! it is so challenging to understand how the health system works. There is limited or no information about where, how, and when to approach the eye clinic, the dental clinic and so on" (IDI-6)*.

Although the majority of the participants believed that they have basic oral healthcare literacy, few said anything about how far their lack of exposure to proper professional dental care in Eritrea, had impacted their overall oral healthcare mentality in Germany. They also expressed their strong beliefs in traditional medicine such as potions, herbs, or prayers, as influences on their oral healthcare perception:

> *"Back in our country [Eritrea], if we experience any kind of illness, we don't simply go to the clinic […]. Our parents and community healers used to give us any traditional herbs, potions, and spells. Then we wait for God to heal us. Likewise, here [in Germany] even though I am not using the herbs and potions […], I simply don't go to the clinic, I just pray at home and wait for God to heal me from my misery" (FGD-1).*

Most participants spoke highly of and trusted their dentists: *"My dentist is so reliable and honest […]. She is always helpful and she treated almost all of the dental problems I had" (IDI-2)*. Some participants, however, disagreed with their health providers' treatment decisions, as well as the bureaucracy involved in dental healthcare for ERNRAS in Germany. They also reported their concerns about the unforeseen forthcoming financial burden associated with dental health care:

> *"Sometimes though, the dentists work on a tooth that you have not complained about and we might not be comfortable with it too. As far as I am concerned, I don't like it" (IDI-13).*

> *"For some of us, it is like we don't even trust some of the dentists in Germany. I think that when they [dentists] are taking out our teeth, they want to do so in their own interest, and to replace ours with artificial teeth, which is not in our interest" (FGD-1).*

> *"I don't trust the dentists too. I have a trust issue! I mean […], the bureaucracy is very tedious […], they tell you to sign here, and there […], I don't know what we are sometimes saying. Who knows, later they [dentists] might ask us to pay all (laughter)?" (FGD-1).*

3.5. Acceptability and Ability to Seek

This theme conveys the intercultural and social competencies of oral healthcare providers to accept refugees, and the ability of refugees to seek dental care services [53]. In addressing that, some participants mentioned a lack of interculturally proficient dental care professionals. One participant iterated:

> *"The dentists should try to understand our difficulties in learning the new culture here [Germany]. In our country [Eritrea], we have a different background and practice for tooth care. We don't know much about the new way of dental care in Germany […], but we used to treat dental pain with herbs. Thus, the doctors should show some kindness and teach us calmly the correct way […]. My dentist expects me to comply to whatever*

he said, and he is very rigid and strict […]. I really didn't understand his instructions and he once yelled at me too (sigh)" (FGD-1).

The majority of the participants are satisfied with the services that they obtain at the dental clinics: *"As my former dentist is so cooperative, I also take many of my fellow Eritreans, who don't understand about their dental health, to her and get the treatment" (IDI-1).* A few, however, have commented on communication and conduct issues of some dental professionals: *"My former dentist had a very arrogant receptionist. I wasn't really comfortable with her. I was discouraged from going to the clinic as I couldn't stand her discriminating look; I take only pain killers and stay at home" (IDI-2).*

Some participants reported a great deal of uncertainty in their capacity to seek dental care. They mentioned that they either were not confident or not independent: *"I once wanted to visit a dental clinic but I couldn't. I honestly had no enough confidence to talk about my complaint" (IDI-6).* Furthermore, despite understanding the need for regular dental visits, some participants admitted to negligence or indifference. They believed that this was deeply rooted because their upbringing in Eritrea most often did not emphasise the significance of regular dental check-ups and care. One participant also alluded to the widespread and serious suffering on his migration journey (Sahara-Libya-Mediterranean Sea) and using this to relativise and justify his non-use of dental care:

"As far as I am concerned, the reason behind my hesitation in visiting a dental clinic, despite experiencing marked dental problems, is that I had been through a very bad experience on my way to Europe. I saw and witnessed a lot of awful distress and health problems along my way in Sahara, Libya or at sea [Mediterranean]. Comparing to those, I consider my teeth problem as a simple discomfort and I just resist the pain until it resolves itself" (FGD-1).

The majority of the participants explained why they opt out of regular dental visits. They usually related this with the presumptive or experienced fear and apprehension of dental instruments or physical dental pain.

"I chose not to go back after six months because I hate the machines that trim the teeth. Do you know how annoying are the rotating machines and the other sharp instruments that they [dentists] use? For example, one day, I had experienced a severe headache because of the instruments that they had stuck into my teeth; honestly, I hated it. Now that I am treated, thanks God it's over […]. It has been three years since I have experienced any kind of dental problem, and I never been in a dental surgery after that too" (IDI-15).

3.6. Availability and Accommodation and Ability to Reach

This theme relates to the availability of services that enable refugees or asylum-seekers to access dental care, as well as their abilities to reach the dental care facilities [53]. Communication is found to hamper ERNRAS access to oral healthcare. This was either a result of a language barrier or the non-availability of a translator. Participants confirmed that language problems were the most significant challenge in accessing dental care or support: *"It is the language problem; I can't tell a dentist what is really happening to me, and that is why I didn't go to them [dentists]" (IDI-6).* In addition, nearly all of the participants were concerned about the unavailability of interpreter service and believed that visiting a dentist without a translator could be a source of both misinformation and non-compliance with instruction: *"For example, I had severe dental pain, and I was waiting for artificial teeth. I had to go for several successive appointments, and I asked for a translator, but they [dental team] couldn't find me one. Thus, I missed several instructions from the dentists" (IDI-3).* Furthermore, dependency on an interpreter and the issue of privacy and confidentiality was also mentioned as a barrier by some participants: *"I might find a translator who can help me translate, but I also don't want to share my health problems with people of my own community as he or she might publicise my health issues" (IDI-14).*

The majority of ERNRAS expressed satisfaction with dental services in Germany. However, some were discontented over rigid clinic working hours, long waiting lists or times, inflexibility of dental appointments, and the long-distances or mobility issues as a hindrance to access to dental care:

> *"Most of the appointments that you get are on weekdays […], where most of us are busy at work or school […]. They [dentists] won't see you at weekends. So, if we need further visits, we couldn't miss work or classes so often […]. Thus, we often miss follow-up appointments" (FGD-1).*

> *"I can say that there is some problem, especially for those who reside in villages, where train transport is unpredictable […]. Pregnant mothers have some access problems. My friend's wife was once caught up in such a difficulty" (FGD-1).*

3.7. Affordability and Ability to Pay

This theme describes the financial ability of refugees or asylum seekers to devote enough funds and time to expend on dental care services and their ability to generate capital to finance the services [53]. Although some participants mentioned the free-of-fee primary dental care services, which are covered by insurance, the majority, however, made it clear that cost is a significant impediment to obtaining dental services. They reported that most of the dental treatments except regular check-ups, teeth cleaning, and tooth filling, are out-of-pocket or require co-payment. Indirect costs like those of transport, dental products, and opportunity costs were also mentioned by some participants as a detrimental factor in accessing dental services:

> *"In my opinion, comparing with the other services, dental care is expensive, and it always requires several consecutive appointments so that you need to skip work, pay for trains, and dental products like tooth brush, paste or mouthwash" (IDI-1).*

Participants acknowledge the complexity of health insurance eligibility and entitlement procedures, i.e., how, where, and when to approach or access dental care services. Many participants experienced that eligibility for free dental services depended on factors such as age and refugee status (asylum application decisions). They also remarked on the impact of employment status when seeking dental care:

> *"I haven't had enough money to get the treatment [orthodontic treatment], because I have no work or income" (IDI-4).*

3.8. Appropriateness and Ability to Engage

Here the compatibility of the dental service with the needs of refugees and their involvement in decision-making and treatment decisions were explored [53]. Most participants were content with their dentists' diagnosis, management, and communication competence: *"My doctor is so good and tells you everything about your oral health. She effectively treated all my dental problems and also cleaned my teeth" (IDI-2).* Some, however, have perceived, and complained about, the technical and interpersonal inadequacy and incompetence of the service providers: *"I can say that my former dentist could have done more […]. Not only did he treat my complaint badly, but he also forced me to go along with his decisions. That's why I always complain about him, the treatments that he gave me were neither appropriate nor satisfying" (IDI-3).*

Some participants indicated that some of their providers (dentists) were not only uncooperative and bad-tempered, but also difficult to build relationships with. One participant complained about a significant amount of money that she was forced to pay for unsuccessful treatment: *"I couldn't express my feelings. I wanted to sue the dentist […], but I don't know how everything works, and I don't know the court and how to take people to court too. I was so frustrated and depressed (sigh)" (IDI-12).* Many participants though admitted to their own lack of adherence to regular dental visits and appointments: *"I have never been in the dental clinic for the last two years, after I had received a dental treatment that actually relieved me from the pain that I had. To speak from experience […], once my dentist strongly advised me to visit a dentist every six months. She told me that I am entitled to two check-ups a year and teeth cleaning [i.e., scaling and polishing]. Still, I am not adhering to her advices (laughter)" (FGD-2).*

When participants were probed about their role in decision-making regarding their treatment options, many of them reported having limited enthusiasm to engage in their treatment decisions. This, they believe, affected their motivation to become involved in dental care and commit to finishing their treatment:

> *"My dentist once informed me that my tooth was decayed and suggested to extract it, and I simply agreed. Then when he [the dentist] attempted the extraction, it took him six hours. Since the tooth was decayed only on the upper part not at root, I should have asked him to restore it. It was my mistake. I was looking for a temporary solution but it cost me a lot and my left cheek was really numb for the following six months" (IDI-3).*

As results have been summarized in Figure 2, the oral healthcare attitude attributes of Eritrean refugees and asylum-seekers not only make proper self and dentist dental care difficult but also appear to negatively affect directly or indirectly the accessibility and utilisation of oral healthcare services as much as the supply (structural) side barriers do.

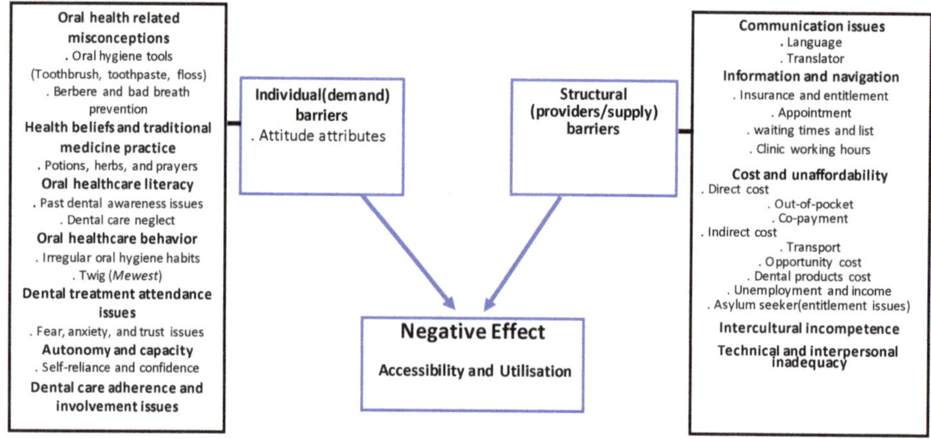

Figure 2. The effect of individual and structural barriers on oral healthcare accessibility and utilization of ERNRAS.

4. Discussion

This qualitative research identifies the major oral health concerns and barriers to dental care services among Eritrean refugees and asylum-seekers living in Heidelberg, Germany. In addressing those concerns, the results of our study indicate that the participants defined good oral health as the absence of any condition that involves problems with teeth, gums, jaws, cheeks, lips, or breath. This finding is consistent with a qualitative study carried out among newly arrived refugees in Canada, where the assessment of participants on what represents good oral health comprised absence of swelling, missing, broken, decayed or painful teeth [56]. As it has been documented in several studies among African refugees [28,57,58], ERNRAS also acknowledged the physical, mental, and social benefits of regular oral healthcare. The WHO pointed out that the oral cavity is not an isolated organ, and consequences of poor oral health are not only limited to the teeth but also affect general health [36]. In line with that report, the current study revealed that most of the participants were conscious of good oral health as part of holistic health, including the need to avoid the possible life-threatening consequences of untreated oral diseases. Compared to findings reached by Keboa and colleagues [28], our study identified the psychosocial concerns of bad breath and further new insight into a custom followed by some ERNRAS, mainly to prevent oral malodour by routine consumption of a traditional pepper (Berbere). The metal chelating activity (to stabilize and remove harmful metals), antioxidant properties, and enhancing effect on the carbohydrate-hydrolysing enzyme (hypoglycemic activity) of spice blend Berbere has been scientifically established [59]. However, the anti-halitosis element perceived and hypothesized by ERNRAS is yet to be researched.

When refugees are settling in Western countries, they are introduced to a higher-sugar diet than they were accustomed to in their COO [60]. Our findings confirm this dietary transition into the consumption of readily available 'carb-heavy' foods (concentrated with sugar and fat) in Germany and are perceived as a leading cause of their dental diseases. The nutritional transition and its negative impacts on the overall oral health of refugees and asylum seekers had been identified in several studies [15,29,30,61].

Dental care is the preservation of a healthy oral cavity, and it relates to regular personal and professional oral healthcare [36]. However, our study found that the majority of participants do not comply regularly with all the recommended dental care practices [15,24,28,30,61,62]. Almost all of the participants accepted toothbrushes and mouthwash, while none of them acknowledged dental floss as a significant oral hygiene tool. This result builds on the existing evidence, where a study reported that refugees were barely utilizing dental floss as a complementary device to clean their interdental areas [15]. In addition, unlike in Germany, most ERNRAS disclosed their utilisation of twig (Mewets) as a sole oral hygiene tool when they were in Eritrea. A study in Eritrea has shown the antimicrobial and anti-cariogenic (caries prevention) effect of Mewets [63]. The study might suggest the change in practice (from Mewets to a toothbrush) reflects the belief or attitude change. However, considering findings in similar studies of East African refugees [29,61], a more plausible explanation is the inconvenience and unavailability of the right tree in Germany. Regarding reasons for dental attendance, the results demonstrate that the majority only seek treatment in times of dental emergency or pain. This finding mirrors several studies amongst East African refugees living in the U.S.A [61], Australia [30], and Canada [28].

Misperception related to oral hygiene tools and methods among many East African refugees has been highlighted in previous studies [29,61]. Nevertheless, the findings from the current study go beyond previous reports, showing that some ERNRAS have negative perceptions about regular use of a toothbrush, toothpaste, and dental floss. However, they also believed that discontinuation of the routing toothbrushing could lead to halitosis (bad breath). The plausible explanation of this erroneous assumption is that users were unable to notice their halitosis before they started using toothpaste; they might not have known that an oral malodour already existed. Since our sense of smell is a learned behaviour [64], they start to differentiate bad breath from normal breath following the discontinuation or skipping of regular toothbrushing. Revisiting the belief that some ERNRAS rely on the hypothetical concept of Berbere's advantage in preventing bad breath, brushing is not only replaced as an important means to prevent oral malodour, but also considered to be detrimental to oral health.

This study found that ERNRAS had or have difficulty in obtaining information, locating oral healthcare services, and navigating the health system in Germany. A similar conclusion was reached on recently-arrived refugees in Finland [25], and East African refugees in Australia [58]. As well, participants with limited oral health literacy and undesirable health beliefs might also hamper their dental care-seeking behaviour as a study conducted in Eritrea by Andegiorgish and colleagues shows [65]. Furthermore, since stress and psychological insecurity are endemic among refugees [13,26], a dentist and trust-based dental treatment seem to be largely unachievable [58]. Consistent with previous findings [26,57], fear of dental pain, anxiety, or past negative dental experiences of oneself, friends, or family members influenced the care-seeking behaviour of ERNRAS. Additionally, the current study goes beyond that some ERNRAS present with certain trust issues towards provider's possible dental malpractice and unpredicted future financial implications of the current dental treatments.

The reported lack of interculturally competent professionals was a significant finding in this study. The perceived experienced discrimination, lack of empathy, and misconduct were also identified in related studies [58,66]. This underlines the importance of a dentist who understands and accepts his or her patient's diverse cultural beliefs and background. Furthermore, in line with similar research [67,68], this study found that apart from personal dental care disregard and negligence, reduced autonomy and self-reliance are also some of the participants' hurdles for seeking dental care.

This study suggests that communication features, language barriers, or unavailability of translators, negatively affect participants' accessibility. This finding mirrors other studies in Europe [8,69], and in Germany [70]. It is understood to be a major barrier to seeking oral healthcare services for the majority of ERNRAS. Moreover, our findings confirm that the patent's dependence on translators not only interferes with an effective conversation, diagnosis, and follow-up but also raises concerns on confidentially and privacy [71,72].

This study established that ERNRAS faces difficulties in understanding the working hours and appointment procedures of dental clinics. This correlates reasonably well with a study conducted by Mattila et al. [25], and further supports existing findings on the challenges of refugees' to secure and attend dental appointments, cope with long waiting times, and make the right treatment choices [30,58,67].

Many of the participants of this study report either postponing or avoiding dental treatments because of direct or indirect costs (transport, dental products, and opportunity costs). Evidence from other studies in Germany (direct payment) [16], Australia (co-payment and indirect payment) [67], and North America (direct and indirect payment) [28], all suggested that financial difficulties discourage refugees from seeking dental treatment. Analogously, a lack of health insurance and entitlement, as well as unemployment seems to pose challenges for asylum-seekers whose legal asylum status were still being processed. Age and refugee status are factors that affect the health entitlement of refugees to publicly funded health insurance. Minors (children and young people under the age of eighteen) and refugees fully covered, whereas asylum-seekers, during their first 18 months in Germany have restricted access (*§4 and § 6 Asylum-Seekers' Benefits Act (AsylBLG)*) [32], that in turn complicate assess and delay integration into the German health system. This result reflects findings from Australia [58], and Germany [73,74]. These show that eligibility criteria based on refugee status not only complicated access to health care but also resulted in delayed care, affect treatment outcomes, and increase expenditures.

This research identified access and follow-up issues of some participants related to professional inadequacies. This includes perceived failure to competently convince, agree on, and accommodate patients' treatment demands. A similar pattern of results was obtained in a systematic review compiled by Keboa and colleagues [28]. As indicated by Hobbs [58], the evidence that we found confirmed that a poor relationship between participants and providers acts as a barrier to seeking dental care. Furthermore, in line with research conducted in Australia [30], access to dental care was inhibited because participants were found to show less adherence to dental treatment. Many of them were found to be poorly motivated and to be less engaged or involved in their own oral health decision-making, including the choice of alternative treatments.

4.1. Study Strengths and Limitations

To back up the credibility of our research, we applied the concept of trustworthiness by Lincoln and Guba (1985) [75] with its four components: Credibility, Dependability, Confirmability, and Transferability. Credibility, which relates to the researchers' ability accurately to identify and describe the study participants [75], was reached through triangulation (both IDI and FGD), prolonged and persistent observations of participants until thematic saturation, and member checking with three participants to verify and provide feedback for their transcript and interpretation of the findings. Dependability, which refers to the consistency of data with time and conditions [75], was appreciated by auditing the rich and thick data set (transcript) against the recorded audio by another Tigrinya-speaking co-author (T.G.) and experienced supervising researcher (C.B.). Confirmability, which closely refers to the objectivity of the research [75], was realised by reflecting on our own preconceptions, bias, backgrounds, and beliefs and thus ensuring that researchers have not influenced the findings in the process of extracting information or analysing the collected data. Yet, being conscious of all these points, and as one of the principal researchers (YSK) is a registered dentist in Eritrea and a Tigrinya speaker, we had continually to reflect on our roles in the research process and the possibility of influencing the response of participants. We embrace this professional perspective and remain as part of the research but at the same time did justice to the shared experiences of our respondents without evaluating them solely from a medical perspective. Transferability concerns how far it is possible to apply methods and findings to other similar study contexts [75]. It was reached by explicitly describing our sample, participants characteristics, methodology and study setting.

The COVID-19 pandemic and subsequent contact restrictions that were in place in Germany during the data-collection phase (March–May 2020) posed some challenges for the research. Due to that, we had to change the initially planned face-to-face into online interviews. Online communication deters the possibility of using the whole spectrum of non-verbal cues and interferes with building a robust relationship between interviewer and interviewee. We addressed that by employing follow-up and probing questions to keep respondents focused throughout and across all the interviews and discussions. We also transcribed, from the recorded video, all the verbal expressions, facial expressions, and emotional intonations of the respondents.

We highlighted a lack of back-translation of the transcripts from English back to Tigrinya as another limitation of this study. Inter-rater reliability could have been approached had another translator back translated the transcripts prior to analysis. We had insufficient funds for a second translator, however, the translations were reviewed for accuracy by both Tigrinya and English-speaking co-author (T.G.).

4.2. Practical Implications

The findings suggest that Eritrean refugees and asylum-seekers are not sufficiently well-informed about their overall oral healthcare, nor do they make enough use of oral healthcare services. Acknowledging refugees' poor oral health status and limited access globally, it is considered that dental services should be included within primary healthcare, and efforts should be made to provide comprehensive dental screening at the first point of entry into host countries.

With regards to issues related to navigating the health system and misconceptions about oral hygiene, authorities and health workers should develop and deliver oral health education, promotion, and outreach activities, to improve awareness, utilisation, and accessibility of dental services. In addition, accessible and understandable information should be provided on scientifically supported oral hygiene measures, preventative dental care, and how to access the German health system. Oral healthcare providers should also build trust with their clients by cultivating a friendly patient-provider relationship, understanding culturally sensitive information, and demonstrating intercultural competency.

Enhancing oral health literacy in the community, and appropriate dental public health strategies would probably also benefit oral health care among refugees and asylum-seekers. Policy-makers should re-define the current framework (*AsylBLG*) of eligibility criteria for asylum seekers to access dental care only for emergencies, or painful and acute conditions, and that in event involves lengthy administration procedures [32].

Finally, as oral health care of refugees is still a neglected field, we recommend further research that focuses on oral healthcare professionals' experience and perspective of dental care services.

5. Conclusions

Eritrean refugees and asylum seekers have a fairly realistic perception and understanding of oral health. However, the majority have poor dental care behaviour or practice, whilst a few have certain misconceptions of the conventional oral hygiene tools. This study uncovered that the majority of the participants were mainly concerned about the psychosocial attributes of poor oral health rather than its functional implications. Since their arrival in Germany, the participants have been found to be influenced by the global dietary transition. This is not only a contributing factor to the rising burdens of non-communicable diseases (NCDs) worldwide [76], but is also believed to be the leading cause of their poor oral health. This study has shed light on participants' reported barriers to oral health-care services. Along with the individual's or client's own ability barriers, communication barriers remain the main hurdle at all stages of accessing oral healthcare. That includes problems with identifying and navigating oral health services. Additionally, it interferes with initiation or building up patient-doctor relationships. It further interferes with what to do when patients arrive at the dental surgery, how to decide on payment arrangements, and whom patients should trust. To address the oral health burdens of ERNRAS, it is prudent to consider language-specific, inclusive, and culturally and professionally appropriate oral health care services. Only then the Universal Health Coverage (UHC) of the Sustainable Developmental Goals 3 (SDG 3) could be achieved.

Author Contributions: Conceptualization, Y.S.K., C.B., J.B.-M.; methodology, Y.S.K., C.B., S.Z., V.K., T.G.; formal analysis, Y.S.K.; writing—original draft preparation, Y.S.K.; writing—review and editing, C.B., S.Z., V.K., J.B.-M., A.J., T.G.; supervision, C.B. All authors have read and agreed to the published version of the manuscript.

Funding: This research received no external funding.

Institutional Review Board Statement: The study was conducted according to the guidelines of the Declaration of Helsinki, and approved by the Ethical Commission of the Medical Faculty of Heidelberg, Germany (protocol number: S-207/2020, 16 April 2020).

Informed Consent Statement: Informed consent was obtained from all subjects involved in the study.

Data Availability Statement: The data presented in this study are available on request from the corresponding author.

Acknowledgments: We thankfully acknowledge for all those who provided assistance in recruiting study participants, especially Yonas Shishay and Abraham Tuemzgi. My heartfelt gratitude goes to all Eritrean refugees and asylum seekers residing in Heidelberg for their generous participation in this work. The principal author (YSK), would like to forward his sincere appreciation to Open Society Foundation (OSF) for the Civil Society Leadership Award (CSLA) and German Academic Exchange Services (DAAD) for granting him this priceless and exceptional opportunity to study Master's degree at Heidelberg Institute of Global Health.

Conflicts of Interest: The authors declare no conflict of interest.

Appendix A

Table A1. Wide-ranging questions of the IDI and FGD.

No.	
1.	What is the first thing that comes to your mind when you hear about oral health?
2.	What is good oral health to you? What about oral healthcare?
3.	How concerned are you about your oral health?
4.	What is your opinion on the relationship of poor oral health and general health?
5.	Thinking as ERNRAS, how would you describe your overall oral health status?
6.	What do you think the risk factors for the poor oral health among ERNRAS?
7.	Can you talk about the oral hygiene tools you are using? How often are you using?
8.	what is your opinion on 'when a dentist should be visited?'
10.	Can you tell me what are the main factors hindering ERNRAS from demanding oral healthcare services in Germany?
11.	what is your opinion on how oral health issues of refugees should be managed at the individual, community, governmental or policy levels?

Appendix B

Table A2. Themes, sub-themes, and exemplar quotes.

Themes and Sub-themes	Pertinent Findings According to Participants (ERNRAS)	Quotes
Perception of Oral Healthcare		
Perceived definition of oral health	• Oral health is related to possessing of white teeth, pleasing smile with no broken, decayed, or crowded tooth and also no bleeding gums or oral malodour.	*"we shouldn't have a dental cavity, oral ulcers, bad mouth odour or no broken or crooked tooth" (FGD-1).*
Social acceptance and Self-esteem	• Sound and satisfactory oral health status believed to enhances self-esteem and social approval, and opposite is true in case of poor oral health. • Routine consumption of a traditionally prepared spice blended pepper (Berbere), is believed to prevent bad breath and its burdens.	*"If we don't have good teeth, we don't have a girlfriend (laughter)" (FGD-1).* *"After I knew from my friends about the smell of my mouth […], it wasn't good news […], I soon lost my confidence and couldn't stand talking to people." (IDI-14).* *"Then, I wouldn't stop eating Berbere so that I could stay free of bad breath" (IDI-8).*
Understanding of Oral Health Determinants		
Awareness of oral health as holistic health	• They acknowledged the close relationship between oral health and the overall health	*"Tooth has to be cleaned and kept healthy so that we can eat nutritious foods and we live longer" (IDI-11).*
Perceived risks of poor oral health	• lifestyle change and dietary transition from low-sugar food in Eritrea to high-sugar content in Germany was believed as the main cause of their current poor oral health conditions.	*"Here [Germany], we, Eritreans, are consuming a lot of sweet, soft, and packed food, unlike the food we used to eat in our country, which was hard to chew and less sweet. I believe this is the reason for this poor oral health" (IDI-5).*
Dental Care Behaviour		
Personal dental care	• All practicing some form of oral hygiene habits that varies from, once or twice a day to an irregular basis using toothbrush, paste, mouth wash or twig (Mewets). • They either not using or unaware of dental floss. • A shift from the habit of using twig (Mewets) to toothbrush and is believed to be due to unavailability of the right tree in Germany.	*"I always clean my teeth in the morning, right after I eat my breakfast, and sometimes in the evening—before I went to bed" (IDI-3).* *"I don't have any idea about this thread, and I have never used one in my life" (FGD-2).* *"Not always and only, but I sometimes use Mewets if I can find a good tree" (IDI-15).*
Professional dental care	• The Majority attend dental clinic in times of dental emergency or pain. Only 2 (8%) regularly (twice a year), another 2 (8%) never attended, and 17 (68%) at least once in their lifetime.	*"I only go to a dentist for an essential treatment; for example, I once went to a dental clinic for a severe dental pain" (IDI-3).* *"I check my tooth every six months; it doesn't matter whether I have a problem or not." (IDI- 2)*
Misperception of oral healthcare practice	• Some believed that continuous use of toothbrush and paste, could harm their teeth, a reason for introducing and widening spacing between teeth, and intermittent use of toothbrush as a risk factor for bad breath. • Flossing perceived to damage gums and initiation of flossing addiction that in turn exacerbates the negative effect of flossing.	*"From my understanding […], and from what I heard, the chemical in the toothpaste is destructive to our teeth." (FGD-1).* *"I guess it might damage our teeth or gum. So, I don't have any plans to use it" (IDI-14)*

Table A2. *Cont.*

Themes and Sub-themes	Pertinent Findings According to Participants (ERNRAS)	Quotes
Approachability and Ability to Perceive		
Information about availability and navigation of the oral healthcare system	• ERNRAS lack clear information to locate dental and navigate the oral healthcare system in Germany. • They lack information on eligibility (service fee exemption), dental care entitlement and health insurance.	*"I don't know where and how to find it [dental clinic" (IDI-6).* *"I don't know even whether a regular visiting is free. I am just hearing now that I could go to a dentist on a twice a year basis"(IDI-12).*
Oral health literacy and beliefs	• Few lacks clear understanding regarding the vital significance of oral health care. • Awareness issues while they were in Eritrea influenced the poor oral healthcare outcome and mentality in Germany. • They have beliefs in homeopathic medicines such as potions, herbs, and saltwater rinse, praying, tolerating, or fighting dental pain affected their perception.	*"In our case [Eritreans], […] we were neither screened nor taught to take care of our teeth when we were in our country. Then we grew up known nothing, and it is costing us a lot to learn to take good care of our teeth. We are simply detached of the reality" (FGD-2).*
The level of trust and Expectations from a dentist	• Perceived negatively for some of their dentists' treatment decisions, the bureaucracy of the dental clinic, and worried about future financial implications.	*"When they [dentists] are taking out our tooth, they want to use it [tooth] for their interest and replace ours with artificial tooth"(FGD-1).*
Acceptability and Ability to Seek		
Interculturally competent professionals	• Some dental care professionals fail to consider the uniqueness of ERNRAS regarding their cultural views, practices, and beliefs towards oral healthcare acts as a barrier to access care.	*"I can say my first dentist could have done more […]. Not only he doesn't want to hear my opinion, but also wanted me to follow his instructions only. That's why the treatment he provided that time couldn't satisfy me" (IDI-3).*
Lack of communication and professional value	• Barriers related to dental care professionals' failure to communicate patiently or productively, and also poor professional conduct and racial discrimination.	*"There is one staff, she doesn't really hear you what you want to say, I don't know why, she is either racist or arrogant" (IDI-4).*
Autonomy and capacity to seek oral healthcare	• ERNRAS were unable to seek dental care due to their limited capacity and doubtful self-reliance related to language barrier.	*"Sometimes, even though we are in a great misery and needed treatment, we don't go to the dentist due to lack of confidence the language barrier puts us into" (FGD-1).*
Disregard or negligence	• Participants fail to seek dental care and regular check-ups related to their negligence, disregard, and unfavourable previous dental experience.	*"Dental care never been my priority […], unless I have a serious pain, I don't care to visit a dentist for a minor discomfort" (IDI-6).*
Fear, anxiety and past dental experience	• ERNRAS fail to seek or discontinue dental treatment due to apprehension related to physical pain, anxiety, and past personal or friends' negative dental experience.	*"Dental appointment is good, [..] but I would never go to my dentist for a regular check-up unless I have pain; I am scared of the machines"(FGD-2).*

Table A2. Cont.

Themes and Sub-themes	Pertinent Findings According to Participants (ERNRAS)	Quotes
Availability and Accommodation and ability to reach		
Language issues and Availability of translator	• Communication attributes hampers access to dental care to ERNRAS: language difficulty, unavailability of translator, and interpreter dependency and privacy concerns.	*"It is the language problem I have; I can't tell a dentist what is really happening to me, and that is why I didn't go to the dentist" (IDI-6).*
Existence of dental services, hours of opening and appointment	• Access was hindered by delay in appointment, long waiting list, inconvenient or inflexible clinic working hours.	*"Sometimes the long waiting time and all […], we [refugees] don't visit unless we have serious problem or pain" (IDI-11).*
Living environment and mobility	• Eritrean refugees, particularly those living in the small towns or villages, raised the impact of long distance in reaching dental clinic.	*"Well, earlier my dental clinic wasn't that far, now that I have changed my residence area to small village, it's a bit far from my dental clinic. I am not going to the dental clinic, maybe because of this" (IDI-5).*
Affordability and Ability to pay		
Direct and Indirect cost	• Unaffordability of direct payments (out-of-pocket, co-payment) and indirect costs (transport, dental product, and opportunity cost) were raised as barriers by most of the participants.	*"And, the other day that I didn't visit my dentist might have been due to the instinct I developed to avoid paying money to the dentist as it is crazy expensive"(IDI-3).*
Entitlement based on age, refugee and employment status	• Adult asylum seekers were unable to access dental care as they were not entitled to (except in cases of emergency). • Unemployed (inadequate income) ERNRAS unable to pay dental services that require fee or co-payment and fail to access dental care.	*"I wanted to check and clean my teeth. However, I couldn't get the treatment, both cleaning and filling my teeth, [..] because I wasn't entitled to that treatment as I was an asylum seeker" (FGD-2).* *"Dental treatment is so expensive that I can't really afford it since a I am not working right now" (IDI-3).*
Appropriateness and ability to engage		
Adequacy and quality of the dental care providers	• ERNRAS claimed that their accessibility to dental care was also hampered by inadequacy and incompetence of the oral healthcare providers.	*"I always think that, at my first visit, the dentist could have given me more information. He [dentist] should have checked my teeth very well and he should have filled it instead of removing" (IDI-3).*
Provider-patient relationships	• Poor relations with the dental team discourage ERNRAS from accessing dental care.	*"Some of the nurses in the reception shows you bad attitude and are not friendly" (IDI-2).*
Adherence and involvement of ERNRAS in dental treatments	• Participants' fail to engage in decision-making and adherence to the recommended dental care. • ERNRAS were inadequately motivated, committed, and less involved in their comprehensive oral healthcare decision-making.	*"Most of us [refugees] are not registered with the town's refugee collaboration community, where we could have participated and seek help when we face difficulty in understanding and accessing the dental services" (IDI-7).*

References

1. UNHCR. Global Trends—Forced Displacement in 2018. Available online: https://www.unhcr.org/globaltrends2018/ (accessed on 30 March 2020).
2. UNHCR. Convention and Protocol Relating to the Status of Refugees. Available online: https://www.unhcr.org/protection/basic/3b66c2aa10/convention-protocol-relating-status-refugees.html (accessed on 18 April 2021).

3.   UNHCR. UNHCR Master Glossary of Terms. Available online: https://www.refworld.org/docid/42ce7d444.html (accessed on 18 April 2021).

4.   Statistics on Migration to Europe. Available online: https://ec.europa.eu/info/strategy/priorities-2019-2024/promoting-our-european-way-life/statistics-migration-europe_en (accessed on 27 January 2021).

5.   UNHCR—Global Trends 2019: Forced Displacement in 2019. Available online: https://www.unhcr.org/globaltrends2019/ (accessed on 21 January 2021).

6.   World Population Prospects. Population Division, United Nations. Available online: https://population.un.org/wpp/Maps/ (accessed on 27 January 2021).

7.   BAMF. Bundesamt für Migration und Flüchtlinge. Available online: https://www.BAMF.de/SharedDocs/Anlagen/DE/Statistik/Asylgeschaeftsbericht/201812-statistik-anlage-asyl-geschaeftsbericht.html?nn=284730 (accessed on 21 January 2021).

8.   Kohlenberger, J.; Buber-Ennser, I.; Rengs, B.; Leitner, S.; Landesmann, M. Barriers to Health Care Access and Service Utilization of Refugees in Austria: Evidence from a Cross-Sectional Survey. *Health Policy Amst. Neth.* **2019**, *123*, 833–839. [CrossRef] [PubMed]

9.   Hamilton, T.G. The Healthy Immigrant (Migrant) Effect: In Search of a Better Native-Born Comparison Group. *Soc. Sci. Res.* **2015**, *54*, 353–365. [CrossRef] [PubMed]

10.   Shawyer, F.; Enticott, J.C.; Block, A.A.; Cheng, I.-H.; Meadows, G.N. The Mental Health Status of Refugees and Asylum Seekers Attending a Refugee Health Clinic Including Comparisons with a Matched Sample of Australian-Born Residents. *BMC Psychiatry* **2017**, *17*, 76. [CrossRef] [PubMed]

11.   Ghiabi, E.; Matthews, D.C.; Brillant, M.S. The Oral Health Status of Recent Immigrants and Refugees in Nova Scotia, Canada. *J. Immigr. Minor. Health* **2014**, *16*, 95–101. [CrossRef] [PubMed]

12.   Davidson, N.; Skull, S.; Calache, H.; Murray, S.; Chalmers, J. Holes a Plenty: Oral Health Status a Major Issue for Newly Arrived Refugees in Australia. *Aust. Dent. J.* **2006**, *51*, 306–311. [CrossRef]

13.   Høyvik, A.C.; Lie, B.; Grjibovski, A.M.; Willumsen, T. Oral Health Challenges in Refugees from the Middle East and Africa: A Comparative Study. *J. Immigr. Minor. Health* **2019**, *21*, 443–450. [CrossRef] [PubMed]

14.   Makan, R.; Gara, M.; Awwad, M.A.; Hassona, Y. The Oral Health Status of Syrian Refugee Children in Jordan: An Exploratory Study. *Spec. Care Dentist.* **2019**, *39*, 306–309. [CrossRef] [PubMed]

15.   Solyman, M.; Schmidt-Westhausen, A.-M. Oral Health Status among Newly Arrived Refugees in Germany: A Cross-Sectional Study. *BMC Oral Health* **2018**, *18*, 132. [CrossRef] [PubMed]

16.   Goetz, K.; Winkelmann, W.; Steinhäuser, J. Assessment of Oral Health and Cost of Care for a Group of Refugees in Germany: A Cross-Sectional Study. *BMC Oral Health* **2018**, *18*, 69. [CrossRef] [PubMed]

17.   Report on the Health of Refugees and Migrants in the WHO European Region: No Public Health without Refugee and Migrant Health. 2018. Available online: https://www.euro.who.int/en/publications/abstracts/report-on-the-health-of-refugees-and-migrants-in-the-who-european-region-no-public-health-without-refugee-and-migrant-health-2018 (accessed on 28 January 2021).

18.   Sivakumar, V.; Jain, J.; Haridas, R.; Paliayal, S.; Rodrigues, S.; Jose, M. Oral Health Status of Tibetan and Local School Children: A Comparative Study. *J. Clin. Diagn. Res. JCDR* **2016**, *10*, ZC29–ZC33. [CrossRef] [PubMed]

19.   Deora, S.; Saini, A.; Bhatia, P.; Abraham, D.V.; Masoom, S.N.; Yadav, K. Periodontal Status among Tibetan Refugees Residing in Jodhpur City. *J. Res. Dent.* **2016**, *3*, 722–729. [CrossRef]

20.   Sugerman, D.E.; Hyder, A.A.; Nasir, K. Child and Young Adult Injuries among Long-Term Afghan Refugees. *Int. J. Inj. Contr. Saf. Promot.* **2005**, *12*, 177–182. [CrossRef]

21.   Singh, H.K.; Scott, T.E.; Henshaw, M.M.; Cote, S.E.; Grodin, M.A.; Piwowarczyk, L.A. Oral Health Status of Refugee Torture Survivors Seeking Care in the United States. *Am. J. Public Health* **2008**, *98*, 2181–2182. [CrossRef] [PubMed]

22.   Cote, S.; Geltman, P.; Nunn, M.; Lituri, K.; Henshaw, M.; Garcia, R.I. Dental Caries of Refugee Children Compared with US Children. *Pediatrics* **2004**, *114*, e733–e740. [CrossRef] [PubMed]

23.   van Berlaer, G.; Carbonell, F.B.; Manantsoa, S.; de Béthune, X.; Buyl, R.; Debacker, M.; Hubloue, I. A Refugee Camp in the Centre of Europe: Clinical Characteristics of Asylum Seekers Arriving in Brussels. *BMJ Open* **2016**, *6*, e013963. [CrossRef]

24.   Keboa, M.T.; Hiles, N.; Macdonald, M.E. The Oral Health of Refugees and Asylum Seekers: A Scoping Review. *Glob. Health* **2016**, *12*, 59. [CrossRef] [PubMed]

25.   Mattila, A.; Ghaderi, P.; Tervonen, L.; Niskanen, L.; Pesonen, P.; Anttonen, V.; Laitala, M.-L. Self-Reported Oral Health and Use of Dental Services among Asylum Seekers and Immigrants in Finland-a Pilot Study. *Eur. J. Public Health* **2016**, *26*, 1006–1010. [CrossRef]

26.   Lamb, C.E.F.; Whelan, A.K.; Michaels, C. Refugees and Oral Health: Lessons Learned from Stories of Hazara Refugees. *Aust. Health Rev. Publ. Aust. Hosp. Assoc.* **2009**, *33*, 618–627. [CrossRef] [PubMed]

27.   Geltman, P.L.; Adams, J.H.; Cochran, J.; Doros, G.; Rybin, D.; Henshaw, M.; Barnes, L.L.; Paasche-Orlow, M. The Impact of Functional Health Literacy and Acculturation on the Oral Health Status of Somali Refugees Living in Massachusetts. *Am. J. Public Health* **2013**, *103*, 1516–1523. [CrossRef]

28.   Keboa, M.; Nicolau, B.; Hovey, R.; Esfandiari, S.; Carnevale, F.; Macdonald, M.E. A Qualitative Study on the Oral Health of Humanitarian Migrants in Canada. *Community Dent. Health* **2019**, *36*, 95–100. [CrossRef] [PubMed]

29.   Willis, M.S.; Bothun, R.M. Oral Hygiene Knowledge and Practice Among Dinka and Nuer from Sudan to the U.S. *Am. Dent. Hyg. Assoc.* **2011**, *85*, 306–315.

30. Riggs, E.; Gibbs, L.; Kilpatrick, N.; Gussy, M.; van Gemert, C.; Ali, S.; Waters, E. Breaking down the Barriers: A Qualitative Study to Understand Child Oral Health in Refugee and Migrant Communities in Australia. *Ethn. Health* **2015**, *20*, 241–257. [CrossRef] [PubMed]

31. S Dhull, K.; Dutta, B.; M Devraj, I.; Samir, P. Knowledge, Attitude, and Practice of Mothers towards Infant Oral Healthcare. *Int. J. Clin. Pediatr. Dent.* **2018**, *11*, 435–439. [CrossRef]

32. Asylum Act (AsylG). Available online: https://www.BAMF.de/SharedDocs/Links/EN/G/gesetze-im-internet-asylvfg.html?nn=285460 (accessed on 22 January 2021).

33. Davidson, N.; Skull, S.; Chaney, G.; Frydenberg, A.; Isaacs, D.; Kelly, P.; Lampropoulos, B.; Raman, S.; Silove, D.; Buttery, J.; et al. Comprehensive Health Assessment for Newly Arrived Refugee Children in Australia. *J. Paediatr. Child Health* **2004**, *40*, 562–568. [CrossRef] [PubMed]

34. Calvasina, P.; Muntaner, C.; Quiñonez, C. Factors Associated with Unmet Dental Care Needs in Canadian Immigrants: An Analysis of the Longitudinal Survey of Immigrants to Canada. *BMC Oral Health* **2014**, *14*, 145. [CrossRef]

35. Quach, A.; Laemmle-Ruff, I.L.; Polizzi, T.; Paxton, G.A. Gaps in Smiles and Services: A Cross-Sectional Study of Dental Caries in Refugee-Background Children. *BMC Oral Health* **2015**, *15*, 10. [CrossRef] [PubMed]

36. Petersen, P.E. The World Oral Health Report 2003: Continuous Improvement of Oral Health in the 21st Century—The Approach of the WHO Global Oral Health Programme. *Community Dent. Oral Epidemiol.* **2003**, *31*, 3–24. [CrossRef]

37. Rautemaa, R.; Lauhio, A.; Cullinan, M.P.; Seymour, G.J. Oral Infections and Systemic Disease—an Emerging Problem in Medicine. *Clin. Microbiol. Infect.* **2007**, *13*, 1041–1047. [CrossRef] [PubMed]

38. Li, X.; Kolltveit, K.M.; Tronstad, L.; Olsen, I. Systemic Diseases Caused by Oral Infection. *Clin. Microbiol. Rev.* **2000**, *13*, 547–558. [CrossRef] [PubMed]

39. Jansson, H.; Lindholm, E.; Lindh, C.; Groop, L.; Bratthall, G. Type 2 Diabetes and Risk for Periodontal Disease: A Role for Dental Health Awareness. *J. Clin. Periodontol.* **2006**, *33*, 408–414. [CrossRef] [PubMed]

40. Dietrich, T.; Webb, I.; Stenhouse, L.; Pattni, A.; Ready, D.; Wanyonyi, K.L.; White, S.; Gallagher, J.E. Evidence Summary: The Relationship between Oral and Cardiovascular Disease. *Br. Dent. J.* **2017**, *222*, 381–385. [CrossRef] [PubMed]

41. Pani, S.C.; Al-Sibai, S.A.; Rao, A.S.; Kazimoglu, S.N.; Mosadomi, H.A. Parental Perception of Oral Health-Related Quality of Life of Syrian Refugee Children. *J. Int. Soc. Prev. Community Dent.* **2017**, *7*, 191–196. [CrossRef]

42. Abu-Awwad, M.; AL-Omoush, S.; Shqaidef, A.; Hilal, N.; Hassona, Y. Oral Health-Related Quality of Life among Syrian Refugees in Jordan: A Cross-Sectional Study. *Int. Dent. J.* **2020**, *70*, 45–52. [CrossRef] [PubMed]

43. Ogunbodede, E.O.; Mickenautsch, S.; Rudolph, M.J. Oral Health Care in Refugee Situations: Liberian Refugees in Ghana. *J. Refug. Stud.* **2000**, *13*, 328–335. [CrossRef]

44. UNHCR. Access to Healthcare. Available online: https://www.unhcr.org/access-to-healthcare.html (accessed on 18 April 2021).

45. Pottie, K.; Greenaway, C.; Feightner, J.; Welch, V.; Swinkels, H.; Rashid, M.; Narasiah, L.; Kirmayer, L.J.; Ueffing, E.; MacDonald, N.E.; et al. Evidence-Based Clinical Guidelines for Immigrants and Refugees. *CMAJ* **2011**, *183*, E824–E925. [CrossRef] [PubMed]

46. Polkinghorne, D.E. Phenomenological Research Methods. In *Existential-Phenomenological Perspectives in Psychology: Exploring the Breadth of Human Experience*; Valle, R.S., Halling, S., Eds.; Springer US: Boston, MA, USA, 1989; pp. 41–60. ISBN 978-1-4615-6989-3.

47. The 2018 Migration Report. Available online: https://www.BAMF.de/SharedDocs/Anlagen/EN/Forschung/Migrationsberichte/migrationsbericht-2018.html?nn=403964 (accessed on 29 January 2021).

48. Initial Distribution of Asylum-Seekers (EASY). Available online: https://www.BAMF.de/EN/Themen/AsylFluechtlingsschutz/AblaufAsylverfahrens/Erstverteilung/erstverteilung-node.html (accessed on 17 April 2021).

49. Heidelberg.de. Arrival Center for Baden-Württemberg. Available online: https://www.heidelberg.de/english/Home/Life/arrival+center+for+baden-wuerttemberg.html (accessed on 29 January 2021).

50. Heidelberg.de. Refugees in Heidelberg. Available online: https://www.heidelberg.de/english/Home/Life/refugees+in+heidelberg.html (accessed on 29 January 2021).

51. Goodman, L.A. Snowball Sampling. *Ann. Math. Stat.* **1961**, *32*, 148–170. [CrossRef]

52. Braun, V.; Clarke, V. Using Thematic Analysis in Psychology. *Qual. Res. Psychol.* **2006**, *3*, 77–101. [CrossRef]

53. Levesque, J.-F.; Harris, M.F.; Russell, G. Patient-Centred Access to Health Care: Conceptualising Access at the Interface of Health Systems and Populations. *Int. J. Equity Health* **2013**, *12*, 18. [CrossRef]

54. Aboul-Enein, B.H. The Miswak (Salvadora Persica L.) Chewing Stick: Cultural Implications in Oral Health Promotion. *Saudi J. Dent. Res.* **2014**, *5*, 9–13. [CrossRef]

55. Araya, Y. Contribution of Trees for Oral Hygiene in East Africa. *Ethnobot. Leafl.* **2007**, *2007*, 8.

56. Prowse, S.; Schroth, R.J.; Wilson, A.; Edwards, J.M.; Sarson, J.; Levi, J.A.; Moffatt, M.E. Diversity Considerations for Promoting Early Childhood Oral Health: A Pilot Study. *Int. J. Dent.* **2014**, *2014*. [CrossRef] [PubMed]

57. Hilton, I.V.; Stephen, S.; Barker, J.C.; Weintraub, J.A. Cultural Factors and Children's Oral Health Care: A Qualitative Study of Carers of Young Children. *Community Dent. Oral Epidemiol.* **2007**, *35*, 429–438. [CrossRef] [PubMed]

58. Hobbs, C. Talking about Teeth—Exploring the Barriers to Accessing Oral Health Services for Horn of Africa Community Members Living in Inner West Melbourne. Available online: https://www.semanticscholar.org/paper/Talking-about-Teeth-exploring-the-barriers-to-oral-Hobbs/a8ba1491d521e1cfbf8e6a3026a0b0a16d5043b4 (accessed on 22 January 2021).

59. Loizzo, M.R.; Lecce, G.D.; Boselli, E.; Bonesi, M.; Menichini, F.; Menichini, F.; Frega, N.G. In Vitro Antioxidant and Hypoglycemic Activities of Ethiopian Spice Blend Berbere. *Int. J. Food Sci. Nutr.* **2011**, *62*, 740–749. [CrossRef] [PubMed]

60. Holmboe-Ottesen, G.; Wandel, M. Changes in Dietary Habits after Migration and Consequences for Health: A Focus on South Asians in Europe. *Food Nutr. Res.* **2012**, *56*, 18891. [CrossRef]

61. Adams, J.H.; Young, S.; Laird, L.D.; Geltman, P.L.; Cochran, J.J.; Hassan, A.; Egal, F.; Paasche-Orlow, M.K.; Barnes, L.L. The Cultural Basis for Oral Health Practices among Somali Refugees Pre- and Post-Resettlement in Massachusetts. *J. Health Care Poor Underserved* **2013**, *24*, 1474–1485. [CrossRef] [PubMed]

62. Riggs, E.; Yelland, J.; Shankumar, R.; Kilpatrick, N. 'We Are All Scared for the Baby': Promoting Access to Dental Services for Refugee Background Women during Pregnancy. *BMC Pregnancy Childbirth* **2016**, *16*, 12. [CrossRef]

63. Jeevan, J.K.; Yacob, T.; Abdurhman, N.; Asmerom, S.; Birhane, T.; Sebri, J.; Kaushik, A. Eritrean Chewing Sticks Potential against Isolated Dental Carries Organisms from Dental Plaque. *Arch. Clin. Microbiol.* **2017**, *8*, 58. [CrossRef]

64. Herz, R.S. Odor-Associative Learning and Emotion: Effects on Perception and Behavior. *Chem. Senses* **2005**, *30*, i250–i251. [CrossRef]

65. Andegiorgish, A.K.; Weldemariam, B.W.; Kifle, M.M.; Mebrahtu, F.G.; Zewde, H.K.; Tewelde, M.G.; Hussen, M.A.; Tsegay, W.K. Prevalence of Dental Caries and Associated Factors among 12 Years Old Students in Eritrea. *BMC Oral Health* **2017**, *17*, 169. [CrossRef]

66. Mariño, R.; Wright, C.; Schofield, M.; Minichiello, V.; Calache, H. Factors Associated with Self-Reported Use of Dental Health Services among Older Greek and Italian Immigrants. *Spec. Care Dentist.* **2005**, *25*, 29–36. [CrossRef]

67. Davidson, N.; Skull, S.; Calache, H.; Chesters, D.; Chalmers, J. Equitable Access to Dental Care for an At-Risk Group: A Review of Services for Australian Refugees. *Aust. N. Z. J. Public Health* **2007**, *31*, 73–80. [CrossRef] [PubMed]

68. Erdsiek, F.; Waury, D.; Brzoska, P. Oral Health Behaviour in Migrant and Non-Migrant Adults in Germany: The Utilization of Regular Dental Check-Ups. *BMC Oral Health* **2017**, *17*, 84. [CrossRef] [PubMed]

69. Sjögren Forss, K.; Mangrio, E. Scoping Review of Refugees' Experiences of Healthcare. *Eur. J. Public Health* **2018**, *28*. [CrossRef]

70. Borgschulte, H.S.; Wiesmüller, G.A.; Bunte, A.; Neuhann, F. Health Care Provision for Refugees in Germany—One-Year Evaluation of an Outpatient Clinic in an Urban Emergency Accommodation. *BMC Health Serv. Res.* **2018**, *18*, 488. [CrossRef]

71. Lamb, C.F.; Smith, M. Problems Refugees Face When Accessing Health Services. *New South Wales Public Health Bull.* **2002**, *13*, 161–163. [CrossRef] [PubMed]

72. Newton, J.T.; Thorogood, N.; Bhavnani, V.; Pitt, J.; Gibbons, D.E.; Gelbier, S. Barriers to the Use of Dental Services by Individuals from Minority Ethnic Communities Living in the United Kingdom: Findings from Focus Groups. *Prim. Dent. Care* **2001**, *8*, 157–161. [CrossRef]

73. Bozorgmehr, K.; Razum, O. Effect of Restricting Access to Health Care on Health Expenditures among Asylum-Seekers and Refugees: A Quasi-Experimental Study in Germany, 1994–2013. *PLoS ONE* **2015**, *10*, e0131483. [CrossRef]

74. Humphris, R.; Bradby, H. Health Status of Refugees and Asylum Seekers in Europe. Available online: https://oxfordre.com/publichealth/view/10.1093/acrefore/9780190632366.001.0001/acrefore-9780190632366-e-8 (accessed on 18 February 2021).

75. Lincoln, Y.S.; Guba, E.G.; Pilotta, J.J. Naturalistic Inquiry Beverly Hills, CA: Sage Publications, 1985, 416 pp. *Int. J. Intercult. Relat.* **1985**, *9*, 438–439. [CrossRef]

76. Terzic, A.; Waldman, S. Chronic Diseases: The Emerging Pandemic. *Clin. Transl. Sci.* **2011**, *4*, 225–226. [CrossRef]

International Journal of
Environmental Research and Public Health

Review

Dental Caries among Refugees in Europe: A Systematic Literature Review

Sneha Bhusari , Chiamaka Ilechukwu, Abdelrahman Elwishahy , Olaf Horstick, Volker Winkler and Khatia Antia *

Heidelberg Institute of Global Health, Heidelberg University Hospital, 69120 Heidelberg, Germany; sneha.bhusari@uni-heidelberg.de (S.B.); nkorika.ilechukwu@alumni.uni-heidelberg.de (C.I.); abdelrhman.elwishahy@alumni.uni-heidelberg.de (A.E.); olaf.horstick@uni-heidelberg.de (O.H.); volker.winkler@uni-heidelberg.de (V.W.)
* Correspondence: khatia.antia@uni-heidelberg.de; Tel.: +49-152-2785-7798

Received: 10 November 2020; Accepted: 16 December 2020; Published: 18 December 2020

Abstract: Oral health is one of the most neglected aspects of refugee health. The study aimed to systematically review evidence on prevalence of dental caries and dental care services provided to refugees in Europe. Following PRISMA guidelines, we searched PubMed, Cochrane, WHOLIS, Web of Science, Medline Ovid, and Google Scholar identifying studies on dental caries among refugees in Europe after the 2015 refugee crisis. From 3160 records, fourteen studies were included in the analysis. Eight studies on oral health showed caries prevalence of between 50% and 100%, while it ranged from 3% to 65% in six general health studies. Caries prevalence was proportional to age and inversely associated with education, whereas gender and country of origin showed no significant association. Nowhere is oral health part of general health assessment on arrival and is complaint based. Primary focus on resettlement, language, cultural, and economic barriers emerged as explanatory models for limited access. Our study identified a high prevalence of caries and limited access to dental health services as main challenges. Integrating oral health check-ups may contribute in shifting towards preventive oral care. Further research is urgently needed to better understand the dental needs of refugees in Europe.

Keywords: caries; decay; Decayed Missing and Filled index (DMF) and dental health; refugee; asylum seeker

1. Introduction

Antonio Guterres, the Secretary-General of the United Nations, described the European situation in 2015 as "primarily a refugee crisis, not only a migration phenomenon" [1]. In 2015, the largest movement of people for 20 years was seen, with more than 3.5 million refugees in Europe [2]. The International Organization for Migration (IOM) defines a refugee as "a person who is outside the country of his nationality and is unable to avail himself of the protection of that country" [3]. During refugees' state of unrest, the most valuable assets become necessities such as clean water, food, nutrition, shelter, sanitation, and protection, while medical assessment and healthcare are neglected [4]. Moreover, refugees are faced with language barriers, unfamiliar surroundings, new laws, rules and regulations [3].

Refugees are always at a risk for innumerable issues regarding health [5]. Of all the conditions faced, necessary or emergency health issues are addressed in European countries [6]. Major areas of health focused upon include non-communicable and communicable diseases, maternal and child health, occupational health and mental health [7]. Even though oral health is a key indicator of overall health, well-being and quality of life [8], it is not part of this essential list [7]. Moreover, the exclusion of dental assessment within basic care makes refugees more vulnerable [4] and the lack of active involvement of a dentist curtails the importance of oral health [9].

Oral and or dental diseases are correlated with non-communicable diseases (NCDs) [9]. They can result in malnutrition due to alterations in diet, and phonation problems, especially in the older age group [10]. There is also higher body dissatisfaction [11] and simple acts of smiling, communicating and eating can be affected negatively [12]. Hence, oral health not only affects one's general health, but also has an impact on mental health. Dental caries is the leading oral health problem, with high prevalence, affecting a large population in the majority of the countries, including Europe [12,13]. The basic motive for seeking oral health is mainly pain based [12]. Ordinarily, oral health, and in particularly caries, is one of the most neglected aspects of health irrespective of region, culture, education or the socioeconomic status of an individual, and more so in low and middle-income countries. The overall burden is decreasing due to public health measures, but prevalence still remains high [12].

Considering war-affected regions, attention to oral health can be even worse or non-existent. Such populations suffer the most, not only with the general requirement for oral care, but also with need based (i.e., pain based) oral care, and prioritizing oral health becomes increasingly difficult for refugees as other priorities are pre-eminent [6]. In light of this situation, the prevalence of dental caries is expected to be high among refugees in general and in Europe in particular. Lack of proper education, information and awareness of oral health, lack of inclination to maintain good oral health, overall neglect of oral health and financial limitations, coupled with geographical constraints, war or devastating surroundings, migration, resettlement in foreign lands, language barriers and lack of stability have resulted in an increase in dental caries (along with other oral problems) [14]. This lack of provision is the main area of concern about, and hindrance to, obtaining health data and achieving good health care.

The aim of this study was to find out the prevalence of dental caries among refugees in the European region. The objectives were twofold: first, to synthesize the evidence of prevalence of dental caries among refugees in the European region after the 2015 crisis by evaluating the Decayed Missing and Filled index (DMF); and second, to evaluate the dental care services provided to the refugees in Europe and their needs and shortcomings

2. Materials and Methods

This study followed the reporting guidelines of Preferred Reporting Items for Systematic Reviews and Meta-Analyses (PRISMA) [15]. We included all types of quantitative and qualitative study. There was no restriction regarding the language, age, gender, country of origin, education, or socioeconomic status of the participants. However, we included only studies focusing on caries among refugees or asylum seekers after 2015 (the European Migrant Crisis), but not on oral conditions of periodontium or oral mucosa. We also excluded studies with the word "migration", "migrant" or "immigrant" from the search. The word 'Europe' was a broad term; therefore, we dropped the term 'Europe' during the database search and manually searched for European studies. We performed the search in English using the key words Refugee or Asylum seeker in combination with Caries, Decay, DMF or Dental Health through the following databases: PubMed, Cochrane, World Health Organization Library Information System (WHOLIS), Web of Science, Medline Ovid and Google Scholar. The search was finalized on 21st November 2020. The database searches as well as the screening procedure were run independently by the first and the second author. Conflicts were resolved upon agreement by focusing on the eligibility criteria and the aims set for this review. Removal of duplicates was carried out at a later stage. Table 1 denotes the PICO criteria used for this study.

Table 1. PICO and eligibility criteria for this systematic literature review.

Criteria	Inclusion	Exclusion
Population	Refugee and Asylum seeker	Migration, migrant, immigrant
Indicator	European region	Other regions
Comparison	No specific comparators set	No specific comparators set
Outcome	Caries, Decay, DMF, Dental health	Other oral conditions of periodontium or oral mucosa

The decayed, missed and filled index, known as the DMF index, is the measure of the prevalence of caries; it identifies the number of teeth with dental caries including its effects on an individual [16]. The DMF index has been a simple, rapid, universally accepted and widely used tool for several decades to determine coronal caries experience, since it requires a minimal inventory: natural light, plain mouth mirror and a fine probe. The calculation of DMF is performed by obtaining the number of decayed, missed and filled teeth or surfaces [17,18]. However, the DMF index does not distinguish the reason for loss of tooth (MT) [16]. We extracted information regarding Decayed teeth (DT), Missing teeth (MT), and Filled teeth (FT), and the average DMF index from the included studies to look mainly at the experience of caries, where only the DT factor was focused on.

We evaluated risk of bias using the quality assessment tool of The U.S. National Institute of Health (NIH) [19]. This provided separate assessment criteria for different types of studies under one domain. We used two groups: Quality Assessment Tool for Observational Cohort and Cross-Sectional Studies, and Quality Assessment Tool for Case Series Studies.

3. Results

Our search yielded 3160 records, 1717 from the five databases and 1443 from Google scholar for which we screened the first 200 hits per combination until no further relevant studies were found. 205 articles remained after title and abstract screening, from which twenty full texts were evaluated against the eligibility criteria. Finally, fourteen studies were included in this systematic literature review. A detailed description of the screening process can be seen in Figure 1.

From the final fourteen articles, only one was a qualitative study [20] while others were: ten cross-sectional [6,21–29], one cohort study [30] and two case reports [2,13]. All except two studies [13,28] had a comparison group. By the Quality Assessment Tool, only one was graded as 'fair' [30] while all other studies were 'good'. The main host countries were Belgium, Finland, Germany, Greece, Norway, Spain, Sweden and the UK, while refugees originated from a wide range of countries with a majority coming from Afghanistan, Iraq and Syria. Less frequently, refugees came from Asia, Africa, Europe and the Middle East as listed in Table 2.

Table 2. Characteristics of Included Studies by hierarchy of evidence.

Author	Year	Study Type	Sample Size	Age in Years	Host Country	Country of Origin
[30] Hermans et al.	2017	Dynamic cohort	2291	18–38	Greece	Afghanistan, Pakistan and Syria
[21] Solyman and Schmidt-Westhausen	2018	Cross-sectional	386	18–60	Germany	Iraq and Syria
[22] Kakalou et al.	2018	Cross-sectional: Descriptive	6688	0–75+	Greece	Afghanistan, Iraq and Syria, other regions: Africa, Asia and the Middle East
[6] Høyvik et al.	2019	Cross-sectional: Comparative	132	18–47	Norway	Africa and The Middle East
[23] Goetz et al.	2018	Cross-sectional: Pilot	102	16–64	Germany	Afghanistan, Armenia, Chechnya, Eritrea, Iran, Iraq, Somalia, Syria and Yemen
[24] Riatto et al.	2018	Cross-sectional	156	5–13	Spain	Syria
[25] Pavlopoulou et al.	2017	Cross-sectional: Prospective	300	0–14	Greece	Afghanistan, Bangladesh, DR Congo, Eritrea, Iran, Kenya, Lebanon, Pakistan, Somalia and Sudan

Table 2. *Cont.*

Author	Year	Study Type	Sample Size	Age in Years	Host Country	Country of Origin
[26] van Berlaer et al.	2016	Cross-sectional: Descriptive	3907	0–75+	Belgium	Afghanistan Iraq, Morocco, Palestine and Syria
[20] Mattila et al.	2016	Cross-sectional: Pilot	38	17–53	Finland	Afghanistan, Hungary, Iran, Iraq, Morocco, Russia, Slovakia, China, Somalia, South Sudan, Sweden, Syria, Thailand, Turkey and Vietnam
[27] Al-Ani et al.	2020	Cross-sectional	544	3–75+	Germany	Mainly from Afghanistan, Iraq and Syria, Others nationalities: African countries, Arabian countries, Asia and Eastern Europe
[29] Hjern and Kling	2019	Cross-sectional	639	6–15	Sweden	Afghanistan and Syria
[28] Freiberg et al.	2020	Retrospective observational	568	20–34	Germany	Afghanistan, Iran, Somalia and Syria
[13] Zaheer et al.	2017	Case report	NS	NS	Greece	Afghanistan, Kurdistan, Iraq and Syria
[2] Williams et al.	2016	Case report	NS	NS	European mainland and the UK	Afghanistan, Albania, Eritrea, Iran, Iraq and Syria

Note: NS—Not specified.

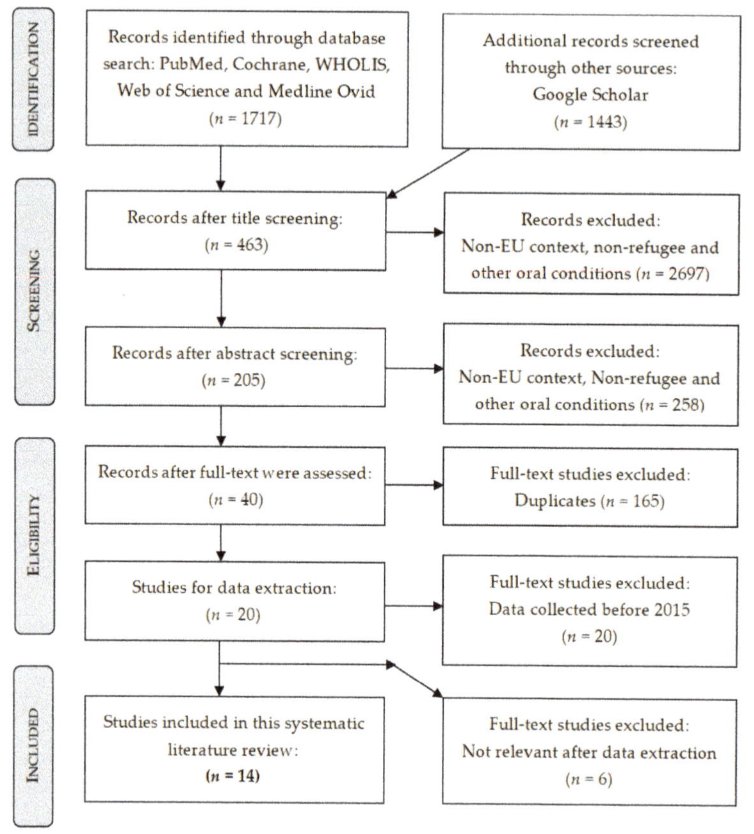

Figure 1. Preferred Reporting Items for Systematic Reviews and Meta-Analyses PRISMA flow chart depicting the selection process [15].

All studies addressed information on oral or health status, four studies [6,13,29,30] focused on the healthcare needed, three studies [20,25,28] examined the treatment provided and two studies [13,30] examined necessary improvements. The study populations in three studies [24,25,29] were children, in five studies children and adults [22,26–28,30] and in four studies only adults [6,20,21,23]. All study samples consisted of more men than women. Questionnaires on both medical and travel history along with present living and medical conditions including clinical assessments were the primary sources of data in all studies except one [20], which used self-reporting as a source.

Only eight studies [6,13,20,21,23,24,27,28] had oral health as the focus while others concentrated on oral checkups along with general health assessments. The prevalence of caries was higher in the oral health focused studies as shown in Table 3. Only one study [24] reported a low prevalence of caries, which was explained by the fact that children from wealthier families had better access to oral health services in the country of origin. From the above mentioned eight studies, only five studies [6,21,23,24,27] used the DMF index as a part of their analysis, while others recorded oral issues based on complaint. Four studies [6,21,23,27] out of the five reported a very high DMF severity. These five studies also showed an expanded version of the Decayed Missing Filled Teeth (DMFT) index with individual components reported, namely Decayed Teeth (DT), Missing Teeth (MT) and Filled Teeth (FT). As seen in Tables 4–6, the DT is observed to be high for all the studies except one [23], where the author suggests the high to moderate social status of the sample population to be the reason for higher MT and FT as compared with DT. DMF is denoted as an index for permanent teeth, and dmf for deciduous teeth.

Table 3. DMF and Caries prevalence.

Study	Focus	Dentist Involved	Instruments Mentioned	DMF	Prevalence of Caries %	Reliability Tested	Guideline
[30] Hermans et al.	GH	NR	NR	NR	2.9 *	NR	SPHERE
[21] Solyman and Schmidt-Westhausen	OH	Yes	Yes	Yes [a]	87.5	Yes	WHO
[22] Kakalou et al.	GH	NR	NR	NR	4.6	NR	ICD-10
[6] Høyvik et al.	OH	Yes	Yes	Yes [a]	89.4	Yes	astdd
[23] Goetz et al.	OH	Yes	Yes	Yes [a]	NR	1 Dentist	ICDAS (STROBE)
[24] Riatto et al.	OH	Yes	Yes	Yes [b]	(50–75)	Yes	WHO
[25] Pavlopoulou et al.	GH	NR	NR	NR	24.7	NR	NR
[26] van Berlaer et al.	GH	NR	NR	NR	8.1	NR	ICD-10
[20] Mattila et al.	OH	Yes	NR	NR	AS 57	NR	NR
[27] Al-Ani et al.	OH	Yes	Yes	Yes [c]	Age groups: 0–3 (49) 6–11 (14) 13–17 (28) 18–34 (10) 35–44 (16) 45–64 (21)	Yes	WHO
[29] Hjern and Kling	GH	NR	Yes	NR	48.1	NR	NR
[28] Freiberg et al.	OH	Yes	NR	NR	98.7	NR	(BEMA)
[13] Zaheer et al.	OH	Yes	NR	NR	100	NR	NR
[2] Williams et al.	GH	NR	NR	NR	65	NR	NR

Note: GH—General health; OH—Oral health; AS—Asylum seeker; NR—Not Reported; * For 30 Patients; [a] All participants were adults; [b] All participants were children; [c] Participants were children and adults. For deciduous (up to 6 years—permanent (up to 12 years) teeth respectively. SPHERE—Global movement started in 1997 to improve quality of humanitarian assistance, Humanitarian Charter and Minimum Standards in Humanitarian Response; WHO—World health organization; ICD-10—International Classification of Diseases, 10th revision; astdd—Association of State and Territorial Dental Directors; ICDAS—International Caries Detection and Assessment System; BEMA—standard of evaluation of dental services and forms within the statutory health insurance in Germany.

Table 4. Detailed DMF Index reported in the included (dental) studies: All participants were adults.

Study	Average/Mean							
	DMFT		DT		MT		FT	
[21] Solyman and Schmidt-Westhausen, 2018	6.4		4.0		1.5		0.9	
[6] Høyvik et al., 2019	7.4		4.3		1.4		1.7	
	ME	A	ME	A	ME	A	ME	A
	10.7	5.7	5.2	3.9	1.6	1.3	3.9	0.5
[23] Goetz et al., 2018	6.9		2.9		3.9		3.8	

Note: ME—Middle East; A—Africa.

Table 5. [24] Riatto et al., 2018.

DMF/Dmf for Different Age Groups: All Participants Were Children						
DMF-dmf/Age	5–7	8–10	11–13	6	12	5–13
DMFT	0.1	0.7	1.8	0.1	1.6	0.8
DT	0.1	0.7	1.5	0.1	1.4	0.7
MT	0	0	0.1	0	0	0
MT	0	0	0.2	0	0.2	0.1
dft	3.2	2.2	0.9	3.2	0	2.2

Table 6. [27] Al-Ani et al., 2020.

Detailed DMF Index: Participants Were Children and Adult							
Age group	d	m	f	dmf	D	M	F	DMF
3	2.54	0.05	0.03	2.62	-	-	-	-
6–7	4.21	0.47	0.55	5.22	0.12	0	0.02	0.13
8–11	2.50	0.53	0.57	3.60	0.42	0.02	0.26	0.70
12	0.62	0.08	0.15	0.85	1.12	0.06	0.82	2
13–17	-	-	-	-	1.93	0.23	0.72	2.87
18–34	-	-	-	-	3.72	1.46	2.24	7.43
35–44	-	-	-	-	3.13	3.22	4.21	10.55
45–64	-	-	-	-	3.64	7.63	3.64	14.92

Six studies found an association between caries and socio-demographic variables [6,21,22,24,27,30]. Age was directly correlated while education was inversely proportional to caries prevalence [30]. Caries was inversely proportional in deciduous dentition age while directly proportional to permanent dentition age [24]. All studies showed that men had a higher prevalence compared to women. No country of origin specific effects were observed [22] but Høyvik et al. [6] suggest that the differences in caries prevalence are related to the origin of the refugee population when comparing two sets of refugees from the Middle East and Africa. None of the included studies had access to the pre-arrival oral health status of the sample population.

All except four studies [13,20,22,24] showed the need for oral screening and all except five studies [2,6,22,25,30] concluded the need for a preventive focus. Freiberg et al. [28] suggested that regular check-ups have a potential to improve refugees' health literacy and raise awareness of the benefits of such preventive measures. The utilization of an existing Primary Health Center (PHC) to incorporate oral health care need was suggested in six studies [20–22,25,26,30]. Furthermore, seven studies [6,20–23,25,26] pointed to the economic burden on both the refugees and on the host country while dealing with easily preventable oral complications. General referral systems seemed to be in place according to four studies [23,25,26,30] while two studies [13,20] directly provided necessary interventions. Specifics about utility of referral systems were not discussed in any of these studies. Six studies [6,21,24,27–29] emphasized the need for interventions. Moreover, Al-Ani et al. [27] encouraged all European migrant receiving countries to strengthen their dental capacity, as refugees' dental care needs are expected to further increase in the near future. Accessibility, cariogenic diet and poor oral hygiene were seen to be the main causes for disease pattern in all the included studies. The study of Hjern and Kling [29] argued that children are especially vulnerable, as they are affected by the caries-promoting food culture of their families. Finally, five studies [2,20,21,23,27] raised the issue of 'Health as a human right'. One [2] study stressed the importance of clinicians to carry out a dual role by providing care and advocating for dental needs. Language and cultural barriers [2,6,13,20,21,25,27,28], selection bias, mainly due to self-reporting or voluntary treatment-seeking behavior, among other reasons [6,20–23,26], lack of diagnostic tools and resources [6,13,21,22,26], small sample size [20,23,24], missed other oral health details [6,21–24,26], lack of representativeness [21,25,28] generalizability [22,25], crude methods used and insufficient data quality [27] were some of the limitations reported in the included studies.

Further Results

Our study found that refugees are at increased risk of developing oral diseases (mainly dental caries) when compared to the local populations (See Table 3). Filled Teeth were more frequent among the local populations and also among other migrants in comparison to refugees, whereas Decayed Teeth were more common among refugees (See Table 3). This clearly shows that the local population has better access to and utilization of available dental treatment. Missing Teeth were similarly distributed among all three groups (See Table 3). The authors explained this by the fact that refugees originated mainly from war-affected regions, where the priority for curative treatment is completely absent [24]. Availability of health services seems to be scarce, along with other necessities such as clean drinking water, a hygienic environment and other cleaning and sanitation products [23]. Moreover, children tend to suffer more since they are not provided with the essential oral health services and practices, which may have long-lasting negative effects [23,29]. Our findings show the need for oral health assessment tools such as overhead light, mouth mirror, probe/explorer and intra oral *x*-ray/orthopantomogram to aptly collect the data [6,13,21,22,26]. The studies emphasized the lack of human and material resources [6,20,23,25]. A shift from curative to conservative to preventive care is highly recommended [13,20,21,23,24,26–29].

Effects of oral health on refugees' general health is an important aspect addressed by several studies; e.g., Høyvik et al. [6] state that dental problems have a substantial effect on social, physical and psychological well-being; missing teeth can be detrimental to self-confidence. Especially, reduced social and psychological well-being can delay the acceptance and amalgamation process and, therefore, lead to social isolation and mental issues resulting in increased overall health problems [23]. Other factors not directly associated, but important, such as dental fear, anxiety or post-traumatic stress disorder (PTSD), also need appropriate planning and time for treatment [2,6,13,20,26,30]. Additionally, the unavailability of orientation from the host country [21] and of proper oral care is one aspect highlighted by all studies except one [30].

Studies included in our analysis emphasize the health needs and oral health seeking behavior of refugees. Findings suggest less motivation and orientation regarding oral health care and prevention among refugees when compared to the local population [21]. Refugees' priorities tend to be more towards resettlement [6]. Additionally, studies suggest that refugees in the transition phase are provided mainly with emergency care. Refugees tend to have similar access to dental services as the local population only once their refugee status is accepted [2,20,21,25,26,28]. Language, cultural and economic barriers, social isolation, the unfamiliarity of the health care system of the host country, laws, regulations and restrictions can further limit access to needed dental care.

4. Discussion

Prevalence of caries and dental treatment needs are high among refugees and the burden is increasing with the ever-growing influx of this population. The complex process of integration entails challenges, which also puts a burden on the host country. The unavailability of oral screening at reception sites leads to missing detectable oral health problems, which should be treated as early as possible to improve treatability. Consistent with our results, other studies from the USA [31–33], Canada [34] and from Australia [35,36] show high prevalence of caries, poor oral hygiene and similar unmet treatment needs among the refugee population. Moreover, a lack of information on pre-arrival oral status makes comparison and assessment difficult.

We found that more data is available on the general health needs of refugees in Europe while data on oral health is scarce. Additionally, the lack of oral assessments and inconsistencies within insurance systems, such as lack of uniformity and harmonization in cost coverage which depends on per capita spending on health care [37], add to the barriers in achieving good oral health. As a result of these large fluctuations and the diversity of refugees, the challenges faced are not homogenous [6]. The journey, and later the waiting time to become an officially accepted refugee by the host country, exacerbate preexisting conditions [21]. This not only increases suffering, but also incurs unnecessary costs [6]. This, in turn, puts excessive pressure on the individual as well as the host country's health system [21]. Language, cultural barriers and the unfamiliarity of the health system further amplify this. The European refugee crisis is a persistent issue gripping refugees and host countries alike and brings in challenges on a daily basis. In spite of all the advancements and resources available at the disposal of European countries with a good health care system in place, inclusion and integration of refugees and asylum seekers still remains challenging [6,7,23].

Studies that primarily focused on refugee oral health examined oral hygiene practices, periodontal health, DMF of teeth and knowledge and self-perception regarding oral health. A dentist and necessary dental equipment were also available for the assessment, making it easier to detect problems. However, equipment for screening, such as dental *x*-rays, was not available which might have led to an underestimation of prevalence. Studies that focused on general health had no dentists in their study teams and dental equipment was not mentioned. Hence, only complaints about oral/teeth problems or pain were registered, which is likely to have led to an underestimated prevalence of oral problems.

Studies included in this review clearly show the substantial effect of oral health on general health and especially on mental health and well-being. Some non-migration focused studies investigated the link between oral and general health, e.g., Kitamoto et al. [38] and Patini [39] suggest an association between oral microbiota and systemic diseases.

Dental fear and anxiety are other important aspects emphasized by the included studies. Especially, children seem to be more vulnerable to pain associated with dental treatment. Some authors examine this issue and emphasize the importance of local anesthesia (LA) in achieving pain-free treatment [40]. However, according to the authors, anxiety and stress associated with local anesthetic injection makes pain-free treatment challenging [41,42].

Little emphasis has been given to oral health research among refugees in European countries during the last 25 years [6]. Further research is needed; however, based on available data, targeted interventions should be implemented [6]. Early detection of oral health conditions can be considered as the most effective way to address the complex problem of oral health. Immediate oral assessment of refugees at the point of entry or registration for consecutive dental screening [21] can prove vital [6]. Communication in the native language can also help avoid any misunderstanding and delays [21].

Consistent with our results, studies from other non-European regions suggest that targeted services will help access major oral health care challenges even with limited resources [6]. Riatto et al. [24] state that a structured assessment of the refugees' situation with respect to the amount of dental care received, economic capability, knowledge and awareness, and access to oral health care services will be needed to plan and arrange necessary services for oral health care. Canada is the only country with specific guidelines for oral services for refugees [43].

Several limitations of this systematic review should be mentioned. No language restrictions were set during the search but the search terms were in English. Due to its simplicity and popularity, we focused on the DMF index as a quantitative measure of caries prevalence. We only concentrated on the D factor of the DMF index. As MT can also be due to multiple reasons other than caries (such as trauma, periodontal issues, etc.), there can be a risk of bias. However, our search and screening procedure did not bring up other measures to quantify caries among refugees; only one study used Index of Restoration (IR) [24]. Due to the lack of comparability, we decided not to perform a formal meta-analysis. Due to unavailability of data, we could not compare pre and post-arrival oral health conditions. Lastly, the element of human error and bias cannot be neglected, which may have caused the loss of some information or a steering of the conclusions.

Despite these limitations, our study provides a comprehensive analysis of the available data on dental caries and provided oral health care among the refugee population in the European region after 2015. This systematic literature review adds to the existing literature on the specific needs and associations required for further planning. Moreover, it brings dental and oral health into focus. Concentrating on caries may help to discretely tackle a major condition, provide required treatment and precisely fulfill unmet needs for better oral health.

5. Conclusions

Our systematic literature review shows a high burden of dental caries, with increasing severity among refugees in Europe. Factors such as pre-existing poor oral health, limited access to treatment, language and cultural barriers, and lack of orientation and unfamiliarity with the host country's health care system might be major reasons and lead to low oral health-seeking behavior. Additionally, dietary behavior and changes owing to migration, low oral hygiene practices and lack of preventive measures in the host countries lead to worse oral health over time. Further research focusing on refugees in Europe is needed to better understand, plan and install preventive measures. Setting priorities now with the available data is urgently needed to improve oral health among refugees in Europe.

Key Points

(1)   High prevalence of caries and limited access to dental health services are the main challenges refugees and asylum seekers face in Europe.

(2)   Further research is urgently needed to better understand the dental health needs of refugees in Europe.

(3)   The necessity of oral health check-up irrespective of need will help make the shift from curative to preventive oral health care.

Author Contributions: Conceptualization, K.A., V.W. and O.H.; methodology, K.A., V.W. and O.H.; formal analysis, S.B., A.E., C.I. and K.A.; resources, A.E., C.I.; writing—original draft preparation, S.B.; writing—review and editing, S.B., K.A., C.I., A.E., V.W., O.H.; visualization, S.B.; K.A.; supervision, K.A.; project administration, V.W., O.H.; funding acquisition, K.A., V.W. All authors have read and agreed to the published version of the manuscript.

Funding: For this systematic review no funding has been received, and no ethical approval is needed, since all relevant information is in the public domain. No research has been performed involving human participants and/or animals.

Conflicts of Interest: The authors have no conflict of interest to declare.

References

1.   Clayton, J. *UNHCR Chief Issues Key Guidelines for Dealing with Europe's Refugee Crisis UNHCR*; The UN Refugee Agency. 2015. Available online: https://www.unhcr.org/55e9793b6.html (accessed on 13 June 2019).

2.   Williams, B.; Cassar, C.; Siggers, G.; Taylor, S. Medical and social issues of child refugees in Europe. *Arch. Dis. Child* **2016**, *101*, 839–842. [CrossRef] [PubMed]

3.   Perruchoud, R. Glossary on Migration. In *International Migration Law*; IOM (International Organization for Migration): Geneva, Switzerland, 2004; p. 233.

4.   United Nations High Commissioner for Refugees. A Handy Guide to UNHCR Emergency Standards and Indicators UNHCR2000 [updated 26 July 2019]. Available online: https://www.refworld.org/docid/3dee456c4.html (accessed on 29 July 2019).

5.   Efird, J.; Bith-Melander, P. *Refugee Health: An Ongoing Commitment and Challenge*; Multidisciplinary Digital Publishing Institute: Basel, Switzerland, 2018.

6.   Høyvik, A.C.; Lie, B.; Grjibovski, A.M.; Willumsen, T. Oral health challenges in refugees from the Middle East and Africa: A comparative study. *J. Immigr. Minority Health* **2019**, *21*, 443–450. [CrossRef] [PubMed]

7.   Rechel, B. *Migration and Health in the European Union*; McGraw-Hill Education: London, UK, 2011.

8.   World Health Organization. Oral Health WHO 2019. Available online: https://www.who.int/news-room/fact-sheets/detail/oral-health (accessed on 23 June 2019).

9.   Glick, M.; Da Silva, O.M.; Seeberger, G.K.; Xu, T.; Pucca, G.; Williams, D.M.; Kess, S.; Eiselé, J.-L.; Séverin, T. FDI Vision 2020: Shaping the future of oral health. *Int. Dent. J.* **2012**, *62*, 278–291. [CrossRef] [PubMed]

10. Maamari, A. Geriatric odontology. *Revue Med Brux.* **2018**, *39*, 322–324.

11. Banu, A.; Șerban, C.; Pricop, M.; Urechescu, H.; Vlaicu, B. Dental health between self-perception, clinical evaluation and body image dissatisfaction—A cross-sectional study in mixed dentition pre-pubertal children. *BMC Oral Health* **2018**, *18*, 74. [CrossRef] [PubMed]

12. Petersen, P.E. Oral health. *Int. Encycl. Public Health.* **2008**, *4*, 677–685.

13. Zaheer, K.; Sheikh, S.; Khalid, O.; Rizvi, Z. Refugees: Addressing the burden of oral disease through prevention. *Br. Dent. J.* **2017**, *223*, 121. [CrossRef]

14. Petersen, P.E. The World Oral Health Report 2003: Continuous improvement of oral health in the 21st century—The approach of the WHO Global Oral Health Programme. *Community Dent. Oral Epidemiol.* **2003**, *31*, 3–24. [CrossRef]

15. Moher, D.; Liberati, A.; Tetzlaff, J.; Altman, D.G. Preferred reporting items for systematic reviews and meta-analyses: The PRISMA statement. *Ann. Intern. Med.* **2009**, *151*, 264–269. [CrossRef]

16. Shulman, J.D.; Cappelli, D.P. Epidemiology of dental caries. In *Prevention in Clinical Oral Health Care*; Elsevier: Amsterdam, The Netherlands, 2008; pp. 2–13.

17. Ejcecada, L. *Caries Process and Prevention Strategies: Epidemiology*; American Dental Association: Niagara Falls, NY, USA, 2014.

18. World Health Organization. *Oral Health Surveys: Basic Methods*; World Health Organization: Geneva, Switzerland, 2013.

19. National Institute of Health. Study Quality Assessment Tools U.S. Department of Health & Human Services, National Heart, Lung and Blood Institute. 2019. Available online: https://www.nhlbi.nih.gov/health-topics/study-quality-assessment-tools (accessed on 12 June 2019).

20. Mattila, A.; Ghaderi, P.; Tervonen, L.; Niskanen, L.; Pesonen, P.; Anttonen, V.; Laitala, M. Self-reported oral health and use of dental services among asylum seekers and immigrants in Finland—A pilot study. *Eur. J. Public Health* **2016**, *26*, 1006–1010. [CrossRef]

21. Solyman, M.; Schmidt-Westhausen, A.-M. Oral health status among newly arrived refugees in Germany: A cross-sectional study. *BMC Oral Health* **2018**, *18*, 132. [CrossRef] [PubMed]

22. Kakalou, E.; Riza, E.; Chalikias, M.; Voudouri, N.; Vetsika, A.; Tsiamis, C.; Choursoglou, S.; Terzidis, A.; Karamagioli, E.; Antypas, T.; et al. Demographic and clinical characteristics of refugees seeking primary healthcare services in Greece in the period 2015–2016: A descriptive study. *Int. Health* **2018**, *10*, 421–429. [CrossRef] [PubMed]

23. Goetz, K.; Winkelmann, W.; Steinhäuser, J. Assessment of oral health and cost of care for a group of refugees in Germany: A cross-sectional study. *BMC Oral Health.* **2018**, *18*, 69. [CrossRef] [PubMed]

24. Riatto, S.G.; Montero, J.; Pérez, D.R.; Castaño-Séiquer, A.; Dib, A. Oral Health Status of Syrian Children in the Refugee Center of Melilla, Spain. *Int. J. Dent.* **2018**, *2018*, 1–7. [CrossRef]

25. Pavlopoulou, I.D.; Tanaka, M.; Dikalioti, S.; Samoli, E.; Nisianakis, P.; Boleti, O.D.; Tsoumakas, K. Clinical and laboratory evaluation of new immigrant and refugee children arriving in Greece. *BMC Pediatr.* **2017**, *17*, 132. [CrossRef]

26. Van Berlaer, G.; Carbonell, F.B.; Manantsoa, S.; De Béthune, X.; Buyl, R.; Debacker, M.; Hubloue, I. A refugee camp in the centre of Europe: Clinical characteristics of asylum seekers arriving in Brussels. *BMJ Open* **2016**, *6*, e013963. [CrossRef]

27. Al-Ani, A.; Takriti, M.; Schmoeckel, J.; Alkilzy, M.; Splieth, C.J.C.O.I. National oral health survey on refugees in Germany 2016/2017: Caries and subsequent complications. *Clin. Oral Investig.* **2020**, 1–7. [CrossRef]

28. Freiberg, A.; Wienke, A.; Bauer, L.; Niedermaier, A.; Führer, A.J. Dental Care for Asylum-Seekers in Germany: A Retrospective Hospital-Based Study. *Int. J. Environ. Res. Public Health* **2020**, *17*, 2672. [CrossRef]

29. Hjern, A.; Kling, S. Health Care Needs in School-Age Refugee Children. *Int. J. Environ. Res. Public Health* **2019**, *16*, 4255. [CrossRef]

30. Hermans, M.P.; Kooistra, J.; Cannegieter, S.C.; Rosendaal, F.R.; Mook-Kanamori, D.O.; Nemeth, B. Healthcare and disease burden among refugees in long-stay refugee camps at Lesbos, Greece. *Eur. J. Epidemiol.* **2017**, *32*, 851–854. [CrossRef]

31. Geltman, P.L.; Adams, J.H.; Cochran, J.; Doros, G.; Rybin, D.; Henshaw, M.; Barnes, L.L.; Paasche-Orlow, M. The impact of functional health literacy and acculturation on the oral health status of Somali refugees living in Massachusetts. *Am. J. Public Health* **2013**, *103*, 1516–1523. [CrossRef] [PubMed]

32. Cote, S.; Geltman, P.; Nunn, M.; Lituri, K.; Henshaw, M.; Garcia, R.I. Dental caries of refugee children compared with US children. *AAP News J. Pediatr.* **2004**, *114*, e733–e740. [CrossRef] [PubMed]

33. Willis, M.S.; Bothun, R.M. Oral hygiene knowledge and practice among Dinka and Nuer from Sudan to the US. *J. Am. Dent. Hyg. Assoc.* **2011**, *85*, 306–315.

34. Ghiabi, E.; Matthews, D.C.; Brillant, M.S. The oral health status of recent immigrants and refugees in Nova Scotia, Canada. *J. Immigr. Minority Health* **2014**, *16*, 95–101. [CrossRef]

35. Davidson, N.; Skull, S.; Calache, H.; Murray, S.; Chalmers, J. Holes a plenty: Oral health status a major issue for newly arrived refugees in Australia. *Aust. Dent. J.* **2006**, *51*, 306–311. [CrossRef]

36. Marino, R.; Wright, F.; Minas, I. Oral health among Vietnamese using a community health centre in Richmond, Victoria. *Aust. Dent. J.* **2001**, *46*, 208–215. [CrossRef]

37. Widström, E.; Eaton, K.A. Oral healthcare systems in the extended European Union. *J. Oral Health Prev. Dent.* **2004**, *2*, 155–194.

38. Kitamoto, S.; Nagao-Kitamoto, H.; Hein, R.; Schmidt, T.; Kamada, N. The Bacterial Connection between the Oral Cavity and the Gut Diseases. *J. Dent. Res.* **2020**, *99*, 1021–1029. [CrossRef]

39. Patini, R. *Oral Microbiota: Discovering and Facing the New Associations with Systemic Diseases*; Multidisciplinary Digital Publishing Institute: Basel, Switzerland, 2020.

40. Monteiro, J.; Tanday, A.; Ashley, P.F.; Parekh, S.; Alamri, H. Interventions for increasing acceptance of local anaesthetic in children and adolescents having dental treatment. *Cochrane Database Syst. Rev.* **2020**, *2*, CD011024. [CrossRef]

41. Patini, R.; Staderini, E.; Cantiani, M.; Camodeca, A.; Guglielmi, F.; Gallenzi, P. Dental anaesthesia for children—Effects of a computer-controlled delivery system on pain and heart rate: A randomised clinical trial. *Br. J. Oral Maxillofac. Surg.* **2018**, *56*, 744–749. [CrossRef]

42. Akmal, N.; Ganapathy, D.; Visalakshi, R. Pain perception in children toward dental anesthesia–A survey. *Drug Invent. Today.* **2019**, *12*, 1475–1481.

43. Pottie, K.; Greenaway, C.; Feightner, J.; Welch, V.; Swinkels, H.; Rashid, M.; Narasiah, L.; Kirmayer, L.J.; Ueffing, E.; Macdonald, N.E.; et al. Evidence-based clinical guidelines for immigrants and refugees. *Can. Med. Assoc. J.* **2011**, *183*, E824–E925. [CrossRef] [PubMed]

Publisher's Note: MDPI stays neutral with regard to jurisdictional claims in published maps and institutional affiliations.

International Journal of
***Environmental Research
and Public Health***

Article

Health Patterns among Migrant and Non-Migrant Middle- and Older-Aged Individuals in Europe—Analyses Based on Share 2004–2017

Nico Vonneilich [1,*], Daniel Bremer [2], Olaf von dem Knesebeck [1] and Daniel Lüdecke [1]

[1] Institute of Medical Sociology, University Medical Center Hamburg-Eppendorf, 20246 Hamburg, Germany;
o.knesebeck@uke.de (O.v.d.K.); d.luedecke@uke.de (D.L.)

[2] Department of Medical Psychology & Center for Health Care Research, University Medical Center
Hamburg-Eppendorf, 20246 Hamburg, Germany; da.bremer@uke.de

* Correspondence: n.vonneilich@uke.de

Abstract: Introduction: European populations are becoming older and more diverse. Little is known about the health differences between the migrant and non-migrant elderly in Europe. The aim of this paper was to analyse changes in the health patterns of middle- and older-aged migrant and non-migrant populations in Europe from 2004 to 2017, with a specific focus on differences in age and gender. We analysed changes in the health patterns of older migrants and non-migrants in European countries from 2004 to 2017. Method: Based on data from the Survey of Health, Ageing and Retirement in Europe (6 waves; 2004–2017; *n* = 233,117) we analysed three health indicators (physical functioning, depressive symptoms, and self-rated health). Logistic regression models for complex samples were calculated. Interaction terms (wave * migrant * gender * age) were used to analyse gender and age differences and the change over time. Results: Middle- and older-aged migrants in Europe showed significantly higher rates of depressive symptoms, lower self-rated health, and a higher proportion of limitations on general activities compared to non-migrants. However, different time trends were observed. An increasing health gap was identified in the physical functioning of older males. Narrowing health gaps over time were observed in women. Discussion: An increasing health gap in physical functioning in men is evidence of cumulative disadvantage. In women, evidence points towards the hypothesis of aging-as-leveler. These different results highlight the need for specific interventions focused on healthy ageing in elderly migrant men.

Keywords: depression; self-rated health; functional limitations; older age; migrant status; health inequalities; trend analysis; Europe

Citation: Vonneilich, N.; Bremer, D.; von dem Knesebeck, O.; Lüdecke, D. Health Patterns among Migrant and Non-Migrant Middle- and Older-Aged Individuals in Europe—Analyses Based on Share 2004–2017. *Int. J. Environ. Res. Public Health* **2021**, *18*, 12047. https://doi.org/10.3390/ijerph182212047

Academic Editor: Paul B. Tchounwou

Received: 29 September 2021
Accepted: 15 November 2021
Published: 16 November 2021

Publisher's Note: MDPI stays neutral with regard to jurisdictional claims in published maps and institutional affiliations.

1. Introduction

In the last decades, European countries have witnessed immigration flows, leading to more diverse populations in Europe [1]. The share of migrant populations in European countries is rising; over the last thirty years it rose from around 7% to about 12% in 2020, with considerable variation between countries [2].

Early research detected the health advantages of first-generation immigrants [3], who initially showed a better health status than non-migrants in host countries [4,5]. Health selection in the countries of origin, e.g., for labour immigration, and the process of migration itself, which often requires good health, played an important role in this. But the health of migrants deteriorates as they settle in host-countries [6], due to socioeconomic disadvantages, increased risks in the workplace, or even experiences of discrimination, among other factors. Migrant and non-migrant populations face different health risks in European countries over the course of their lives [7–9].

Given the ongoing population ageing and increasing longevity in many European countries [1,10], it is necessary to focus on health differences between middle- and older-aged migrant and non-migrant populations, aged 50 years and older. Studies indicate

145

poorer self-rated health, higher rates of depression, worse physical functioning, and lower life expectancy in older migrant populations [11–14]. However, most of these studies do not observe trends over time, nor do they consider variations regarding gender or age. When looking at health patterns in migrant and non-migrant elderly, a steeper rate of health decline was observed in older migrants [13].

When considering the health disparities between older migrant and non-migrant populations and their development over time, different hypotheses have been discussed: aging-as-leveler, persistent inequality, and cumulative disadvantage. The aging-as-leveler hypothesis states that health differences between population groups decrease in late life, when physical and mental limitations increase in all groups. The persistent inequality hypothesis states that, regardless of age, inequalities remain stable. And the hypothesis of cumulative disadvantage states that across the lifespan, socioeconomic disadvantages cumulate and lead to an increase in health risks and health inequalities between migrant and non-migrant elderly populations [15]. Evidence exists for the accumulation of disadvantage [13,16,17], for aging-as-leveler [18], and for persistent inequality [19], with most of the research focusing on black, Hispanic and non-Hispanic white populations in the US.

Earlier research analysing trends over time in the health of migrant and non-migrant elderly has often focused on the health of migrant and non-migrant populations in general, without specifying age or gender [13,14]. As migration background is one among other relevant factors associated with health, such as gender, age, or socioeconomic status [5,20], these factors are taken into account in the present study. Earlier studies indicated gender differences in health disparities according to migration [20–22].

Based on SHARE data, the health patterns of migrant and non-migrant middle- and older-aged individuals in different European countries were investigated. Moreover, time trends over 14 years were examined. Three health outcomes were assessed, as differences between groups might vary depending on the indicator used and in order to provide a more holistic and complex picture of health [15]: physical functioning, depression, and self-rated health (SRH). Therefore, the aim of the present trend analysis was to analyse health patterns in migrant and non-migrant middle- and older-aged individuals in Europe, focusing on individuals aged 50 years and older over a time period of 14 years, with a specific focus on gender and age differences.

2. Methods

2.1. Data

The analyses were based on data from SHARE, the Survey of Health, Ageing, and Retirement in Europe [23–28]. The sample comprised 233,117 observations from 28 European countries and Israel across 6 out of 7 waves (2004, 2006, 2010, 2013, 2015, and 2017), including Israel (Austria, Belgium, Bulgaria, Croatia, Cyprus, Czech Republic, Denmark, Estonia, Finland, France, Germany, Greece, Hungary, Ireland, Israel, Italy, Latvia, Lithuania, Luxembourg, Malta, Netherlands, Poland, Portugal, Romania, Slovakia, Slovenia, Spain, Sweden, and Switzerland). Data from the third wave were excluded, since this was a retrospective SHARELIFE survey. Based on the population registers, SHARE used probability samples within the countries and included non-institutionalized adults aged 50 years or older and, if available, their partners. Further exclusion criteria were being incarcerated, moved abroad, unable to speak the language of questionnaire, deceased, hospitalized, moved to an unknown address, or not residing at the sampled address [29,30].

2.2. Measures

The health patterns were assessed using three indicators that were dichotomised for analytical reasons. Physical functioning was measured using the global activity limitation index (GALI) [31]. The goal of this index is to assess perceptions in both general and more specific populations regarding long-standing, health-related limitations in common daily activities [31]. The GALI was measured by asking: "For the past 6 months at least, to what extent have you been limited because of a health problem in activities people usually do?".

Possible answers were "severely limited", "limited, but not severely" and "not limited". The first two limitations were combined for the binary variable with the categories 0 ("not limited") and 1 ("limited").

As a measure of mental health, the EURO-D was used to assess depressive symptoms [32]. The EURO-D was developed to allow the cross-country measurement and comparison of depressive symptoms. It includes 12 relevant symptoms of depression. The score ranged from 0 ("not depressed") to 12 ("very depressed"). A binary variable for "depression" was categorised as 0 ("not depressed") if the EURO-D score was lower than 4, otherwise it was categorised as 1 ("case of depression"), as proposed by the authors of the EURO-D scale [32].

The SRH was measured by asking, "Would you say your health is … ", offering five options as potential answers ("very good", "good", "fair", "bad" and "very bad"). The variable was dichotomised, with 0 indicating "good or very good health", and 1 "less than good health".

Migrant background, age, and gender were used as the main predictors. The respondents' age and gender were recorded at each wave. The respondents were considered as migrants if they were not born in the country of data collection. Wave is a continuous variable, ranging from 1 to 7, reflecting the respective year of data collection, where "1" relates to the year 2004, while "7" stands for 2014.

The household size and educational level of the respondents were used as further covariates. Educational levels were based on the International Standard Classification of Education [33] and recoded into three levels: "low (lower/upper secondary)", "mid (post-secondary)" and "high (tertiary)".

2.3. Analyses

For each of the three dichotomised health indicators, a logistic regression model for complex samples was calculated, using quasi-binomial links to properly account for survey-weighting, disproportional sampling, and selective mortality. The country variable was used to define the strata in the survey design; hence, the regression models accounted for the fact that the respondents were clustered within different countries. Robust Horvitz–Thompson standard errors were reported [34]. Models without interaction terms were calculated first, to investigate the overall associations between the dependent variables and the predictors of interest, migration, gender and age. The interactions between wave, migration, gender, and age were analysed to explore changes over time and to stratify the results by the aforementioned variables (migration, gender, and age). Age was standardized (z-score). Predictors that vary over time, particularly age, usually exert an effect within subjects (e.g., the individual outcome level changes when a person gets older) and between subjects (e.g., outcome levels differ between younger and older people). This results in biased regression coefficients, because both the within- and between-effect are captured in *one* coefficient. This is called "heterogeneity bias" [35]. To avoid this bias, age was separated into its within- and between-component. Since differences between age groups were the main focus of interest, the between-effect of age was used in the interactions. Educational status, household size, and the within-effect of age were included as further covariates. The associations between the interaction of the focal predictors and the three indicator variables are presented as predicted probabilities. The trends for each combination of the focal predictors' categories (migrant yes/no, female/male gender, and middle/older-aged persons) are shown, resulting in figures with four panels per health indicator. Since age was a continuous variable, the values at one standard deviation below and above the average age in the sample were chosen as representative values through which to distinguish between younger and older respondents, following the suggestions from Aiken and West [36]. Thus, the figures represent model predictions as if an individual was at the age of 60 or 80, respectively. To answer the question of whether the change over time for both migrants and non-migrants and whether the comparison between these trends was statistically significant, contrast analyses and pairwise comparisons for the predicted probabilities were conducted to test for the differences between the stratified groups. The related *p*-values are shown in the figures. All the analyses were performed using the R statistical package (R Core Team, Vienna, Austria) [37], including the packages "survey" [38], "ggeffects" [39] and "emmeans" [40].

3. Results

Our analyses were based on data from 233,117 cases (55% female, 45% male, see Table 1). Overall, 10.4% were classified as migrants (10,923 men, 13,351 women). In the sample, 39.3% of the respondents were 70 years or older. The mean age of the non-migrants was 67.0 years and in migrants it was 66.6 years. The following significant differences between migrant and non-migrant populations were observed: the non-migrant populations were older (higher share of people aged 70 years and older) and had a higher share of lower education (45% upper secondary level or lower). Migrants across all waves under study reported a significantly higher share of GALI and a significantly higher share of poor SRH.

3.1. Associations between Health Indicators, Age, Gender and Migrant Status

In model 1 (Table 2), the regression models showed a 24% increased risk of reporting limited GALI for migrant populations, after controlling for wave, age (between and within subjects), gender, educational status, and household size. The age between and gender were not significantly associated with limited GALI. Similar results were found for depression (model 2). The migrant populations demonstrated a 19% increased risk for depressive symptoms compared to non-migrants. Model 3 demonstrates that migrants had a 28% higher chance of reporting poor SRH.

3.2. Subgroup Analyses and Time Trends

To analyse the interactions between the three focal variables (migrant background, age, and gender) with regard to three health indicators (limited GALI, depression, and poorer SRH), three logistic regressions were conducted, including several interaction terms (see Supplementary Table S1).

3.2.1. Global Activity Limitation Index

Figure 1 presents the predicted probabilities of respondents who were limited in daily activities for migrant background, age, and gender over time in four line graphs (Figure 1a–d). The left-handed graphs show middle-aged males (a) and older-aged males (b). The right-handed graphs show middle-aged females (c) and older-aged females (d). When looking at the predicted probability models for GALI across the six waves, it can be observed that significantly more middle-aged migrant men reported limitations in global activities than non-migrant men, but that these differences decreased over time (Figure 1a). In terms of older males (on average 80 years of age), the trends in GALI over time contrasted significantly, as a higher proportion of migrant men reporting limitations in global activities over time was observed (Figure 1b). A significant increase in the predicted probability of GALI over time was found in the non-migrant men. In females, different trends regarding GALI over time were found. The gap between migrant and non-migrant elderly women narrowed both in the middle- and in the older-age groups (Figure 1c,d). In both age groups, GALI in non-migrant women significantly increased over time, leading to a reduction in the health gap in GALI observed in the early waves of SHARE. Generally, no significant trends were found for pairwise comparisons of trends in patterns of GALI over time between the migrant and non-migrant populations.

Table 1. Descriptive statistics of SHARE data ($n = 233,117$; across all countries and waves).

Variable	Wave 1 (2004)		Wave 2 (2006)		Wave 4 (2010)		Wave 5 (2013)		Wave 6 (2015)		Wave 7 (2017)		Total		p-Value for Difference
	Non-Migrants ($n = 19,585$)	Migrants ($n = 2554$)	Non-Migrants ($n = 21,241$)	Migrants ($n = 2334$)	Non-Migrants ($n = 37,043$)	Migrants ($n = 4136$)	Non-Migrants ($n = 40,535$)	Migrants ($n = 5452$)	Non-Migrants ($n = 40,975$)	Migrants ($n = 4955$)	Non-Migrants ($n = 49,464$)	Migrants ($n = 4843$)	Non-Migrants (all waves, $n = 208,843$)	Migrants (all waves, $n = 24,274$)	
Mean Age (SD)	64.8 (10.4)	64.9 (10.2)	64.8 (10.4)	64.3 (10.4)	65.7 (10.5)	65.4 (9.7)	67.8 (10.5)	67.6 (10.3)	68.4 (10.8)	68.0 (10.2)	69.5 (10.4)	69.7 (10.0)	67.0 (10.7)	66.6 (10.3)	0.007 [a]
Age (70 and older), %	31.9	32.0	31.4	27.7	34.9	32.2	40.3	40.0	42.8	40.1	46.1	48.5	38.5	36.9	0.013 [b]
Female Gender, %	53.9	54.0	53.5	56.5	55.7	52.5	55.6	53.7	55.9	58.2	57.3	57.5	55.5	55.4	0.941 [b]
Mean Household Size (SD)	2.2 (1.0)	2.2 (1.0)	2.1 (1.0)	2.1 (0.9)	2.2 (1.0)	2.1 (0.9)	2.1 (1.0)	2.1 (0.9)	2.1 (1.0)	2.0 (1.0)	2.0 (1.0)	1.90 (1.0)	2.1 (1.0)	2.1 (1.0)	0.279 [c]
Education															<0.001 [b]
lower/upper secondary, %	53.3	38.5	48.9	37.6	45.8	34.5	43.2	33.8	42.2	32.9	39.7	30.2	45.0	34.6	
post-secondary, %	30.7	37.6	34.7	39.1	37.0	40.3	36.9	36.9	38.8	38.6	41.1	39.9	36.9	38.7	
tertiary, %	16.0	24.0	16.4	23.3	17.2	25.2	19.9	29.3	19.0	28.5	19.2	29.8	18.1	26.7	
Health Indicators															
Global Activity Limitation Index, % limited	43.7	51.6	44.9	48.7	49.0	51.3	45.2	47.4	45.9	49.2	47.5	51.0	46.2	49.9	<0.001 [b]
Depression, % depressed	27.6	32.1	27.9	28.4	30.5	30.6	28.4	30.0	29.6	31.6	31.4	29.8	29.0	30.5	0.074 [b]
Self-Rated Health, % less than good	76.3	79.4	78.6	81.7	79.5	80.8	76.3	79.8	77.3	78.1	79.0	80.7	77.9	80.1	0.004 [b]

[a] Design-based t-test; [b] Pearson's Chi^2 (Rao & Scott adjustment); [c] Design-based Kruskal–Wallis test.

Table 2. Logistic regression models with survey design for GALI (model 1), depression (model 2) and self-rated health (model 3) (SHARE, waves 1, 2, 4–7, *n*= 233,117); odds ratios, 95% confidence intervals (CI) and significances (*p*).

Parameter	Limited Global Activity Limitation Index (GALI) (Model 1)			Depression (Model 2)			Poorer Self-Rated Health (Model 3)		
	Odds Ratio	95% CI	*p*	Odds Ratio	95% CI	*p*	Odds Ratio	95% CI	*p*
(Intercept)	0.95	(0.89, 1.01)	0.088	0.53	(0.50, 0.57)	<0.001	4.85	(4.51, 5.21)	<0.001
Wave	1.03	(1.02, 1.03)	<0.001	1.03	(1.02, 1.04)	<0.001	1.02	(1.01, 1.03)	<0.001
Migrant (ref. non-migrant)	1.24	(1.15, 1.34)	<0.001	1.19	(1.09, 1.29)	<0.001	1.28	(1.17, 1.40)	<0.001
Age (between subjects)	0.98	(0.95, 1.01)	0.184	1.03	(1.00, 1.07)	0.086	0.98	(0.95, 1.02)	0.359
Female Gender (ref. male)	1.01	(0.98, 1.05)	0.367	1.00	(0.97, 1.04)	0.863	1.02	(0.98, 1.06)	0.256
Education (ref. lower/upper secondary)									
post-secondary	0.73	(0.70, 0.77)	<0.001	0.57	(0.54, 0.60)	<0.001	0.60	(0.57, 0.64)	<0.001
tertiary	0.52	(0.49, 0.55)	<0.001	0.40	(0.38, 0.43)	<0.001	0.33	(0.31, 0.35)	<0.001
Household Size	1.00	(0.99, 1.02)	0.656	0.98	(0.96, 1.00)	0.055	1.00	(0.98, 1.02)	0.893
Age (within subject)	1.02	(1.00, 1.04)	0.020	0.99	(0.97, 1.02)	0.539	1.02	(1.00, 1.04)	0.080

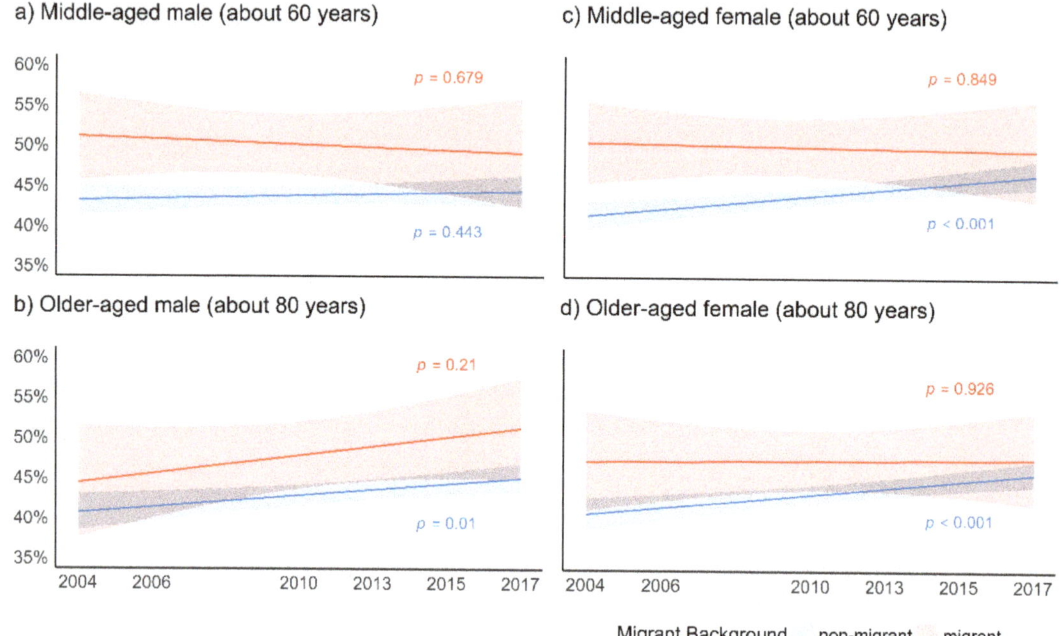

Figure 1. Predicted probabilities of Global Activity Limitation Index (% limited in daily activities) for migrant background, age, and gender (across six waves of SHARE); *p*-values relate to change over time for migrant and non-migrant elderly.

3.2.2. Depression

Figure 2 presents the predicted probabilities of the respondents who suffered from depression on migrant background, age, and gender (Figure 2a–d).

In depression, a significant increase over time was found in non-migrant middle-aged men (Figure 2a). For migrant men, a stagnation over time among the middle-aged and a decrease among the older-aged was identified, although the latter did not show a significant level (Figure 2b). A significant increase in depression over time was also found in middle-aged non-migrant women, but not for migrant women in the same age group (Figure 2c). For middle-aged men and women alike, the gap between migrant and non-migrant populations in depression found in the early waves disappeared over time. In older-aged individuals, the likelihood of depressive symptoms increased over time, particularly in non-migrant men (Figure 2b,d). No significant trends for pairwise comparisons of patterns in depression over time between migrant and non-migrant elderly were identified.

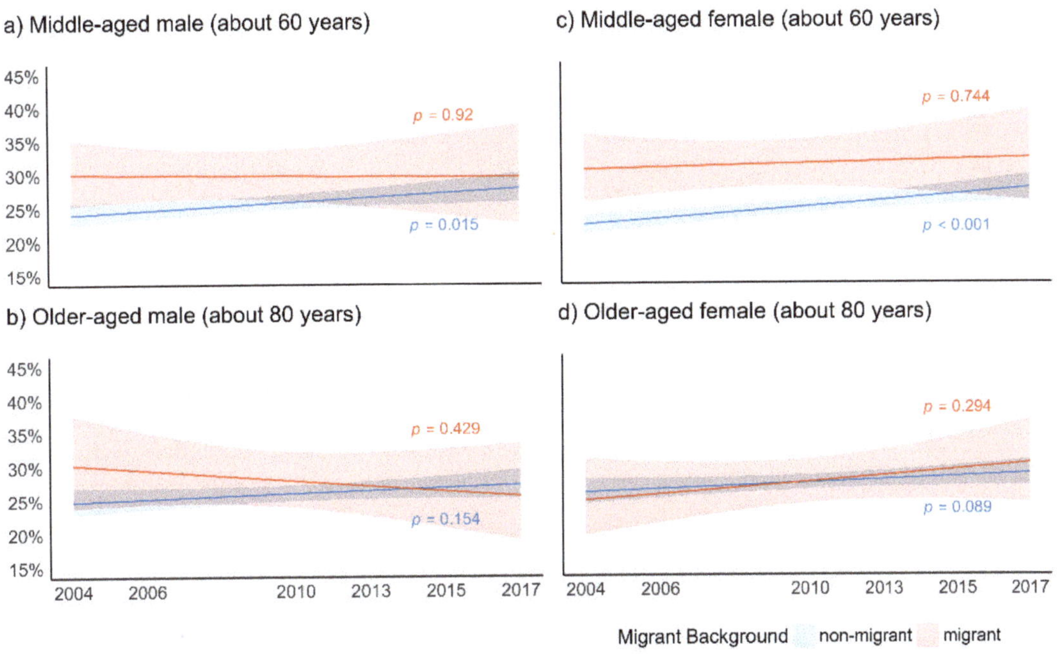

Figure 2. Predicted probabilities of EURO-D (% depressed = four or more symptoms of depression) for migrant background, age and gender (across six waves of SHARE); *p*-values relate to change over time for migrant and non-migrant elderly.

3.2.3. Self-Rated Health

Figure 3 presents the predicted probabilities of respondents who reported poorer SRH for the migrant and non-migrant elderly, distinguished by age and gender (Figure 3a–d).

In men, the gap in SRH between migrant and non-migrant populations increased over time (Figure 3a,b). This trend was particularly pronounced among middle-aged males, reaching significant differences in wave 4 and 5 (2010 and 2013, Figure 3a). A slightly significant increase in probability of reporting poor SRH over time was found among older-aged non-migrant men only. In women, a contrary time trend was observed (Figure 3c,d). In middle-aged females, a significant gap in wave 1 (2004), with higher risks of reporting bad SRH in migrant populations, diminished over time (Figure 3c). No significant differences and very similar amounts of poorer SRH were found in wave 7 (2017). An increase in time of poorer SRH in non-migrant middle-aged women and a decrease in migrant middle-aged women was found (Figure 3c). For older-aged women, the gap was narrowing as time trends of poor SRH were increasing in non-migrant populations (Figure 3d). This increase was more pronounced than in migrant older-aged women, leading to a decrease of the initial gap in SRH. No significant trends for pairwise comparisons in patterns of SRH over time between migrant and non-migrant elderly were identified.

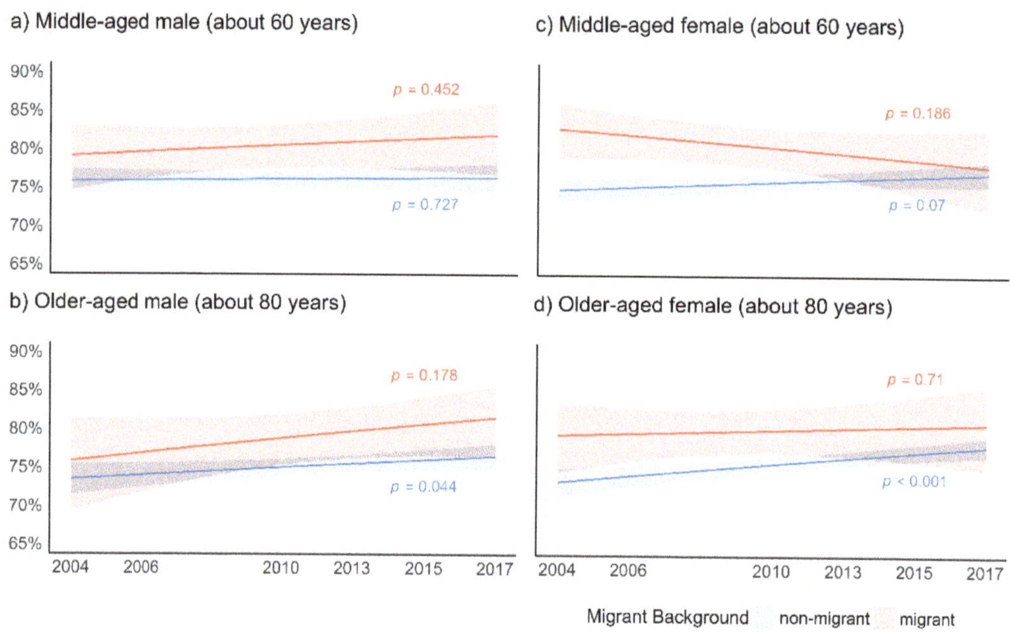

Figure 3. Predicted probabilities of self-rated health (% less than good health) for migrant background, age, and gender (across 6 waves of SHARE; model 3.2); *p*-values relate to change over time for migrant and non-migrant elderly.

4. Discussion

4.1. Summary

Based on the trend analysis of the data from SHARE for 2004 to 2017, this study shows different health patterns between middle and older-aged migrant and non-migrant-populations. A significant health gap for migrant populations exists, with higher risks for reporting worse health outcomes in all three indicators (physical functioning, depressive symptoms, and SRH) for migrant populations. Over time, these gaps develop differently, especially when age and gender are taken into account. In women, the health gap between migrant and non-migrant elderly narrowed significantly across all three health indicators. Moreover, interestingly, there are indications for a reverse health gap, with higher risks of reporting depression in non-migrant elderly women. In men, there is evidence for widening health gaps, such as the SRH in middle-aged men and physical limitations in older-aged men. In depression, differences between migrant and non-migrant men in the all ages under study diminished over the years. The results presented need to be interpreted with caution, as the differences in time trends between migrant and non-migrant populations did not reach significant levels.

While previous studies found some evidence for deteriorating health trends over time, especially in migrant populations [13,41], the analysis revealed a more complex picture of health patterns and trends over time in the health of migrant and non-migrant elderly in Europe by considering age and gender differences (see also [5]). The results show that health differences between the migrant and non-migrant elderly change over time. There is no clear trend of deteriorating health among migrants in Europe [13]. Evidence indicates a widening of health disparities among men, while among women there is evidence indicating a reduction in these disparities.

With regard to theoretical considerations of the development of health disparities over time [15], some results point towards the hypothesis of accumulation. This is especially true for men with regard to physical limitations in older age groups and for SRH. When looking at the health trends in women, there is evidence for the hypothesis of aging-as-leveler, since the health gap between migrant and non-migrant elderly women decreased over time, especially in

GALI and SRH. As other scholars have pointed out, downplaying differences among elderly migrant populations such as age or gender will likely lead to false assumptions regarding the development of health patterns [5,41]. The findings presented here underline these conclusions.

How can these age and gender differences in health patterns be explained? In terms of increasing health gaps in GALI in elderly men, physically demanding jobs and higher risks in the workplace experienced during the lifespan might play an important role [4,42,43], especially for the older cohort of elderly migrant men. Elderly migrant women might be less likely to suffer from physical risks in the workplace. Health gaps in SRH and depression suggest psychosocial explanations, such as social isolation and the experience of discrimination [8,43], and some research suggests that psychological distress in migrant populations might increase with age [44]. In order to explain differences in health patterns between men and women, other scholars have pointed towards the role of welfare states and other contextual factors [45,46], such as health systems, health coverage or health equity concerns in national strategies, which are likely to affect health services and access to health care for certain (marginalised) groups (e.g., due to language barriers) as well as living and working conditions of women with corresponding consequences for health in later life stages.

We find evidence for higher health risks among migrant elderly despite their higher educational status. How can this be explained? Educational status might not reflect the true socioeconomic status of migrant elderly, as the process of migration often occurs after the end of educational training. In the process of migration, recognition of educational attainment is difficult, resulting in a greater likelihood of lower-qualified occupations and, consequently, lower income [47]. It has also been shown that within Europe, tertiary level graduates are more likely to migrate, which is in line with the finding of higher educational attainment within the migrant populations studied here [48].

4.2. Limitations

Some methodological limitations need to be taken into account when interpreting the results. Firstly, SHARE excludes institutionalized women and men [49]. The survey is representative for community-dwelling people 50 years and older with regard to the participating European countries [49]. Since institutionalized individuals show worse health patterns, this could have led to more positive health trends in the analyses. Furthermore, panel attrition of potentially older and more handicapped people can contribute to sample selection, limiting the representativeness of the data and the generalisability of results [29].

Moreover, SHARE includes only those participants that were able to speak the official language of the respective country [49]. Consequently, migrants with weaker language skills are underrepresented in the sample [13]. This effect might partially explain the higher educational status within the migrant populations under study. This could also be also true for specific migrant populations, e.g., undocumented migrants in precarious living conditions and asylum seekers, among others. Conceivably, this limitation would make an underestimation of health differences more likely. Additionally, the SHARE data did not provide detailed information on the migrant background, such as length of stay, migration history, or the nationality of the migrants' parents. When looking at migrant subpopulations, the design of the data is improvable [13]. Consequently, the variable of migrant background represents no more than a proxy for the diversity of migrant experiences [4]. In particular, since the proportion of migrants is comparably low in the overall sample, a further differentiation of migrant populations (such as country of origin, length of stay, or type of immigration) would lead to small subsamples and would not be feasible for the present analysis.

The analyses are to some extent crude as a limited set of control variables, namely age, gender, household size, and education were included, in order to reduce complexity and enhance the interpretation of the results. Other potential explanatory variables were not included as the focus of the present analysis is to analyse changes over time rather than to explain the complex associations between the variables.

Our three health indicators are characterized by self-reporting. GALI and EURO-D are validated and widely used instruments [50]. On the one hand, SRH is based on a general single item and caution is recommended when drawing conclusions from studies using SRH as health indicator [51]. On the other hand, SRH represents a proper summary of health status [52] and it has been shown to be a valid measure of health status, especially among older adults, regardless of varying cultures and social conditions [53–55]. The three health indicators were dichotomised for reasons of comparability and clearness. As a consequence, the analyses are crude to some extent.

Lastly, all the analyses are based on the pooled data of 28 European countries and Israel. This procedure involves the risk of missing country-specific trends caused by different health systems [56], migration patterns [57], and welfare regimes [58]. In this regard, country-specific health patterns in migrant and non-migrant populations were analyzed but did not reveal any clear country-specific patterns (details not shown here).

5. Conclusions

The results of the present analyses highlight the need for interventions focused on healthy ageing for the migrant elderly, with a specific focus on middle- and older-aged migrant men. Especially in SRH and GALI, widening health gaps between non-migrant and migrant elderly men were observed. Migration should be included in healthy ageing policies in Europe [20,58]. Policies for migrant integration can help to reduce health disparities [59]; explicit migrant health policies are needed in all European countries in order to adapt health systems to the specific risks and needs of migrant populations [4]. These policies and interventions should be targeted regarding age and gender of migrants. As the results show, treating migrants as a homogenous group underestimates the differences in health and health patterns over time. Further migrant and health specific panel data are needed. Further differentiation of migrant populations in future studies is needed, for example by oversampling migrant populations, as they could take countries or regions of origin and length of stay in host countries as further potential explanations of health differences into account [4,41]. This also implies greater efforts to include migrant populations, especially those who are potentially harder to reach, in future research and to reduce existing barriers, such as language restrictions.

Supplementary Materials: The following are available online at https://www.mdpi.com/article/10.3390/ijerph182212047/s1, Table S1 includes the results of logistic regression models, including interaction terms. All R scripts to reproduce the analyses are available at osf.io (https://osf.io/7wd8e, accessed on 14 November 2021).

Author Contributions: N.V., D.B., D.L. and O.v.d.K. developed the research questions. N.V. prepared, analysed, and interpreted the data, and drafted and finalized the manuscript. D.B., O.v.d.K. and D.L. made an essential contribution to data analyses and interpretation, drafting the manuscript, and critically revised and approved the final manuscript. All authors have read and agreed to the published version of the manuscript.

Funding: This research received no external funding.

Institutional Review Board Statement: Not applicable.

Informed Consent Statement: Not applicable.

Data Availability Statement: The SHARE data is available for research purpose after registering at the SHARE website (www.share-project.org, accessed on 14 November 2021).

Acknowledgments: This paper uses data from SHARE Waves 1-7 [23–28]. Please see [29] for methodological details. The SHARE data collection was primarily funded by the European Commission through FP5 (QLK6-CT-2001-00360), FP6 (SHARE-I3: RII-CT-2006-062193, COMPARE: CIT5-CT-2005-028857, SHARELIFE: CIT4-CT-2006-028812) and FP7 (SHARE-PREP: N°211909, SHARE-LEAP: N°227822, SHARE M4: N°261982). Additional funding from the German Ministry of Education and Research, the US National Institute on Aging (U01_AG09740-13S2, P01_AG005842, P01_AG08291, P30_AG12815, R21_AG025169, Y1-AG-4553-01, IAG_BSR06-11, OGHA_04-064) and from various national funding sources is gratefully acknowledged (see www.share-project.org, last accessed 31 October 2020).

Conflicts of Interest: The authors declare no conflict of interest.

References

1. Lanzieri, G. Fewer, Older and Multicultural? Projections of the EU Populations by Foreign/National Background. In *Eurostat Methodologies and Working Papers*; European Union: Luxmburg, 2011. [CrossRef]
2. Migration Data Portal. Total Number of International Migrants at Mid-Year 2020. Available online: https://www.migrationdataportal.org/data?m=1&i=stock_abs_&t=2020 (accessed on 24 August 2021).
3. Razum, O.; Zeeb, H.; Akgün, H.S.; Yilmaz, S. Low Overall Mortality of Turkish Residents in Germany Persists and Extends into a Second Generation: Merely a Healthy Migrant Effect? *Trop. Med. Int. Health* **1998**, *3*, 297–303. [CrossRef]
4. Rechel, B.; Mladovsky, P.; Ingleby, D.; Mackenbach, J.P.; McKee, M. Migration and Health in an Increasingly Diverse Europe. *Lancet* **2013**, *381*, 1235–1245. [CrossRef]
5. Gkiouleka, A.; Huijts, T. Intersectional Migration-Related Health Inequalities in Europe: Exploring the Role of Migrant Generation, Occupational Status & Gender. *Soc. Sci. Med.* **2020**, *267*, 113218. [CrossRef] [PubMed]
6. La Parra-Casado, D.; Stornes, P.; Solheim, E.F. Self-Rated Health and Wellbeing among the Working-Age Immigrant Population in Western Europe: Findings from the European Social Survey (2014) Special Module on the Social Determinants of Health. *Eur. J. Public Health* **2017**, *27* (Suppl. S1), 40–46. [CrossRef]
7. Arnold, M.; Razum, O.; Coebergh, J.-W. Cancer Risk Diversity in Non-Western Migrants to Europe: An Overview of the Literature. *Eur. J. Cancer* **2010**, *46*, 2647–2659. [CrossRef] [PubMed]
8. Gkiouleka, A.; Avrami, L.; Kostaki, A.; Huijts, T.; Eikemo, T.A.; Stathopoulou, T. Depressive Symptoms among Migrants and Non-Migrants in Europe: Documenting and Explaining Inequalities in Times of Socio-Economic Instability. *Eur. J. Public Health* **2018**, *28* (Suppl. S5), 54–60. [CrossRef] [PubMed]
9. Lebano, A.; Hamed, S.; Bradby, H.; Gil-Salmerón, A.; Durá-Ferrandis, E.; Garcés-Ferrer, J.; Azzedine, F.; Riza, E.; Karnaki, P.; Zota, D.; et al. Migrants' and Refugees' Health Status and Healthcare in Europe: A Scoping Literature Review. *BMC Public Health* **2020**, *20*, 1039. [CrossRef] [PubMed]
10. Janssen, F.; Bardoutsos, A.; El Gewily, S.; De Beer, J. Future Life Expectancy in Europe Taking into Account the Impact of Smoking, Obesity, and Alcohol. *eLife* **2021**, *10*, e66590. [CrossRef] [PubMed]
11. Aichberger, M.; Schouler-Ocak, M.; Mundt, A.; Busch, M.; Nickels, E.; Heimann, H.; Ströhle, A.; Reischies, F.; Heinz, A.; Rapp, M. Depression in Middle-Aged and Older First Generation Migrants in Europe: Results from the Survey of Health, Ageing and Retirement in Europe (SHARE). *Eur. Psychiatry* **2010**, *25*, 468–475. [CrossRef] [PubMed]
12. Lanari, D.; Bussini, O. International Migration and Health Inequalities in Later Life. *Ageing Soc.* **2012**, *32*, 935–962. [CrossRef]
13. Reus-Pons, M.; Mulder, C.H.; Kibele, E.U.; Janssen, F. Differences in the Health Transition Patterns of Migrants and Non-Migrants Aged 50 and Older in Southern and Western Europe (2004–2015). *BMC Med.* **2018**, *16*, 57. [CrossRef]
14. Reus-Pons, M.; Kibele, E.U.; Janssen, F. Differences in Healthy Life Expectancy between Older Migrants and Non-Migrants in Three European Countries over Time. *Int. J. Public Health* **2017**, *62*, 531–540. [CrossRef]
15. Brown, T.H.; O'Rand, A.M.; Adkins, D.E. Race-Ethnicity and Health Trajectories: Tests of Three Hypotheses across Multiple Groups and Health Outcomes. *J. Health Soc. Behav.* **2012**, *53*, 359–377. [CrossRef] [PubMed]
16. Evandrou, M.; Falkingham, J.; Feng, Z.; Vlachantoni, A. Ethnic Inequalities in Limiting Health and Self-Reported Health in Later Life Revisited. *J. Epidemiol. Community Health* **2016**, *70*, 653–662. [CrossRef]
17. Shuey, K.M.; Willson, A.E. Cumulative Disadvantage and Black-White Disparities in Life-Course Health Trajectories. *Res. Aging* **2008**, *30*, 200–225. [CrossRef]
18. Haas, S.; Rohlfsen, L. Life Course Determinants of Racial and Ethnic Disparities in Functional Health Trajectories. *Soc. Sci. Med.* **2010**, *70*, 240–250. [CrossRef] [PubMed]
19. Levchenko, Y. Aging into Disadvantage: Disability Crossover among Mexican Immigrants in America. *Soc. Sci. Med.* **2021**, *285*, 114290. [CrossRef] [PubMed]
20. Kristiansen, M.; Razum, O.; Tezcan-Güntekin, H.; Krasnik, A. Aging and Health among Migrants in a European Perspective. *Public Health Rev.* **2016**, *37*, 20. [CrossRef] [PubMed]
21. Carnein, M.; Milewski, N.; Doblhammer, G.; Nusselder, W.J. Health Inequalities of Immigrants: Patterns and Determinants of Health Expectancies of Turkish Migrants Living in Germany. In *Health among the Elderly in Germany*; Doblhammer, G., Ed.; New Evidence on Disease, Disability and Care Need; Verlag Barbara Budrich: Opladen, Germany, 2015; Volume 46, pp. 157–190.
22. Milewski, N.; Doblhammer, G. Mental Health among Immigrants: Is There a Disadvantage in Later Life. In *Health among the Elderly in Germany*; Doblhammer, G., Ed.; New Evidence on Disease, Disability and Care Need; Verlag Barbara Budrich: Opladen, Germany, 2015; Volume 46, pp. 191–212.
23. Börsch-Supan, A. Survey of Health, Ageing and Retirement in Europe (SHARE) Wave 1. Published online 2020. Available online: http://www.share-project.org/data-documentation/waves-overview/wave-1.html (accessed on 14 November 2021).
24. Börsch-Supan, A. Survey of Health, Ageing and Retirement in Europe (SHARE) Wave 2. Published online 2020. Available online: http://www.share-project.org/data-documentation/waves-overview/wave-2.html (accessed on 14 November 2021).
25. Börsch-Supan, A. Survey of Health, Ageing and Retirement in Europe (SHARE) Wave 4. Published online 2020. Available online: http://www.share-project.org/data-documentation/waves-overview/wave-4.html (accessed on 14 November 2021).
26. Börsch-Supan, A. Survey of Health, Ageing and Retirement in Europe (SHARE) Wave 5. Published online 2020. Available online: http://www.share-project.org/data-documentation/waves-overview/wave-5.html (accessed on 14 November 2021).

27. Börsch-Supan, A. Survey of Health, Ageing and Retirement in Europe (SHARE) Wave 6. Published online 2020. Available online: http://www.share-project.org/data-documentation/waves-overview/wave-6.html (accessed on 14 November 2021).

28. Börsch-Supan, A. Survey of Health, Ageing and Retirement in Europe (SHARE) Wave 7. Published online 2020. Available online: http://www.share-project.org/data-documentation/waves-overview/wave-7.html (accessed on 14 November 2021).

29. Börsch-Supan, A.; Brandt, M.; Hunkler, C.; Kneip, T.; Korbmacher, J.; Malter, F.; Schaan, B.; Stuck, S.; Zuber, S.; SHARE Central Coordination Team. Data Resource Profile: The Survey of Health, Ageing and Retirement in Europe (SHARE). *Int. J. Epidemiol.* **2013**, *42*, 992–1001. [CrossRef]

30. SHARE. *SHARE Release Guide 1.0.0 of Wave 8*; MEA: Munich, Germany, 2021.

31. van Oyen, H.; Van der Heyden, J.; Perenboom, R.; Jagger, C. Monitoring Population Disability: Evaluation of a New Global Activity Limitation Indicator (GALI). *Soz. Praventivmed.* **2006**, *51*, 153–161. [CrossRef] [PubMed]

32. Prince, M.J.; Reischies, F.; Beekman, A.T.; Fuhrer, R.; Jonker, C.; Kivela, S.L.; Lawlor, B.A.; Lobo, A.; Magnusson, H.; Fichter, M.; et al. Development of the EURO-D Scale—A European, Union Initiative to Compare Symptoms of Depression in 14 European Centres. *Br. J. Psychiatry* **1999**, *174*, 330–338. [CrossRef]

33. UNESCO. *International Standard Classification of Education ISCED 1997*; UNESCO: Paris, France, 1997.

34. Lumley, T.; Scott, A. Fitting Regression Models to Survey Data. *Stat. Sci.* **2017**, *32*, 265–278. [CrossRef]

35. Bell, A.; Jones, K. Explaining Fixed Effects: Random Effects Modeling of Time-Series Cross-Sectional and Panel Data. *Political. Sci. Res. Methods* **2015**, *3*, 133–153. [CrossRef]

36. Aiken, L.S.; West, S.G. *Multiple Regression: Testing and Interpreting Interactions*; Sage Publications, Inc.: Thousand Oaks, CA, USA, 1991.

37. R: A Language and Environment for Statistical Computing. Available online: https://www.gbif.org/tool/81287/r-a-language-and-environment-for-statistical-computing (accessed on 23 September 2021).

38. Lumley, T. *Complex Surveys: A Guide to Analysis Using R*; John Wiley & Sons: Hoboken, NJ, USA, 2011.

39. Lüdecke, D. Ggeffects: Tidy Data Frames of Marginal Effects from Regression Models. *J. Open Source Softw.* **2018**, *3*, 772. [CrossRef]

40. Lenth, R.V. Emmeans: Estimated Marginal Means Aka Least Square Means. R Package Version 1.6.3. Available online: https://github.com/rvlenth/emmeans (accessed on 31 October 2021).

41. Gubernskaya, Z. Age at Migration and Self-Rated Health Trajectories after Age 50: Understanding the Older Immigrant Health Paradox. *J. Gerontol. B Psychol. Sci. Soc. Sci.* **2015**, *70*, 279–290. [CrossRef] [PubMed]

42. Ahonen, E.Q.; Benavides, F.G.; Benach, J. Immigrant Populations, Work and Health—A Systematic Literature Review. *Scand. J. Work Environ. Health* **2007**, *33*, 96–104. [CrossRef]

43. Spallek, J.; Zeeb, H.; Razum, O. What Do We Have to Know from Migrants' Past Exposures to Understand Their Health Status? A Life Course Approach. *Emerg. Themes Epidemiol.* **2011**, *8*, 6. [CrossRef]

44. Honkaniemi, H.; Juárez, S.P.; Katikireddi, S.V.; Rostila, M. Psychological Distress by Age at Migration and Duration of Residence in Sweden. *Soc. Sci. Med.* **2020**, *250*, 112869. [CrossRef]

45. Read, S.; Grundy, E.; Foverskov, E. Socio-Economic Position and Subjective Health and Well-Being among Older People in Europe: A Systematic Narrative Review. *Aging Ment. Health* **2016**, *20*, 529–542. [CrossRef]

46. Sadana, R.; Blas, E.; Budhwani, S.; Koller, T.; Paraje, G. Healthy Ageing: Raising Awareness of Inequalities, Determinants, and What Could Be Done to Improve Health Equity. *Gerontologist* **2016**, *56* (Suppl. S2), S178–S193. [CrossRef]

47. Peixoto, J. Migration and Policies in the European Union: Highly Skilled Mobility, Free Movement of Labour and Recognition of Diplomas. *Int. Migr.* **2001**, *39*, 33–61. [CrossRef]

48. Lulle, A.; Janta, H.; Emilsson, H. Introduction to the Special Issue: European Youth Migration: Human Capital Outcomes, Skills and Competences. *J. Ethn. Migr. Stud.* **2021**, *47*, 1725–1739. [CrossRef]

49. Börsch-Supan, A.; Alcser, K.H. (Eds.) *The Survey of Health, Aging and Retirement in Europe-Methodology*; Mannheim Research Institute for the Economics of Aging (MEA): Mannheim, Germany, 2005.

50. Maskileyson, D.; Seddig, D.; Davidov, E. The EURO-D Measure of Depressive Symptoms in the Aging Population: Comparability across European Countries and Israel. *Front. Polit. Sci.* **2021**, *3*, 90. [CrossRef]

51. Bremer, D.; Lüdecke, D.; Vonneilich, N.; von dem Knesebeck, O. Social Relationships and GP Use of Middle-Aged and Older Adults in Europe: A Moderator Analysis. *BMJ Open* **2018**, *8*, e018854. [CrossRef] [PubMed]

52. Singh-Manoux, A.; Martikainen, P.; Ferrie, J.; Zins, M.; Marmot, M.; Goldberg, M. What Does Self Rated Health Measure? Results from the British Whitehall II and French Gazel Cohort Studies. *J. Epidemiol. Community Health* **2006**, *60*, 364–372. [CrossRef] [PubMed]

53. DeSalvo, K.B.; Bloser, N.; Reynolds, K.; He, J.; Muntner, P. Mortality Prediction with a Single General Self-Rated Health Question. A Meta-Analysis. *J. Gen. Intern. Med.* **2006**, *21*, 267–275. [CrossRef] [PubMed]

54. Miilunpalo, S.; Vuori, I.; Oja, P.; Pasanen, M.; Urponen, H. Self-Rated Health Status as a Health Measure: The Predictive Value of Self-Reported Health Status on the Use of Physician Services and on Mortality in the Working-Age Population. *J. Clin. Epidemiol.* **1997**, *50*, 517–528. [CrossRef]

55. Johnson, R.J.; Wolinsky, F.D. The Structure of Health Status among Older Adults: Disease, Disability, Functional Limitation, and Perceived Health. *J. Health Soc. Behav.* **1993**, *34*, 105–121. [CrossRef] [PubMed]

56. Wendt, C. Mapping European Healthcare Systems: A Comparative Analysis of Financing, Service Provision and Access to Healthcare. *J. Eur. Soc. Policy* **2009**, *19*, 432–445. [CrossRef]

57. Cvajner, M.; Echeverría, G.; Sciortino, G. What Do We Talk When We Talk about Migration Regimes? The Diverse Theoretical Roots of an Increasingly Popular Concept. In *Was ist ein Migrationsregime? What Is a Migration Regime?* Pott, A., Rass, C., Wolff, F., Eds.; Migrationsgesellschaften; Springer Fachmedien: Wiesbaden, Germany, 2018; pp. 65–80. [CrossRef]
58. Esping-Andersen, G. *The Three Worlds of Welfare Capitalism*; Polity Press: Cambridge, UK, 2008.
59. Giannoni, M.; Franzini, L.; Masiero, G. Migrant Integration Policies and Health Inequalities in n Europe. *BMC Public Health* **2016**, *16*, 463. [CrossRef] [PubMed]

International Journal of
*Environmental Research
and Public Health*

MDPI

Article

The Weight of Migration: Reconsidering Health Selection and Return Migration among Mexicans

Aresha M. Martinez-Cardoso [1,*] and Arline T. Geronimus [2]

1   Department of Public Health Sciences, University of Chicago, Chicago, IL 60637, USA
2   Department of Health Behavior and Health Education, University of Michigan, Ann Arbor, MI 48109, USA;
    arline@umich.edu
*   Correspondence: areshamc@uchicago.edu

Abstract: While migration plays a key role in shaping the health of Mexican migrants in the US and those in Mexico, contemporary Mexican migration trends may challenge the health selection and return migration hypotheses, two prevailing assumptions of how migration shapes health. Using data from the Mexican Family Life Survey (2002; 2005), we tested these two hypotheses by comparing the cardiometabolic health profiles of (1) Mexico–US future migrants and nonmigrants and (2) Mexico–US return migrants and nonmigrants. First, we found limited evidence for health selection: the cardiometabolic health of Mexico–US future migrants was not measurably better than the health of their compatriots who did not migrate, although migrants differed demographically from nonmigrants. However, return migrants had higher levels of adiposity compared to those who stayed in Mexico throughout their lives; time spent in the US was also associated with obesity and elevated waist circumference. Differences in physical activity and smoking behavior did not mediate these associations. Our findings suggest positive health selection might not drive the favorable health profiles among recent cohorts of Mexican immigrants in the US. However, the adverse health of return migrants with respect to that of nonmigrants underscores the importance of considering the lived experience of Mexican migrants in the US as an important determinant of their health.

Keywords: Mexican; Hispanic/Latino paradox; stress; migration

Citation: Martinez-Cardoso, A.M.; Geronimus, A.T. The Weight of Migration: Reconsidering Health Selection and Return Migration among Mexicans. *Int. J. Environ. Res. Public Health* **2021**, *18*, 12136. https://doi.org/10.3390/ijerph182212136

Academic Editors: Heiko Becher, Volker Winkler and Hajo Zeeb

Received: 4 October 2021
Accepted: 18 November 2021
Published: 19 November 2021

Publisher's Note: MDPI stays neutral with regard to jurisdictional claims in published maps and institutional affiliations.

1. Introduction

Among people of Mexican descent in the United States (US), nativity and generational status are consistently associated with adiposity, diabetes, and cardiometabolic health. In general, recent Mexican immigrants present with lower levels of cardiometabolic risk factors and adverse outcomes in comparison to their US-born Mexican American counterparts and, in some cases, non-Latino US-born Whites [1]. For immigrants, more time in the US is also associated with increasingly deleterious health outcomes [2,3], namely, obesity, elevated blood pressure, and type-2 diabetes. Advantageous cardiometabolic health among recent Mexican immigrants is one example of what has been coined the Latino/Hispanic or immigrant health paradox, where despite their considerable social and economic vulnerabilities, immigrants have lower rates of mortality, adverse birth outcomes, and cardiovascular mortality risk than US-born Mexican Americans [4–6].

In this research literature, health selection, return migration, and acculturation hypotheses are proposed to explain the disparate health outcomes between Mexican immigrants and US-born Mexicans and between recent and long-stay Mexican immigrants. Health selection arguments suggest that Mexican migration is positively selected on the basis of health and other sociodemographic characteristics associated with health, leading the healthiest individuals to engage in migration [7]. In addition, the return migration/salmon bias hypothesis proposes that immigrants who become unhealthy in the US might return to their country of origin for care, leaving the healthiest immigrants in the US [8]. Therefore, research that compares the health of immigrants and that of US-born may be mis-specified because it fails to account for these in-migration and out-migration selection processes and compares an artificially healthy cohort of immigrants to the US-born population with a more diverse health profile [9].

Moreover, researchers have proposed that acculturation towards the US unhealthy behaviors and social practices might also drive the health disadvantages of US-born and long-stay migrants. Some argue, for example, that health behaviors that are associated with cardiovascular disease may be better in sending countries compared to the US, with more individuals engaging in behaviors that promote health, such as physical activity and healthy dietary practices [10]. Therefore, immigrants may arrive in the US and still engage in cardiometabolic-related behaviors that promote health, such as physical activity and healthy dietary practices [11]. In the US, recent immigrants may also retain positive social networks that promote their health [11]. With more time in the US, however, acculturation arguments suggest that immigrants shift to US norms, which worsens the health behaviors, social, and cultural practices that were protective of immigrants' health [12].

The evidence supporting these hypotheses is mixed, calling into question these potential drivers of immigrant health. For example, whether health selection plays a large role in the calculus of migration as opposed to other drivers such as economic instability and violence is subject to debate [13]. In addition, new evidence demonstrates that the health behaviors of recent immigrants may not, in fact, be better than those of US-born or long-stay migrants [14]. Similarly, the role of return migration due to health may be overestimated, since return migration to Mexico is often due to both involuntary and voluntary factors including family reunification, employment prospects, and deportation [15]. Instead, drawing on scholarship in migration, racialization, and health, a new stream of research argues that the deleterious health of migrants may be linked to chronic exposure to social, political, and environmental stressors over time and across generations [16]. These more contemporary arguments have offered an alternative explanation for the declines in immigrants' health with longer residence in the US, yet traditional health selection and acculturation explanations are replete in the literature.

New methodological approaches also offer innovative ways to evaluate the hypotheses of the immigrant health advantage. Most recently, research has shifted from cross-sectional data in the US to the use of binational data from both sending and receiving countries [17–22]. This approach has enabled comparisons between behaviors and outcomes of migrants and their compatriots who do not migrate, an arguably more appropriate comparison group to assess the possibility of health selection and health-driven return migration.

Our analysis contributes to these debates by using data from the Mexican Family and Life Survey (MxFLS) to explore health selection and return migration among Mexican migrants to the US, with a focus on adiposity, blood pressure, cardiovascular disease, and diabetes. The MxFLS is a recent nationally representative study that measures health and migration in the Mexican population considering future and return migration. As such, our analysis lends new evidence about contemporary migration and health. In addition, we advance the literature findings by using health data that were collected by trained staff, rather than relying only on self-rated health measures, as most surveys do.

First, within the population of Mexico, we compared cardiometabolic health between Mexico–US future migrants (n = 322) and nonmigrants (n = 14,441). We hypothesized that Mexicans who migrated to the US between the waves of the MxFLS would have similar cardiometabolic health profiles as Mexicans who did not migrate, adjusting for demographic characteristics. Next, we compared the cardiometabolic health profiles of Mexico–US return migrants (n = 276) and those in Mexico who never out-migrated (stayers, n = 14,441). We hypothesized that return migrants would have worse cardiometabolic health than stayers. Furthermore, we explored associations between health and time spent in the US, age at migration, and documentation status to test whether these dimensions of migration to the US were associated with poorer health. Finally, we tested whether health behaviors related to physical activity or smoking mediated the relationship between return migration and health.

2. Methods

Data

We use two waves of data from the Mexican Family Life Survey (MxFLS), a nationally representative and longitudinal survey of households in Mexico [23]. The baseline survey in 2002 collected sociodemographic and anthropometric data on 8440 households and 19,809 adults across 150 communities in Mexico. Private dwellings formed the primary sampling units of this survey, and the sample was designed to be nationally, urban-rurally, and regionally representative of the Mexican population [24]. Sampling units with similar geographic and socio-economic characteristics were grouped into a single stratum, and households were selected among the strata. All adults and children in the sampled households were eligible for inclusion in the MxFLS. The second wave of data was obtained in 2005–2006 by successfully recontacting over 90% of the original household sample, including those who migrated within Mexico or emigrated to the US. We leverage the unique migration and migration history measures in the data to create indicators of future migration to the US and return migration from the US (elaborated below). We limited our sample to adult respondents with measures of their migration history, health, and sociodemographic variables of interest (*n* = 14,763).

3. Measures

3.1. Migration Indicators

The MxFLS includes a variety of questions about local and international migration, which we leveraged for our analysis. The MxFLS tracked individuals across wave 1 in 2002 and wave 1 in 2005, identifying individuals who lived in Mexico in wave I but moved to the US in wave 2. For the first research question testing health selection, we use this variable to classify *future migrants*, respondents who migrated to the US between waves 1 in 2002 and wave 2 in 2005, and *nonmigrants*, respondents who remained in Mexico during both waves of the study. In total, we identified 322 future migrants and 14,441 nonmigrants.

For the second research question that tested the association between return migration from the US and health, we leveraged a series of variables that asked the participants to detail all the places they had moved to both within and outside Mexico since the age of 12. Using this variable, we identified *return migrants*, respondents who migrated to the US and returned to Mexico, and *stayers*, respondents who never migrated to the US. We classified respondents as return migrants if they reported living in the US for a period of 12 months or longer (*n* = 276). Stayers included respondents who reported that they had never lived in the US (*n* = 14,487). In addition, for the return migrant group, we computed the total number of years spent in the US and age at migration. Finally, we created a variable for documentation status during migration to the US, classifying individuals as undocumented or documented at migration. For those with multiple migration trips to the US, information from the last migration trip was used for the age at migration and documentation status variable.

3.2. Cardiometabolic Health Indicators and Health Behaviors

A trained survey staff collected anthropometric health data of the respondents at baseline. Using these data, we computed measures for waist circumference and mean arterial blood pressure (MAP) (SBP + 3 *DBP/3). For sensitivity analysis, we also created measures of body mass index (BMI) (weight (kg)/height(m)^2) and hypertension. We created dichotomous indicators of elevated levels for each of these measures (elevated waist circumference, obesity, elevated MAP, and elevated blood pressure) using appropriate clinical cutoffs [25,26]. In addition, we created dichotomous measures of diabetes status and cardiometabolic disease (history of heart disease, heart attack, cholesterol/arteriosclerosis, or stroke) based on self-reported data.

Health behaviors included physical activity and smoking status. Using information on the frequency and duration of physical activity, we created a dichotomous indicator of whether the respondents met the criteria for the recommended amount of physical activity

(>150 min/week). The respondents were also classified as current, former, or nonsmokers based on a series of questions about their smoking history.

3.3. Controls

We controlled for several demographic variables in our models that are associated with the selected health outcomes, including age, gender, marital status, current employment status, highest level of education completed, health insurance status, and household assets. Finally, we included controls for urban/rural residence, since previous research suggests that residents of urban/rural regions have distinct migration patterns.

4. Analysis

The analysis was completed using Stata 15. The datasets generated during the current study are available from the corresponding author on reasonable request. Determination of exempt status was obtained from the University of Michigan and University of Chicago BSD/UCMC Institutional Review Boards. Descriptive statistics were generated to assess the quality of the data and the proportion of missing cases. Cases that were missing data on variables of interest were dropped from the analysis. The final analytical sample included 14,763 respondents. Pearson's chi-square tests and t-tests were used to compare demographic and health characteristics between future migrants and non-migrants as well as between return migrants and stayers at Wave 1.

Next, mixed-effects models were used to examine the association between health at baseline and future migration and between return migration and health. Mixed-effects models were used because data in the MxFLS are clustered on two levels: respondents were nested within families/households, and families/households were grouped within communities. Mixed-effects models allow us to make appropriate statistical inferences while accounting for the multilevel nature of the data. Odds ratios (OR) and 95% CI are reported for dependent variables that were modeled using mixed-effects logistic regression, while beta coefficients are reported for dependent variables that were modeled using mixed-effects linear regression.

First, we tested whether health at wave 1 was associated with Mexico–US migration at wave 2. In these models, Mexico–US migration was the outcome variable, while health variables at baseline were modeled as covariate variables. That is, we tested whether each of the health indicators significantly predicted whether a respondent was a *future migrant* or a *nonmigrant*. To account for potential collinearity among the health variables, each health variable was entered by itself in separate models.

Second, we tested whether return migration from the US was associated with each of the health variables at wave 1. In these models, each health indicator at wave 1 was the outcome variable, and US–Mexico return migration was modeled as a covariate. We then ran a similar series of models with time in the US as a continuous independent variable; non-migrants were assigned a value of 0, while return migrants were assigned a value corresponding to the year(s) they spent in the US. We also modeled time in the US as a categorical variable based on quintile cutpoints to gauge which level of time in the US was most consequential for health. We also tested if associations between return migration or time in the US and health were mediated by smoking and physical activity behaviors by entering these variables in the model and comparing point estimates and *p*-values. Finally, we ran a subset of models among return migrants to test the association between age at migration, documentation status at migration, and health.

5. Results

5.1. Descriptive Statistics

Table 1, column 1 presents descriptive statistics for the overall sample (*n* = 14,763). There was a larger proportion of women (55%) than of men, and the mean age was 40 years (sd = 16.6). The majority of the respondents were married (67%). In addition, over 60% of the sample had achieved at least a primary school education in Mexico, and half of the respondents (58%) reported that they had worked in the past month.

Table 1. Descriptive Statistics and Bivariate Analysis of Future Migrants vs. Non-migrants and Return Migrants vs. Stayers, MxFLS, Wave 1, *n* = 14,763.

Variable	Total Sample (*n* = 14,763)	Future Migrants (*n* = 322)	Non-Migrants (*n* = 14,441)	Return Migrants (*n* = 276)	Stayers (*n* = 14,487)
	Mean (sd)/%	Mean (sd)/%		Mean (sd)/%	
Age	40(16.6)	29(11.5) [†]	41(16.6)	39(14.5)	40(16.7)
Female	55	47 [†]	56	29 [†]	56
Married	67	51 [†]	68	75 [†]	68
Primary School Education	63	76 [†]	63	67 [*]	63
Currently Working	57	60	57	68 [†]	57
Health Insurance	45	21 [†]	46	36 [†]	46
Household Assets (ref = owns house)	85	86	84	84	85
Urban Region	57	42 [†]	58	58	57
Return Migrant	2	9 [†]	2	–	–
Future Migrant	2	–	–	–	–
Time in the US (years)	–	–	–	4.2 (5.4)	–
Age at Migration	–	–	–	24 (8.5)	–
Undocumented at Migration	–	–	–	75	–
Obese	26	17 [†]	26	29	26
Elevated Waist-Circumference	26	12 [†]	27	25	26
High Blood Pressure	37	27 [†]	37	37	36
Elevated Mean Arterial Press.	38	34	38	42	38
Self-Reported Diabetes	6	2 [†]	6	4	6
Self-Reported CVD	3	3	3	3	3
Current Smoker	14	14	14	18 [†]	14
Recommended Level of Physical Activity	14	18 [†]	14	18 [†]	14

Chi-square tests used to compare categorial variables; *t*-tests used to compare continuous variables; [†] Indicates significance at $p < 0.05$ level, [*] indicates $p < 0.10$.

In terms of health, nearly a quarter of the respondents had an elevated waist circumference, and 26% of the sample was classified as obese, which is consistent with national estimates [27]. Based on blood pressure measures, 38% had an elevated mean arterial pressure, and 37% were classified as having high blood pressure. The prevalence of self-reported diabetes and cardiometabolic disease was 6% and 3%, respectively. Finally, 14% of the respondents were current smokers, and 14% performed at least 150 min of physical activity per week.

5.2. Migrant Groups Bivariate Analysis

Next, we compared the sociodemographic and health characteristics of future migrants vs. those of non-migrants (Table 1, Column 2 and 3) and of return migrants vs. those of stayers (Table 1, column 3 and 4). Mexico–US future migrants were respondents who lived in Mexico during Wave 1 of the study in 2002, then lived in the US during Wave 2 in 2005–2006; non-migrants were respondents who remained in Mexico across both waves of the study. The sample comprised a total of 322 Mexico–US future migrants and 14,441 non-migrants. As compared to non-migrants, future migrants were younger and more likely to be men, unmarried, and uninsured. Future migrants had higher rates of primary school completion and were more likely to reside in rural regions of Mexico. In addition, future migrants had lower levels of elevated waist circumference, obesity, high blood pressure, and self-reported diabetes and were more likely to engage in at least 150 min of physical activity per week.

Finally, 276 individuals in the sample were Mexico–US return migrants—individuals who had previously migrated to the US for a period of at least 12 months but had returned to Mexico—whereas 14,487 were stayers, i.e., individuals who had only lived in Mexico. On average, return migrants reported living in the US for 4 years and migrating at the age of 24 years; 75% reported that they were undocumented when they migrated. Compared to stayers, return migrants included a larger proportion of men and were more likely to be married, currently working, uninsured, current smokers, and meet exercise recommendations. Return migrants were largely comparable to stayers across all health variables, based on the bivariate analysis.

5.3. Health Selection

We estimated mixed-effects regression models that tested whether health at wave 1 was associated with migration to the US by wave 2. All models included the sociodemographic control variables. We entered each of the health indicators as an independent predictor of future migration without any other health variable to avoid multicollinearity issues. A summary of regression coefficients for each health variable, estimated from separate models, is shown in Table 2; full models with control variables are provided in Supplemental Table S1. When entered singly, only elevated waist circumference was significantly associated with migration to the US in wave 2 ($p < 0.10$).

Table 2. Summary of Mixed-Effects Models of the Association between Cardiometabolic Health and Future Migration to the US, Wave 1, $n = 14,763$.

Dependent Variable:	Future Migrant to the US Migrant = 1; Nonmigrant = 0		
Health Indicator:	**OR**	**s.e.**	**95% CI**
Elevated Waist Circumference	0.61 *	0.16	0.37–1.02
Elevated MAP	0.78	0.15	0.53–1.15
Diabetes	1.05	0.55	0.38–2.95
Cardiovascular Disease	2.20	1.20	0.76–6.38
Smoker	1.35	0.36	0.81–2.27
Physical Activity	1.34	0.33	0.84–2.16

Mixed-effects logistic regression models predicting future migration to the US. Each health indicator was entered alone as an independent variable; controls include age, gender, marital status, education, employment status, insurance status, household assets, urbanicity, smoking status, and physical activity. * Indicates significance at $p < 0.10$. Model adjusted for level 1 clustering at the family level and level 2 clustering at the locality level.

Notably, however, the point estimates for the ORs suggested the possibility of other differences, and the confidence intervals were wide, suggesting imprecise estimates that we attributed to the small sample size of future migrants. Examining the estimated ORs at face value only, they were not in a consistent direction. For example, the estimated ORs suggested that future migrants were more likely to be smokers and to report a history of cardiometabolic disease, yet also more likely to be physically active and have a smaller waist circumference. In other words, we found no consistent evidence for positive health selection, net of controls.

For robustness checks, we tested the models in Table 2 with continuous versions of health variables as well as various comparison groups. In addition, we substituted BMI and hypertension for weight circumference and mean arterial pressure, respectively. The results were consistent. Because future migration was associated with a previous migration to the US and we hypothesized that return migration was independently associated with health, we also controlled for return migration status in our models. We checked whether excluding return migrants from the sample changed these findings and found that the results remained consistent.

5.4. Return Migration

Next, we explored whether ever migrating to the US was associated with each of the health variables during wave 1 and modeled the relationship between time in the US and health. In these models, each of the health indicators was separately modeled as the dependent variable, with the return migration indicator included as an independent variable. A summary of the regression coefficients for return migration and time in the US regarding each of the health variables is shown in Table 3.

Table 3. Summary of Mixed-Effects Models of the Association between Migration History to the US and Cardiometabolic Health, MxFLS Wave 1, *n* = 14,763.

Migration History Variable	Return Migration (Return Migrant = 1; Stayer = 0)				Time in the US (Years)			
Health Variable:	OR [1]/b [2]	s.e.	95% CI		OR [1]/b [2]	s.e.	95% CI	
Waist Circumference [1]	1.49 [†]	0.67	0.19	2.80	0.27 [†]	0.10	0.07	0.47
Obese [2]	1.41 [†]	0.22	1.03	1.91	1.04	0.02	0.99	1.08
Elevated MAP [2]	1.08	0.16	0.81	1.44	1.01	0.02	0.97	1.06
High Blood Pressure [2]	0.98	0.14	0.73	1.30	1.02	0.02	0.98	1.06
Diabetes [2]	0.71	0.25	0.36	1.43	1.02	0.04	0.94	1.10
Cardiovascular Disease [2]	1.20	0.42	0.60	2.40	1.06 [†]	0.03	1.003	1.12

Mixed-effects regression models of the association between return migration and health and between time in the US and health; [1] beta coefficients are reported for dependent variables that were modeled using mixed-effects linear regression; [2] odds ratios are reported for dependent variables that were modeled using mixed-effects logistic regression. Controls include age, gender, marital status, education, employment status, insurance status, household assets, urbanicity, smoking status, and physical activity. [†] Indicates significance at $p < 0.05$ level. Models were adjusted for level 1 clustering at the family level and level 2 clustering at the locality level.

Return migration was significantly associated with waist circumference and obesity. Return migrants had 41% increased odds of being obese. Waist circumference was also statistically associated with time spent in the US, suggesting a stress-related distribution of adiposity toward the abdomen among return migrants. On average, return migrants had a waist circumference that was 1.49 cm greater than stayers, and each additional year in the US increased waist circumference by a quarter of a centimeter. The inclusion of physical activity or smoking in the models did not significantly change these findings. When we modeled time in the US based on categorical cutpoints, we found that immigrants who spent 2.16–5 years had a waist circumference that was 3.39 cm greater on average than non-migrants, the largest difference in waist circumference across the time in the US groups (Supplemental Table S2). Those who spent less than 1.16 years in the US were indistinguishable from non-migrants.

However, we found no significant association between return migration and hypertension, mean arterial pressures, or diabetes. Time in the US was also not associated with high blood pressure or self-reported diabetes. Finally, we tested return migration and time in the US on self-reported cardiometabolic disease. In these models, return migration was not significantly associated with cardiometabolic disease; however, time in the US was associated with increased odds of reporting heart disease, stroke, or atherosclerosis (OR = 1.06, 95% CI 1.003–1.126).

In a sub-analysis among return migrants, we explored whether age at migration and documentation status during migration were associated with the health indicators. Neither was. Given the small sample sizes, however, these point estimates from the models proved to be unstable, with large confidence intervals. Therefore, we hesitate to meaningfully interpret these results.

6. Discussion

This analysis drew on a large and multi-thematic dataset of adults in Mexico to understand how health might shape contemporary migration to the US. We also explored how migration to the US might shape health and health behaviors among Mexican migrants who return to their country of origin. In general, we found that a variety of health behaviors, cardiometabolic health risk factors, and cardiometabolic disease were not associated with future migration to the US. Instead, more traditional factors such as age and gender were consistently associated with migration. In addition, having health insurance also reduced the odds of migration. This finding may be due, in part, to the types of employment sectors that provided health insurance to Mexican residents in 2002 (private business and government employees) and to the employment and economic opportunities these individuals had in Mexico compared to uninsured individuals outside of these sectors. We found little and mixed evidence in support of the view that those Mexicans who migrate to the US are positively health-selected. Based on these findings, health was not an important predictor of the respondents' decision to migrate to the US between the waves of the MxFLS. While there is some evidence of selection in migration by demographic factors, a previous analysis with older Mexican cohorts still found evidence of health selection and health advantages among immigrants, net of these demographic controls. Our analysis departs from these findings. As such, it is unlikely that health selection with respect to cardiometabolic health acts as a major driver of the advantageous health profile of recent Mexican migrants to the US These findings are consistent with those of a similar analysis that explored other health outcomes that may be linked to health selection [28].

While we found that migrants and nonmigrants had similar health profiles before migration, health outcomes among return migrants as compared to those of stayers revealed a different story. Return migrants, those who had ever spent a year or longer in the US, were more likely to be overweight and had a larger waist circumference as compared to stayers. Moreover, return migrants who had lived in the US for more time also had higher waist circumferences and self-reported cardiometabolic disease. However, neither return migration nor time in the US were associated with other measured cardiometabolic indicators including hypertension and diabetes. Our mixed findings may be due to the fact that the health outcomes associated with adiposity take a shorter time to manifest as compared to diabetes, hypertension, and cardiometabolic disease. In a similar analysis with a more restricted sample of male Mexican migrants, Ullman et al. also found return migrants had higher levels of obesity, with no differences in hypertension or diabetes [22]. Similarly an analysis of children in the MxFLS found that children in Mexico within migrant households also had higher levels of obesity [29]. If followed for more time, we might expect that return migrants may in fact display more adverse chronic disease profiles than stayers.

In summary, we found that while the health of Mexico–US migrants was on par with that of their compatriots who did not migrate, the health of US–Mexico return migrants was worse than that of stayers on some indicators. These results have multiple potential explanations. The acculturation hypothesis suggests that return migrants could have adopted worse health behaviors in the US that drove these differences. However, we tested whether physical activity and smoking mediated these associations and found limited support for an assimilation explanation. Other health behaviors, such as diet, were not measured in the study and may have played a role in these outcomes. However, several studies have called into question the acculturation argument, showing that health behaviors among migrants are already poor upon arriving to the US, and few change meaningfully enough to drive health differences (an important exception being smoking) [30,31]. Return migrants in our sample could also have returned to Mexico because they were sick or unhealthy, as suggested by the salmon bias hypothesis. However, scholars speculate that return migration due to health is for serious illnesses such as cancer or disability, rather than the health indicators that emerged in the analysis, i.e., obesity and elevated waist circumference.

Instead, our findings lend support to the notion that the worse health of returning Mexican migrants compared to those who remain in Mexico may reflect the embodiment of the stressful social environment Mexican migrants endure in the US [16,32]. Migrants experience an incredible amount of stress in the process of migrating and adapting to a new place [33]. Mexican migrants, in particular, contend with being defined as marginalized and racialized immigrants in the US and the explicit and implicit experiences of discrimination and othering as a result [34]. These chronic exposures to stress due to the migration experience could over-activate stress responses in the body, which have been linked to adiposity [35,36]. In particular, chronic stress has been specifically indicated in the progression of central adiposity, including waist circumference and visceral fat deposits [37,38]. Our findings that return migration from the US and more time in the US were associated with higher waist circumference point to the potential of this stress exposure hypothesis, warranting future research in this area. For policy and practice, our findings underscore the importance of developing strategies to protect the health of Mexican immigrants beyond traditional health behavior interventions. For example, Mexican consulates have developed *Ventanillas de Salud* programs as one approach to address immigrants social and health vulnerabilities in the US [39]. At a population level, local and state-level immigration policies have also proven to be an important lever to protect immigrants' health [40]. In addition, given the growing population of return migrants in Mexico, health practitioners in Mexico should monitor the health of Mexicans with previous migration histories in the US as at a potential risk for poor health.

We interpret our findings and their contributions while also noting some important limitations of our analysis and data. The cross-sectional nature of the return migration data limits our ability to infer causation between return migration and health. Those who were return migrants could have had higher levels of adiposity before they migrated to the US as compared to stayers. On average, however, those who were future migrants had smaller waist circumferences than stayers, although it is unclear if these findings are consistent for earlier immigrant cohorts. In addition, given the nature of the data, we were unable to compare the health of return migrants with that of Mexicans who remained in the US; the inclusion of this group in future analysis would help strengthen our findings about the association between migration to the US and adverse levels of adiposity. Looking only at Mexican immigrants who remained in the US, Kaestner et al. found that the length of time in the US was associated with a higher allostatic load, an indicator of stress-mediated wear and tear, net of age, diet, insurance, and other socioeconomic and health behaviors [32]. Our findings related to increased adiposity among return migrants are consistent with such findings and their implications. While our analysis augments the current literature by comparing the health of migrants to that of those who remained in Mexico, future studies that explicitly compare Mexican non-migrants, return migrants, and migrants in the US with larger samples would lend more clarity to the mechanisms driving health differences among these groups. Finally, the age of our data is one drawback, as migration mechanisms and cohorts may have changed between now and the data collection period; yet, the MxFLS is among the most high-quality nationally representative datasets that measure both health and migration in the Mexican population.

7. Conclusions

The Mexico–US migration flow represents one of the largest global migration flows, and Mexican migrants account for the largest immigrant-origin group in the US [41]. Furthermore, Mexican migrants comprise 35% of the Mexican-origin population in the US and 20% of the overall Latino population [42,43]. This analysis therefore elucidates some of the potential mechanisms and drivers of Latino health in the US and provides an important examination of the Latino health paradox using data on Mexicans. While the current literature is mixed on how health might shape individuals' propensity to migrate, how the US shapes migrants' health, and the various mechanisms that influence these processes, we demonstrated that health does not appear to be a major selection factor for future migration among Mexican adults. However, upon returning to Mexico, adults with migration histories in the US fare worse in weight, specifically, in having a stress-related distribution of adiposity, especially if they lived in the US long. Mexicans' experience of being "othered" in the US, arguably, has become more severe in light of contemporary efforts toward criminalizing immigration. Future work should explore and examine the unique social and health environments faced by Mexican migrants to the US in light of these findings. More work is also needed to understand how reintegration into one's country of origin shapes the health of Mexican return migrants.

Supplementary Materials: The following are available online at https://www.mdpi.com/article/10.3390/ijerph182212136/s1, Table S1: Mixed-Effects Model of the Association between Cardiometabolic Health and Future Migration to the US, MxFLS Wave 1, n = 14,763. Table S2: Mixed Effects Model of the Association between Time in the US and Waist Circumference, MxFLS Wave 1, n = 14,763.

Author Contributions: Conceptualization, A.M.M.-C.; Formal analysis, A.M.M.-C.; Methodology, A.M.M.-C. and A.T.G.; Writing–original draft, A.M.M.-C.; Writing–review & editing, A.M.M.-C. and A.T.G. All authors have read and agreed to the published version of the manuscript.

Funding: This research was supported in part by an NIA training grant (T32AG000221), an NICHD training grant (T32HD007339), and an NICHD center grant (P2CHD041028) to the Population Studies Center at the University of Michigan. The content is solely the responsibility of the authors and does not necessarily represent the official views of the National Institutes of Health.

Institutional Review Board Statement: Determination of exempt status was obtained from the University of Michigan and University of Chicago BSD/UCMC Institutional Review Boards.

Informed Consent Statement: Not applicable. This is a secondary data analysis of public data. Informed consent was obtained from all individual participants included in the original study.

Data Availability Statement: All codes for data cleaning and analysis associated with the current submission are available from the corresponding author, AMC, upon request.

Conflicts of Interest: The authors declare no conflict of interest that are relevant to the content of this article.

Previous Publication: A draft of this work was previously presented at the Population Association of American Annual Meeting, 2019 [44].

References

1. Acevedo-Garcia, D.; Bates, L.M. Latino Health Paradoxes: Empirical Evidence, Explanations, Future Research, and Implications. In *Latinas/os in the United States: Changing the Face of América*; Rodríguez, P.H., Sáenz, P.R., Menjívar, A.P.C., Eds.; Springer: Berlin/Heidelberg, Germany, 2008; pp. 101–113. ISBN 978-0-387-71941-2.
2. Antecol, H.; Bedard, K. Unhealthy assimilation: Why do immigrants converge to American health status levels? *Demography* **2006**, *43*, 337–360. [CrossRef]
3. Barcenas, C.H.; Wilkinson, A.V.; Strom, S.S.; Cao, Y.; Saunders, K.C.; Mahabir, S.; Hernández-Valero, M.A.; Forman, M.R.; Spitz, M.R.; Bondy, M.L. Birthplace, Years of Residence in the United States, and Obesity Among Mexican-American Adults. *Obesity* **2007**, *15*, 1043–1052. [CrossRef]
4. Gutman, M.P.; Frisbie, W.P.; DeTurk, P.; Blanchard, K.S. Dating the origins of the epidemiologic paradox among Mexican Americans. In Proceedings of the Annual Meeting of the Population Association of America, Washington, DC, USA, 26–27 March 1997. Available online: http://liberalarts.utexas.edu/prc/_files/pdf/workingpapers/97-98-07.pdf (accessed on 14 November 2016).
5. Markides, K.S.; Eschbach, K. Aging, Migration, and Mortality: Current Status of Research on the Hispanic Paradox. *J. Gerontol. Ser. B* **2005**, *60*, S68–S75. [CrossRef] [PubMed]
6. Ruiz, J.M.; Steffen, P.; Smith, T.B. Hispanic Mortality Paradox: A Systematic Review and Meta-Analysis of the Longitudinal Literature. *Am. J. Public Health* **2013**, *103*, e52–e60. [CrossRef]
7. Akresh, I.R.; Frank, R. Health Selection Among New Immigrants. *Am. J. Public Health* **2008**, *98*, 2058–2064. [CrossRef] [PubMed]
8. Abraido-Lanza, A.F.; Dohrenwend, B.P.; Ng-Mak, D.S.; Turner, J.B. The Latino mortality paradox: A test of the "salmon bias" and healthy migrant hypotheses. *Am. J. Public Health* **1999**, *89*, 1543–1548. [CrossRef] [PubMed]
9. Palloni, A.; Morenoff, J.D. Interpreting the Paradoxical in the Hispanic Paradox. *Ann. N. Y. Acad. Sci.* **2001**, *954*, 140–174. [CrossRef] [PubMed]
10. Kennedy, S.; Kidd, M.P.; McDonald, J.T.; Biddle, N. The Healthy Immigrant Effect: Patterns and Evidence from Four Countries. *J. Int. Migr. Integr.* **2015**, *16*, 317–332. [CrossRef]
11. Jasso, G.; Massey, D.S.; Rosenzweig, M.R.; Smith, J.P. Immigrant Health: Selectivity and Acculturation. In *Critical Perspectives on Racial and Ethnic Differences in Health in Late Life*; Anderson, N.B., Bulatao, R.A., Cohen, B., Eds.; National Academies Press: Washington, DC, USA, 2004. Available online: https://www.ncbi.nlm.nih.gov/books/NBK25533/ (accessed on 14 November 2017).
12. Lara, M.; Gamboa, C.; Kahramanian, M.I.; Morales, L.S.; Bautista, D.E.H. Acculturation and latino health in the united states: A Review of the Literature and its Sociopolitical Context. *Annu. Rev. Public Health* **2005**, *26*, 367–397. [CrossRef]

13. Ro, A.; Fleischer, N.L.; Blebu, B. An examination of health selection among U.S. immigrants using multi-national data. *Soc. Sci. Med.* **2016**, *158*, 114–121. [CrossRef]
14. González, H.M.; Tarraf, W.; Mosley, T.; Wright, C.; Rundek, T.; Lutsey, P.L.; Gouskova, N.; Grober, E.; Pirzada, A.; Daviglus, M. Life's Simple 7 s of Cardiovascular Health among Hispanic/Latinos: The Hispanic Community Health Study/Study of Latinos (HCHS/SOL). *Circulation* **2012**, *126*, A17624.
15. Diaz, C.J.; Koning, S.M.; Martinez-Donate, A.P. Moving Beyond Salmon Bias: Mexican Return Migration and Health Selection. *Demography* **2016**, *53*, 2005–2030. [CrossRef] [PubMed]
16. Viruell-Fuentes, E.A.; Miranda, P.Y.; Abdulrahim, S. More than culture: Structural racism, intersectionality theory, and immigrant health. *Soc. Sci. Med.* **2012**, *75*, 2099–2106. [CrossRef] [PubMed]
17. Beltrán-Sánchez, H.; Palloni, A.; Riosmena, F.; Wong, R. SES Gradients Among Mexicans in the United States and in Mexico: A New Twist to the Hispanic Paradox? *Demography* **2016**, *53*, 1555–1581. [CrossRef] [PubMed]
18. Langellier, B.A.; Martínez-Donate, A.P.; Gonzalez-Fagoaga, J.E.; Rangel, M.G. The Relationship Between Educational Attainment and Health Care Access and Use Among Mexicans, Mexican Americans, and U.S.–Mexico Migrants. *J. Immigr. Minor. Health* **2020**, *22*, 314–322. [CrossRef]
19. Lu, Y. Test of the 'healthy migrant hypothesis': A longitudinal analysis of health selectivity of internal migration in Indonesia. *Soc. Sci. Med.* **2008**, *67*, 1331–1339. [CrossRef]
20. Ro, A.; Fleischer, N. Changes in health selection of obesity among Mexican immigrants: A binational examination. *Soc. Sci. Med.* **2014**, *123*, 114–124. [CrossRef]
21. Rubalcava, L.N.; Teruel, G.M.; Thomas, D.; Goldman, N. The Healthy Migrant Effect: New Findings From the Mexican Family Life Survey. *Am. J. Public Health* **2008**, *98*, 78–84. [CrossRef]
22. Ullmann, S.H.; Goldman, N.; Massey, D.S. Healthier before they migrate, less healthy when they return? The health of returned migrants in Mexico. *Soc. Sci. Med.* **2011**, *73*, 421–428. [CrossRef]
23. Rubalcava, L.; Teruel, G. Mexican Family Life Survey, Second Round. 2006. Available online: www.envih-mxfls.org (accessed on 12 December 2019).
24. Berumen. *"Sample Design". Description of MxFLS-2 Samples*; University Iberoamericana: Mexico City, Mexico, 2007.
25. World Health Organization. *Waist Circumference and Waist–Hip Ratio: Report of a WHO Expert Consultation*; WHO: Geneva, Switzerland, 2008. Available online: http://www.who.int/nutrition/publications/obesity/WHO_report_waistcircumference_and_waisthip_ratio/en/ (accessed on 26 March 2016).
26. World Health Organization. *Prevention of Cardiovascular Disease*; World Health Organization: Geneva, Switzerland, 2007. ISBN 978-92-4-154726-0. Available online: http://www.who.int/cardiovascular_diseases/guidelines/Full%20text.pdf (accessed on 12 December 2019).
27. Barquera, S.; Campos-Nonato, I.; Hernández-Barrera, L.; Flores, M.; Durazo-Arvizu, R.; Kanter, R.; Rivera, J.A. Obesity and central adiposity in Mexican adults: Results from the Mexican National Health and Nutrition Survey 2006. *Salud Pública Méx.* **2009**, *51*, S595–S603. [CrossRef]
28. Bostean, G. Does Selective Migration Explain the Hispanic Paradox? A Comparative Analysis of Mexicans in the U.S. and Mexico. *J. Immigr. Minor. Health Cent. Minor. Public Health* **2013**, *15*, 624–635. [CrossRef] [PubMed]
29. Vilar-Compte, M.; Bustamante, A.V.; López-Olmedo, N.; Gaitán-Rossi, P.; Torres, J.; Peterson, K.E.; Teruel, G.; Pérez-Escamilla, R. Migration as a determinant of childhood obesity in the United States and Latin America. *Obes. Rev. Off. J. Int. Assoc. Study Obes.* **2021**, *22* (Suppl. S3), e13240. [CrossRef] [PubMed]
30. Riosmena, F.; Wong, R.; Palloni, A. Migration Selection, Protection, and Acculturation in Health: A Binational Perspective on Older Adults. *Demography* **2013**, *50*, 1039–1064. [CrossRef]
31. Finch, B.K.; Do, D.P.; Frank, R.; Seeman, T. Could "Acculturation" Effects Be Explained by Latent Health Disadvantages Among Mexican Immigrants? *Int. Migr. Rev.* **2009**, *43*, 471–495. [CrossRef]
32. Kaestner, R.; Pearson, J.A.; Keene, D.; Geronimus, A.T. Stress, Allostatic Load, and Health of Mexican Immigrants. *Soc. Sci. Q.* **2009**, *90*, 1089–1111. [CrossRef]
33. Ornelas, I.J.; Perreira, K.M. The role of migration in the development of depressive symptoms among Latino immigrant parents in the USA. *Soc. Sci. Med.* **2011**, *73*, 1169–1177. [CrossRef]
34. Viruell-Fuentes, E.A. It's a lot of work. *Bois Rev. Soc. Sci. Res. Race* **2011**, *8*, 37–52. [CrossRef]
35. Bose, M.; Oliván, B.; Laferrère, B. Stress and obesity: The role of the hypothalamic–pituitary–adrenal axis in metabolic disease. *Curr. Opin. Endocrinol. Diabetes Obes.* **2009**, *16*, 340–346. [CrossRef]
36. Wardle, J.; Chida, Y.; Gibson, E.L.; Whitaker, K.L.; Steptoe, A. Stress and Adiposity: A Meta-Analysis of Longitudinal Studies. *Obesity* **2011**, *19*, 771–778. [CrossRef]
37. Aschbacher, K.; Kornfeld, S.; Picard, M.; Puterman, E.; Havel, P.J.; Stanhope, K.; Lustig, R.H.; Epel, E. Chronic stress increases vulnerability to diet-related abdominal fat, oxidative stress, and metabolic risk. *Psychoneuroendocrinology* **2014**, *46*, 14–22. [CrossRef]
38. Kyrou, I.; Chrousos, G.P.; Tsigos, C. Stress, Visceral Obesity, and Metabolic Complications. *Ann. N. Y. Acad. Sci.* **2006**, *1083*, 77–110. [CrossRef] [PubMed]
39. Vilar-Compte, M.; Gaitán-Rossi, P.; Félix-Beltrán, L.; Bustamante, A.V. Pre-COVID-19 Social Determinants of Health Among Mexican Migrants in Los Angeles and New York City and Their Increased Vulnerability to Unfavorable Health Outcomes During the COVID-19 Pandemic. *J. Immigr. Minor. Health* **2021**. [CrossRef] [PubMed]

40. Perreira, K.M.; Pedroza, J.M. Policies of Exclusion: Implications for the Health of Immigrants and Their Children. *Annu. Rev. Public Health* **2019**, *40*, 147–166. [CrossRef] [PubMed]
41. Radford, J.; Noe-Bustamante, L. *Facts on US Immigrants, 2017: Statistical Portrait of the Foreign-Born Population in the United States*; Pew Research Center: Washington, DC, USA, 2019. Available online: https://www.pewresearch.org/hispanic/2019/06/03/facts-on-u-s-immigrants/ (accessed on 12 December 2019).
42. Gonzalez-Barrera, A.; Lopez, M.H. *A Demographic Portrait of Mexican-Origin Hispanics in the United States*; Pew Hispanic Center: Washington, DC, USA, 2013.
43. US Census Bureau. Current Population Survey: Characteristics of the Foreign-Born Population by Nativity and U.S. Citizenship Status. 2013. Available online: https://www.census.gov/data/tables/2013/demo/foreign-born/cps-2013.html (accessed on 14 April 2017).
44. Martinez-Cardoso, A.; Geronimus, A. Migration to the United States and the Health of Mexico-U.S. Migrants and Nonmigrants. In Proceedings of the Population Association of American Annual Meeting, Austin, TX, USA, 10–13 April 2019.

International Journal of
*Environmental Research
and Public Health*

MDPI

Review

Parent Emigration, Physical Health and Related Risk and Preventive Factors of Children Left Behind: A Systematic Review of Literature

Justina Račaitė [1,*], Jutta Lindert [2,3], Khatia Antia [4], Volker Winkler [4], Rita Sketerskienė [1], Marija Jakubauskienė [1], Linda Wulkau [2,5] and Genė Šurkienė [1]

[1] Department of Public Health, Institute of Health Sciences, Faculty of Medicine, Vilnius University, M.K. Čiurlionio Str. 21, LT-03101 Vilnius, Lithuania; rita.sketerskiene@mf.vu.lt (R.S.); marija.jakubauskiene@mf.vu.lt (M.J.); gene.surkiene@mf.vu.lt (G.Š.)
[2] Department of Social Work and Health, University of Applied Sciences Emden/Leer, Constantiaplatz 4, 26723 Emden, Germany; jutta.lindert@hs-emden-leer.de (J.L.); linda.wulkau@hs-emden-leer.de (L.W.)
[3] WRSC, Brandeis University, Epstein Building, MS 079, 515 South Street, Waltham, MA 02453, USA
[4] Heidelberg Institute of Global Health, Heidelberg University Hospital, Im Neuenheimer Feld 130/3, 69120 Heidelberg, Germany; khatia.antia@uni-heidelberg.de (K.A.); volker.winkler@uni-heidelberg.de (V.W.)
[5] Institute for Epidemiology, Social Medicine and Health Systems Research, Hannover Medical School, Carl-Neuberg-Straße 1, 30625 Hannover, Germany
* Correspondence: justina.racaite@mf.vu.lt; Tel.: +370-63158873

Citation: Račaitė, J.; Lindert, J.; Antia, K.; Winkler, V.; Sketerskienė, R.; Jakubauskienė, M.; Wulkau, L.; Šurkienė, G. Parent Emigration, Physical Health and Related Risk and Preventive Factors of Children Left Behind: A Systematic Review of Literature. *Int. J. Environ. Res. Public Health* **2021**, *18*, 1167. https://doi.org/10.3390/ijerph18031167

Academic Editors: Ko-Ling Chan and David Schwebel

Received: 4 January 2021
Accepted: 22 January 2021
Published: 28 January 2021

Abstract: The aim of our study was to systematically review the literature on physical health and related consequences of internal and international parental migration on left-behind children (LBC). This review followed PRISMA guidelines. We searched the PubMed, Web of Science, Academic Search Complete, PsycINFO, and Cochrane databases and included studies reporting physical health-related outcomes of children affected by parental migration. The quality of the studies was assessed using the Quality Assessment Tool for Observational Cohort and Cross-Sectional Studies. We selected 34 publications from a total of 6061 search results. The study found that LBC suffer from poor physical health as compared with non-LBC. Physical health-related risk factors such as underweight, lower weight, stunted growth, unhealthy food preferences, lower physical activity, smoking, alcohol consumption, injuries, and incomplete vaccination tend to be more prevalent among LBC in China. Studies focussing on international migration argue that having migrant parents might be preventive for undernutrition. Overall, our study showed that children affected by internal or international migration tend to have similar physical health outcomes. Moreover, we identified a lack of evidence on international parental migration that may have influenced the overall impacts. Further studies addressing international migration would contribute to better understand the impacts of migration for LBC.

Keywords: children left behind; parental migration; physical health; children health

1. Introduction

The United Nation's Convention on the Rights of the Child " [recognized] that the child, for the full and harmonious development of his or her personality, should grow up in a family environment, in an atmosphere of happiness, love and understanding" [1]. The attachment theory, formulated by John Bowlby, states that for the successful social and emotional development, every child needs a close relationship with at least one primary caregiver [2]. However, there are multiple reasons for parental absence, such as divorce or death. Moreover, children are sometimes taken from unsafe family environments temporarily or permanently.

The migration of parents is another form of child separation from one or both parents. The International Organization for Migration defines migration as "the movement of a person or a group of persons, either across an international border, or within a State [independent of] its length, composition and causes [including] migration of refugees, displaced persons, economic migrants, and persons moving for other purposes" [3]. The World Migration Report has reported that, in 2019, the number of international migrants globally was 272 million, which is around 3.5% of the world's population [4]. Migration is important for the economic growth and improvement of countries. The overwhelming majority of people choose to migrate internationally for reasons related to work, family, and study. However, events such as a conflict, persecution, and disaster force people to migrate without having any choice [4]. Safe, orderly, and regular international migration is included in the 2030 Agenda for Sustainable Development Goals [5].

Employment migration may cause a number of short- and long-term consequences. One of them is the separation of families where children are left behind in their region or countries of origin. In the available literature, children left behind are defined as individuals below the age of 18, whose parent(s) migrate to other places for work for at least six months [6,7]. Thousands of children are considered to be left behind in many low- and middle-income countries. For example, it has been estimated that in the Philippines, 27%, Ecuador, 36%, and rural South Africa, 40% of children have at least one migrant parent [8].

Health issues of left-behind children are increasingly being discussed in the scientific literature. Some authors have conducted systematic literature reviews focussing mostly on rural–urban migration in China [8–11]. Previous studies in this field have made an important contribution regarding the understanding of how parental migration affects the social environment and psychological well-being of LBC, education, and health [8,12–14]. Following scientific interest, the United Nations Children's Fund (UNICEF) recently drew attention to the vulnerability of these children [15]. This issue has been well explored in China, where migration happens internally, from rural to urban areas. Despite the increased attention, there is a lack of data on the impact of international parental migration on LBC's physical health outcomes. Our study seeks to analyse and synthesize the most recent evidence on health consequences of internal and international parental migration on LBC's physical health and related risk and preventive factors.

2. Materials and Methods

For this study, we followed the Preferred Reporting Items for Systematic Reviews and Meta-Analyses (PRISMA) guidelines [16].

We searched PubMed, PsycINFO, Web of Science, Academic search complete, and Cochrane databases for relevant studies published up to 15 May 2020. Our search was guided by the concept of population, exposure, comparator, and outcomes (PECO). We applied the following search terms: left alone OR left behind and stay at home OR left over AND child* AND parent* AND emigrant* OR migrant household AND physical* health OR overweight OR obesity OR stunting OR vaccination OR breastfeed* OR "physical* activity". In addition, we searched the reference lists of included studies and relevant systematic reviews.

We included studies based on the following criteria: (1) study population children (below age 18), (2) original study, (3) one or both parents live in internal or international migration, (4) quantitative measure of physical health outcomes on children, (5) available in English. Two authors (J.R. and R.S.) independently performed title and abstract screening. Disagreements were solved by discussions with a third opinion (G.Š.). Studies with the following criteria were excluded: published before 1 January 2008; qualitative or experimental studies; mental health, well-being or educational outcomes; and children living in migration together with parents.

Two authors (J.R. and R.S.) extracted the following information from included papers: first author; year of publication; geographical area; study design; sample size and method, age and gender distribution, definitions and measures of exposures and outcomes, results, covariates and limitations. Disagreement was solved by including a third opinion (G.Š.).

Table 1. Characteristics of the studies.

No	First Author, Year of Publication	Country	Study Design	Sample Size (N)	Age (Range, Mean, SD)	Gender Distribution (Male; %)	Outcomes	Quality Rating [1]
1.	Ban, 2017 [26]	China	Cross-sectional	6136	0–35 months	Neither parent migrated 54.3%, father migrated only 52.7%, mother with/without father migrated 54.2%	1. Stunting 2. Breastfeeding 3. Milk feeding	Good
2.	Cebotari, 2018 [25]	Moldova, Georgia	Cross-sectional	Moldova (1601), Georgia (1193)	10–18 years Moldova 14.3 (SD 2.59), Georgia 13.44 (SD 2.40)	Moldova 51.93%, Georgia 53.96%	Child health status	Good
3.	Edelblute, 2019 [17]	Mexico	Cross-sectional	542	5.3 (SD 2.96)	Father present 51.5%, father absence (migration) 48.1%, father absence (other reasons) 42.9%	Maternal ratings of child poor health	Good
4.	Gao, 2013 [27]	China	Cross-sectional	2558	13.8 (SD 1.14)	55%	1. Past 30 days smoking 2. Self-efficacy of smoking	Good
5.	Gao, 2010 [6]	China	Cross-sectional	2986	10–18 years 14.2 (SD 1.4)	51.4%	1. Health-related behaviours 2. Nutritional status	Good
6.	Graham, 2013 [23]	The Philippines, Vietnam	Cross-sectional	The Philippines (480), Vietnam (482)	9–11 years	NA [2]	Stunting	Good
7	Guo, 2017 [28]	China	Longitudinal	6083	12.27 (SD 3.71)	55%	Self-rated health	Good
8.	Hipgrave, 2014 [29]	China	Cross-sectional	2244	6–23 months	56.3%	Haemoglobin concentration	Good
9.	Hu, 2018 [30]	China	Cross-sectional	4479	6–16 years	46.5%	Unintentional injuries	Good
10.	Huang, 2018 [7]	China	Cross-sectional	916	11.6	57%	Children's health conditions	Good
11.	Islam, 2019 [22]	Bangladesh	Cross-sectional	23,402	0–5 years 2	51.3%	1. Stunting 2. Wasting 3. Underweight 4. Nutritional disorders	Good

Table 1. *Cont.*

No	First Author, Year of Publication	Country	Study Design	Sample Size (N)	Age (Range, Mean, SD)	Gender Distribution (Male; %)	Outcomes	Quality Rating [1]
12.	Jayatissa, 2016 [19]	Sri Lanka	Cross-sectional	7500	6–59 months 32.9 (SD 14.7)	50.2%	1. Stunting, wasting and underweight 2. Health status 3. Food intake	Fair
13.	Jiang, 2015 [31]	China	Cross-sectional	1367	12.2 (SD 1.3)	56%	Current alcohol drinking	Good
14.	Lei, 2018 [32]	China	Cross-sectional	5413	1–15 years	Rural without migrant parents 55%;left behind 56%.	1. Child health status 2. Child health status assessed by guardians	Good
15.	Li, 2015 [33]	China	Longitudinal	13,171	9755 (SD 4.9)	53%	1.Sickness 2. Injuries 3. Chronic conditions 4. Acute disease	Good
16.	Luo, 2008 [34]	China	Cross-sectional	1548	0.33–6.92 years 3.51 (SD 1.59)	56.2%	1. Anthropometry 2. Dietary intake 3. Haemoglobin concentration	Fair
17.	Mo, 2016 [35]	China	Cross-sectional	735	1–6 years 4.58 years (SD 55.0 months)	44.1%	Physical health	Good
18.	Nguyen, 2016 [24]	Ethiopia, India, Peru, Vietnam	Cross-sectional	7725	5–8 years	NA[2]	Health status	Good
19.	Ni, 2017 [36]	China	Cross-sectional	1368	12–72 months	85%	1. The full vaccination rate 2. The age-appropriate vaccinationrate	Good
20.	Palos-Lucio, 2015 [18]	Mexico	Cross-sectional	239	9–12 years	Migrant household 51.69%; Non-migrant household 55.46%.	1. Physical activity	Good
21.	Shen, 2009 [37]	China	Cross-sectional	3019	5–17 years	60.3%	Unintentional injuries	Fair

Table 1. *Cont.*

No	First Author, Year of Publication	Country	Study Design	Sample Size (N)	Age (Range, Mean, SD)	Gender Distribution (Male; %)	Outcomes	Quality Rating [1]
22.	Smeekens, 2012 [21]	The Philippines	Cross-sectional	205	13–18 14.58 (SD 1.04)	31.7%	Physical health	Good
23.	Tang, 2019 [38]	China	Cross-sectional	1662	12–15 years	50.2%	1. Physical health status 2. Health behaviours 3. Frequency of not going to school due to sickness 4. Vaccination	Good
24.	Tang, 2016 [39]	China	Cross-sectional	1216	18–24 months	50.2%	Vaccination status	Good
25.	Tao, 2016 [40]	China	Cross-sectional	827	7–15 years	51.2%	1. Nutritive status 2. Food preference	Good
26.	Tian, 2017 [41]	China	Longitudinal	446	11–18 years	60%	1. Growth 2. Nutrition	Good
27.	Tong, 2015 [42]	China	Longitudinal	8662	10.18 (SD 4.77)	83.8%	Childhood illness	Good
28.	Wen, 2016 [43]	China	Longitudinal	2170	7–17 years	At baseline 53.9%	Blood pressure	Good
29.	Wickramage, 2015 [20]	Sri Lanka	Cross-sectional	770	1–17 years	Left-behind group 50.6%; control group:53.2%.	1. Nutritional status 2. Anthropometry 3. Immunization history	Good
30.	Yang, 2016 [44]	China	Cross-sectional	1343	10–14 years	56%	Smoking	Good
31.	Yue, 2020 [45]	China	Longitudinal	1802	6–30 months	53.1%	1. Nutritional status 2. Anaemia status 3. General health 4. Feeding practises	Good
32.	Zhang, 2015 [46]	China	Longitudinal	2555	0–17 years	57.3%	Child growth	Good
33.	Zhang (a), 2015 [47]	China	Longitudinal	975	1–17 years	56.9%	1. Dietary 2. Macronutrient intakes	Good
34.	Zhou, 2015 [48]	China	Cross-sectional	141,000	3–17 years	NA [2]	1. Health 2. Nutrition	Fair

[1] Quality of the studies assessed using National Institute of Health Quality Assessment Tool for Observational Cohort and Cross-Sectional Studies. Possible quality ratings (good, fair, poor). [2] NA, no data available.

Table 2. Main outcomes (only statistically significant outcomes provided).

Group	Outcome (Left-Behind Children)	Sample Size (N)	Covariates [2]	Statistics (OR, Mean, SD, p, 95% CI or Other Statistics If Provided) [1]	Reference
			Internal Migration		
Physical Health	More susceptible to illness	735	Parenting styles, age of child, health literacy	OR 1.628, $p < 0.05$	Mo, 2015 [35]
	Childhood illness	8662	Age, gender, household size, income per capita, grandparents living together, village context, village size	OR 1.29, SE = 0.164, $p < 0.05$	Tong, 2015 [42]
	Pre-hypertension or hypertension	2170	Age, gender, mothers education, fathers education, annual household per capita income	OR 7.77, 95% CI 2.05–29.4, $p < 0.01$	Wen, 2016 [43]
Nutrition, Weight and Height	Lower Height for age z-score	5413	Age, gender	OR −0.165, $p < 0.01$	Lei, 2018 [32]
	Lower Weight for age z-score	5413	Age, gender	OR −0.142, $p < 0.05$	Lei, 2018 [32]
	Malnutrition rate	827	NR/NA	LBC 14.83%, NoN-LBC 7.04%, $p < 0.01$	Tao, 2016 [40]
	Less likely to be ever breastfed	6136	Age, gender, ethnicity, elder siblings, guardian education attainment, the number of household electrical appliances, year of survey	OR 0.30, 95% CI 0.17–0.52	Ban, 2017 [26]
	Less likely to be breastfed	1548	NR/NA	LBC (78.7%), Non-LBC (82.8%), $p < 0.05$	Luo, 2008 [34]
	Shorter breastfeeding	6136	Age, gender, ethnicity, elder siblings, guardian education attainment, the number of household electrical appliances, year of survey	β-3.77, 95% CI −5.01-−2.53	Ban, 2017 [26]
	Duration of breastfeeding (months)	1548	NR/NA	LBC M = 9.48, SD = 3.58; Non-LBC M = 10.70, SD = +3.26; $p < 0.001$	Luo, 2008 [34]
Unhealthy behaviors	Alcohol use	1367	Gender, age, grade, if only child in the family, perceived school performance	OR 2.01, 95% CI 1.28–3.16, $p < 0.05$	Jiang, 2015 [31]
	Higher smoking rate	1343	Gender, grade, if only child in the family, perceived school performance	OR 5.59, 95% CI 2.38–13.15, $p < 0.001$	Yang, 2016 [44]
Injuries	More likely to experience injury	4479	Gender, age, fair physical health, school academic achievement, only child in the family, household income level, parental marital status; maternal education attainment, family conflicts; model school; peer rejection, rural region	OR 1.208, SE 0.104, $p < 0.05$	Hu, 2018 [30]

Table 2. *Cont.*

Group	Outcome (Left-Behind Children)	Sample Size (N)	Covariates [2]	Statistics (OR, Mean, SD, p, 95% CI or Other Statistics If Provided) [1]	Reference
	Higher injury rate	3019	NR/NA	LBC 252.9/1000, 95% CI 233.0–273.0, Non-LBC 119.7/1000, 95% CI 104.9–134.7, $p < 0.0001$	Shen, 2009 [37]
	Lower rates of timely vaccination	1216	NR/NA	LBC55.7%, 95% CI 51.3–60.0, Non-LBC 60.8, 95% CI 57.3–64.0, $p = 0.011$	Tang, 2016 [39]
Immunization	Less likely to receive complete vaccination	1368	NR/NA	LBC 92.7%, Non-LBC 79.9%, $X^2 = 35.2$, $p < 0.001$	Ni, 2017 [36]
	Lower coverage of complete vaccination	1662	NR/NA	LBC 38.7%, Non-LBC 44.2%, $p < 0.036$	Tang, 2019 [38]
International Migration					
	Maternal reported child poor health	542	Data from model with no covariates added	OR 0.33, 95% CI 0.16–0.7, $p < 0.01$	Edelblute, 2018 [17]
Physical Health	Poorer physical health	205	NR/NA	LBC M = 5.09, SD = 0.78; NoN LBC M = 5.43, SD = 0.63; F (1.199) = 9.08; $p < 0.01$	Smeekens, 2012 [21]
Unhealthy behaviors	Lower physical activity	239	Age, gender, body mass index, maternal schooling, paternal schooling, household characteristics	OR −0.56, $p < 0.05$	Palos-Lucio, 2015 [18]

[1] LBC, left-behind children; Non-LBC, not left-behind children; [2] NR/NA, not reported or not applicable.

3.2. Internal Migration and Physical Health Outcomes

In China, five studies found that parental migration had negative consequences on child health [32,33,35,42,43]. As compared with children of non-migrant parents, LBC were more susceptible to illness and had a higher prevalence of acute and chronic diseases [33,35]. Being a child left behind was also strongly and positively associated with the pre-hypertension or hypertension (OR = 7.77, p < 0.01) [43].

Studies included in our analysis reported contradictory results with respect to physical health outcomes of LBC with one or both migrant parents. In China, findings suggest that a mother's absence alone would not affect a child's health, but both a mother's and father's absence together would have a significant negative effect on LBC in rural China [33]. Children raised by a single parent tend to be more susceptible to illness than children raised by both parents [35]. Among LBC, those living with their mother were more likely to be in better health, than those living with their father only [33].

Regarding the age and a gender of LBC, some studies suggested that left-behind adolescents (13–18 years) could have worse outcomes than younger children [7]. Girls may be more vulnerable than boys to the absence of parental care [33].

One study did not find any significant relationship between parent emigration and child health status [28] and one study found no evidence that children living in migration together with their parents had better health than children left behind [7].

3.3. Internal Migration and Risk and Preventive Factors

3.3.1. Nutrition, Weight, and Height

Studies included in this review found that, in China, parental migration negatively affected child nutrition [45]. Firstly, as compared with the control group (children of non-migrant parents), children left behind were less likely to receive age-appropriate breastfeeding and the duration of breastfeeding was significantly shorter [26,34]. Total food intake, as well as intake of meat, fish and eggs were lower among LBC [34]. Moreover, children with both parents absent were most likely to skip breakfast, as well as eat high-fat food and sweetened snacks [6,46]. Higher fat and lower protein diet were more common among left-behind boys [46]. Accordingly, more LBC disliked vegetables (M = 3.66, SD = 0.55) and fruits (M = 3.81, SD = 0.47, p < 0.01) than non-LBC (M = 3.89, SD = 0.27 and M = 3.97, SD 0.83, p < 0.01) [40]. Due to iron-poor food intake, LBC were at a higher risk of developing anaemia, especially at a younger age [29,34]. Additionally, Zhang et al. found gender differences, for example, on the one hand, LB boys in early childhood showed slower height and weight gain as compared with boys living in non-migrant households [47]. On the other hand, Gao et al. suggested that children left behind might have an increased risk of being overweight [6].

Two studies found no negative impact of parental migration for weight and height outcomes [34,41]. One study found high rates of nutrition problems regardless of parental migration status [48].

3.3.2. Unhealthy Behaviours

In general, children with both parents absent were more likely to engage in risky behaviours such as unhealthy diet, physical inactivity, sedentary lifestyle, smoking, and drinking [48].

With respect to addictive behaviours, studies found that alcohol consumption and smoking were higher among children of both migrant parents as opposed to only one (or none) migrant parent [6,31,44]. Other authors claimed that maternal migration increased the risk of adolescent smoking, while paternal migration could even protect children from smoking [27]. In terms of gender, some authors found that LB girls were at a higher risk for smoking and binge drinking [6], whereas other studies reported that more LB boys tended to be smokers and current alcohol users than LB girls [31,44].

3.3.3. Injuries

Some authors have suggested that LBC have a higher risk of getting injured [30,37]. The annual injury rate was more than double among LBC as compared with children living with both parents [37]. When controlling for other variables, LBC were more likely to experience unintentional injuries than residential children [30]. The most frequently reported injuries were falls, contact with sharp instrument, striking by objects or person, bitten, or struck by animals, and injuries caused by nature or environment factors [30].

3.3.4. Immunization

Studies included in this review found a lower coverage of vaccination among LBC as compared with non-LBC [38]. Children of non-migrating parents (95.7%) were more likely to receive complete vaccination as opposed to LBC (79.9%, $p < 0.001$) [36]. Moreover, LBC had significantly lower coverage of timely vaccination [39].

3.4. International Migration and Physical Health Outcomes

Physical health outcomes were analysed in three studies conducted in different regions [17,21,25]. A study from the Philippines found that left-behind adolescents (M = 5.09, SD = 0.78) reported poorer physical health than non-LBC (M = 5.43, SD = 0.63), $p < 0.01$). The same study found that left-behind adolescents (13–18 years) might be more negatively affected than younger children [21].

A Mexican study compared differences between mother and father emigration and found poorer health outcomes among LBC of international migrant fathers then children of non-migrant fathers [17].

One study conducted in Moldova and Georgia did not find any significant association between parent emigration and child health status [25].

3.5. International Migration and Risk and Preventive Factors
3.5.1. Nutrition, Weight, and Height

Regarding weight and height outcomes, findings from Sri Lanka found a lower prevalence of stunting (11.5% vs. 14.8%), wasting (18.1% vs. 21.5%), and underweight (24.3% vs. 26.2%) among LBC as compared with non-LBC, respectively [19]. A study from Ethiopia, India, and Peru found higher weight and height and lower proportion of malnourished children in migrant households as opposed to children from non-migrant households [24]. However, another study conducted in the Philippines and Vietnam did not support this [23].

Two studies from Bangladesh and Sri Lanka found no negative impact of parental migration for weight and height outcomes [20,22].

3.5.2. Physical Activity

There was only one study from Mexico which found lower physical activity among LBC than non-LBC. Children with parental migration experience had 0.56 less physical active time (hours) per day as compared with children from non-migrant households [18].

4. Discussion

In this study, we systematically reviewed the evidence on the effects of internal and international parental migration on their children's physical health outcomes and related risk and preventive factors. By doing so, we provided comparative analysis of the outcomes of the internal rural–urban and international migration. Previous studies in this field analysed the outcomes independently from type of migration [8] or focused on international migration only [14]. Our study was motivated by the substantial research gap in research on international labour migration effects on LBC in many low- and middle-income countries. A predominance of studies focused on internal migration from China have clearly shown the emerging need to shift this paradigm and investigate the issue in a global context [49].

As explained above, all studies analysing internal migration outcomes were conducted in China, while all studies from other regions (Americas, South and Southeast Asia, Africa, and the East European Region) examined international parental migration outcomes on LBC. Our study found that, despite the type of migration/region, LBC suffer from poor general health. Children with migrating parents are at a higher risk of developing poor nutrition, overweight or obesity, addictive behaviours, physical inactivity, lower vaccination coverage, and more frequent injuries than non-LBC. Some authors have explained such findings using the cognitive stress theory developed by Lazarus and Folkman, in 1984 [21]. Our findings are consistent with previous studies [8] showing that LBC's physical health outcomes are not improving over time.

Despite the negative outcomes reported, several authors also discuss the potential benefits of parental migration on their children's physical health. Some studies suggested that remittances could prevent undernutrition and improve access to medical care for LBC [17,19,25]. In contrast, studies that focused on internal migration from China found that remittances were related to a higher risk of overweight [35].

Studies included in this review found that socioeconomic conditions and characteristics of caregivers play an important role for potential outcomes. The following factors were most reported: parents' and caregivers' education, sex of migrant parent, household size, income per-capita, parental marital status, and siblings. The health literacy of a primary caregiver was found to be essential for nutrition, health, and development of a child [35]. Our study findings show that having a migrant mother might be more harmful, than having a migrant father [6,19,33]. Some authors emphasize the negative influence of a culture, for example, traditions of physical punishments in some countries, such as Moldova and Georgia [50]. However, most of the outcomes stay negative and significant after controlling for potential socioeconomic cofounders (Table 2).

In general, studies from China and Mexico found that LBC might be more vulnerable to risky behaviour and an unhealthy lifestyle than non-LBC. LBC affected by internal migration tend to have more risky behaviour such as alcohol consumption, smoking, as well as a high fat and low protein diet [44,47]. A study from Mexico also found lower physical activity among LBC [18]. This shows the need for improving health literacy and health education in schools and among caregivers of LBC.

Several limitations of our study should be mentioned. We included studies published only in English. Most of the included articles focused on Chinese populations, with a few exceptions from the Caribbean region, South America, South Asia, and Europe. Among all the included studies, only nine studies focused on international migration, while all other studies came from China, and therefore addressed only internal migration. Most of the studies were cross-sectional, which did not allow drawing conclusions on causation. Finally, authors noted that according to the International Organization of Migration, labour migration trends have increased significantly in recent years [4]. With this in mind, we decided to included studies published after 1 January 2008 aiming at providing the most recent evidence in this field. However, this could be considered to be a limitation, since we may have missed some relevant studies. Despite these limitations, to the best of our knowledge, this is the first study considering both internal and international migration aspects while examining the effects of parental migration on LBC's physical health.

The importance of migration is growing together with globalization, while millions of children are left behind in their countries of origin. Our findings emphasize the need for preventive actions to address the health of LBC. Various international organizations such as UNICEF have brought attention to this vulnerable groups. Even though scholars have addressed this issue in China, there is an urgent need for more evidence from other labour migration-affected regions of the world. Public health interventions for LBC is needed.

5. Conclusions

This study found that both internal and international parental migration is associated with child outcomes such as physical health, nutrition, weight and height, injuries, and

In this systematic review, we analysed studies from the following two perspectives: (1) type of migration (internal/international) outcomes and (2) physical health outcomes and related risk and preventive factors (weight and height, nutrition, health behaviour, injuries, immunization).

The quality of the studies was assessed using the Quality Assessment Tool for Observational Cohort and Cross-Sectional Studies (National Heart, Lung, and Blood Institute, 2014). Assessment consisted of 14 questions. Studies were defined as "good", "fair", or "poor" according to the number of "yes" answers in the evaluation from 50%, 49–21%, and below 20%, respectively. Two reviewers (J.R. and L.W.) independently rated the studies and disagreements were resolved by consensus including a third opinion (G.Š.).

3. Results

3.1. Study Selection and Characteristics

We identified 6061 studies by searching the databases, from which 1386 were duplicates. We excluded 4597 records after title and abstract screening. Six articles were added after searching the reference lists of included studies. Full-text reading of 84 articles led to the exclusion of 50 manuscripts. We included 34 studies published between 1 May 2008 and 12 March 2020 in the final analysis. Reasons of exclusion are given in Figure 1.

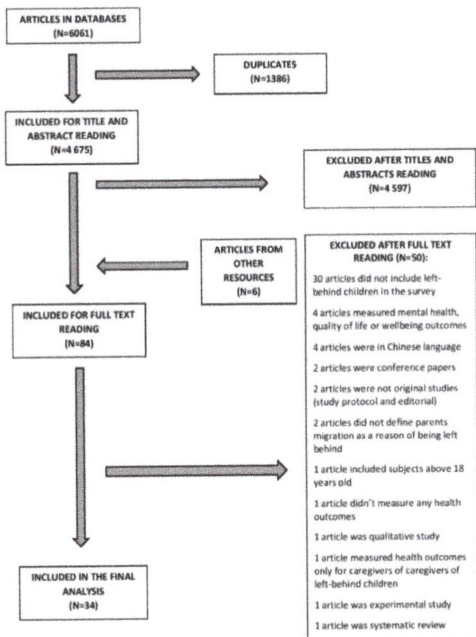

Figure 1. Study selection scheme.

Included studies were cross-sectional (26) or longitudinal (8). The majority of studies (25) were conducted in China. Studies conducted in other countries included the following: two studies from Mexico [17,18]; two studies from Sri Lanka [19,20]; one study from the Philippines [21]; one study from Bangladesh [22]; one study from the Philippines and Vietnam [23]; one study from Ethiopia, India, Peru, and Vietnam [24]; one study from Moldova and Georgia [25]. We evaluated 30 studies as "good" and four studies as "fair". The detailed characteristics of included studies are provided in Table 1. Table 2 describes the main outcome measures.

immunization. In most cases, the consequences for child health are negative, however, in low- and middle-income countries parental migration might also prevent left-behind children from undernutrition. When comparing studies from China and other countries, we found similar outcomes (regardless of internal or international migration). This study highlights the knowledge gap on the topic, especially in Western Asia and the East European Region, and calls for action from governments and international institutions, and the research community to better investigate and address the health needs of children affected by parental migration.

Author Contributions: Conceptualization, J.L., M.J. and J.R.; methodology, G.Š. and J.R.; validation, J.R., R.S. and L.W.; formal analysis, J.R. and R.S.; resources, J.R., K.A. and M.J.; writing—original draft preparation, J.R.; writing—review and editing, J.L., K.A., V.W. and G.Š.; visualization, J.R. and K.A.; supervision, G.Š. All authors have read and agreed to the published version of the manuscript.

Funding: This research received no external funding.

Institutional Review Board Statement: Not applicable.

Informed Consent Statement: Not applicable.

Acknowledgments: Authors wish to thank Vilnius University librarian Julija Lekavičiutė for assistance in article search useful information on databases.

Conflicts of Interest: The authors declare no conflict of interest.

References

1. United Nations. Convention on the Rights of the Child. 1989. Available online: https://www.ohchr.org/en/professionalinterest/pages/crc.aspx (accessed on 30 September 2019).
2. Bowlby, J. *Attachment: Attachment and Loss Volume One (Basic Books Classics)*; Basic Books: New York, NY, USA, 1983.
3. International Organization of Migration. Key Migration Terms. 2015. Available online: https://www.iom.int/key-migration-terms (accessed on 30 September 2019).
4. International Organization of Migration. World Migration Report. 2020. Available online: https://publications.iom.int/system/files/pdf/wmr_2020.pdf (accessed on 8 December 2020).
5. United Nations. The Sustainable Development Agenda. 2016. Available online: https://www.un.org/sustainabledevelopment/sustainable-development-goals/ (accessed on 30 September 2019).
6. Gao, Y.; Li, L.P.; Kim, J.H.; Congdon, N.; Lau, J.; Griffiths, S. The impact of parental migration on health status and health behaviours among left behind adolescent school children in China. *BMC Public Health* **2010**, *10*, 56.
7. Huang, Y.; Song, Q.; Tao, R.; Liang, Z. Migration, Family Arrangement, and Children's Health in China. *Child. Dev.* **2018**, *89*, e74–e90.
8. Fellmeth, G.; Rose-Clarke, K.; Zhao, C.; Busert, L.K.; Zheng, Y.; Massazza, A.; Sonmez, H.; Eder, B.; Blewitt, A.; Lertgrai, W.; et al. Health impacts of parental migration on left-behind children and adolescents: A systematic review and meta-analysis. *Lancet* **2018**, *392*, 2567–2582.
9. Zhao, F.; Yu, G. Parental migration and rural left-behind children's mental health in China: A meta-analysis based on mental health test. *J. Child. Fam. Stud.* **2016**, *25*, 3462–3472.
10. Wang, X.; Ling, L.; Su, H.; Cheng, J.; Jin, L.; Sun, Y.H. Self-concept of left-behind children in China: A systematic review of the literature. *Child. Care Health Dev.* **2015**, *41*, 346–355.
11. Sun, X.; Chen, M.; Chan, K.L. A meta-analysis of the impacts of internal migration on child health outcomes in China. *BMC Public Health* **2016**, *16*, 66.
12. Valtolina, G.G.; Colombo, C. Psychological Well-Being, Family Relations, and Developmental Issues of Children Left Behind. *Psychol. Rep.* **2012**, *111*, 905–928.
13. Dillon, M.; Walsh, C.A. Left Behind: The Experiences of Children of the Caribbean Whose Parents Have Migrated. *J. Comp. Fam. Stud.* **2012**, *43*, 871.
14. Antia, K.; Boucsein, J.; Deckert, A.; Dambach, P.; Račaitė, J.; Šurkienė, G.; Jaenisch, T.; Horstick, O.; Winkler, V. Effects of International Labour Migration on the Mental Health and Well-Being of Left-Behind Children: A Systematic Literature Review. *Int. J. Environ. Res. Public Health* **2020**, *17*, 4335.
15. UNICEF. Working Paper Children "Left-behind". 2019. Available online: https://www.unicef.org/media/61041/file (accessed on 16 January 2020).
16. David Moher, P.; Alessandro Liberati, M.D.; Jennifer Tetzlaff, B.S.; Douglas, G.; Altman, D.S.; the PRISMA Group. Preferred Reporting Items for Systematic Reviews and Meta-Analyses: The PRISMA Statement. *Ann. Intern. Med.* **2009**, *144*, 364–367.
17. Edelblute, H.B.; Altman, C.E. Father Absence, Social Networks, and Maternal Ratings of Child Health: Evidence from the 2013 Social Networks and Health Information Survey in Mexico. *Matern. Child. Health J.* **2018**, *22*, 626–634.

18. Palos-Lucio, G.; Flores, M.; Rivera-Pasquel, M.; Salgado-de-Snyder, V.N.; Monterrubio, E.; Henao, S.; Macias, N. Association between migration and physical activity of school-age children left behind in rural Mexico. *Int. J. Public Health* **2015**, *60*, 49–58.
19. Jayatissa, R.; Wickramage, K. What Effect Does International Migration Have on the Nutritional Status and Child Care Practices of Children Left Behind? *Int. J. Environ. Res. Public Health* **2016**, *13*, 218.
20. Wickramage, K.; Siriwardhana, C.; Vidanapathirana, P.; Weerawarna, S.; Jayasekara, B.; Pannala, G.; Adikari, A.; Jayaweera, K.; Peiris, S.; Siribaddana, S.; et al. Risk of mental health and nutritional problems for left-behind children of international labor migrants. *BMC Psychiatry* **2015**, *15*, 39.
21. Smeekens, C.; Stroebe, M.S.; Abakoumkin, G. The impact of migratory separation from parents on the health of adolescents in the Philippines. *Soc. Sci. Med.* **2012**, *75*, 2250–2257.
22. Islam, M.M.; Khan, M.N.; Mondal, M.N.I. Does parental migration have any impact on nutritional disorders among left-behind children in Bangladesh? *Public Health Nutr.* **2018**, *22*, 95–103.
23. Graham, E.; Jordan, L.P. Does Having a Migrant Parent Reduce the Risk of Undernutrition for Children Who Stay Behind in South-East Asia? *Asian Pac. Migr. J.* **2013**, *22*, 315–348.
24. Nguyen, C.V. Does parental migration really benefit left-behind children? Comparative evidence from Ethiopia, India, Peru and Vietnam. *Soc. Sci. Med.* **2016**, *153*, 230–239.
25. Cebotari, V.; Siegel, M.; Mazzucato, V. Migration and child health in Moldova and Georgia. *Comp. Migr. Stud.* **2018**, *6*, 3.
26. Ban, L.; Guo, S.; Scherpbier, R.W.; Wang, X.; Zhou, H.; Tata, L.J. Child feeding and stunting prevalence in left-behind children: A descriptive analysis of data from a central and western Chinese population. *Int. J. Public Health* **2017**, *62*, 143–151.
27. Gao, Y.; Li, L.; Chan, E.Y.; Lau, J.; Griffiths, S.M. Parental migration, self-efficacy and cigarette smoking among rural adolescents in south China. *PLoS ONE.* **2013**, *8*, e57569.
28. Guo, Q.; Sun, W.K.; Wang, Y.J. Effect of Parental Migration on Children's Health in Rural China. *Rev. Dev. Econ.* **2017**, *21*, 1132–1157.
29. Hipgrave, D.B.; Fu, X.; Zhou, H.; Jin, Y.; Wang, X.; Chang, S.; Scherpbier, R.W.; Wang, Y.; Guo, S. Poor complementary feeding practices and high anaemia prevalence among infants and young children in rural central and western China. *Eur. J. Clin. Nutr.* **2014**, *68*, 916–924.
30. Hu, H.; Gao, J.; Jiang, H.; Xing, P. A comparative study of unintentional injuries among schooling left-behind, migrant and residential children in China. *Int. J. Equity Health* **2018**, *17*, 47.
31. Jiang, S.; Chu, J.; Li, C.; Medina, A.; Hu, Q.; Liu, J.; Zhou, C. Alcohol consumption is higher among left-behind Chinese children whose parents leave rural areas to work. *Acta Paediatr.* **2015**, *104*, 1298–1304.
32. Lei, L.L.; Liu, F.; Hill, E. Labour Migration and Health of Left-Behind Children in China. *J. Dev. Stud.* **2018**, *54*, 93–110.
33. Li, Q.; Liu, G.; Zang, W. The health of left-behind children in rural China. *China Econ. Rev.* **2015**, *36*, 367–376.
34. Luo, J.; Peng, X.; Zong, R.; Yao, K.; Hu, R.; Du, Q.; Fang, J.; Zhu, M. The status of care and nutrition of 774 left-behind children in rural areas in China. *Public Health Rep.* **2008**, *123*, 382–389.
35. Mo, X.; Xu, L.; Luo, H.; Wang, X.; Zhang, F.; Gai Tobe, R. Do different parenting patterns impact the health and physical growth of 'left-behind' preschool-aged children? A cross-sectional study in rural China. *Eur. J. Public Health* **2016**, *26*, 18–23.
36. Ni, Z.L.; Tan, X.D.; Shao, H.Y.; Wang, Y. Immunisation status and determinants of left-behind children aged 12–72 months in central China. *Epidemiol. Infect.* **2017**, *145*, 1763–1772.
37. Shen, M.; Yang, S.; Han, J.; Shi, J.; Yang, R.; Du, Y.; Stallones, L. Non-fatal injury rates among the "left-behind children" of rural China. *Inj. Prev.* **2009**, *15*, 244–247.
38. Tang, D.; Choi, W.I.; Deng, L.; Bian, Y.; Hu, H. Health status of children left behind in rural areas of Sichuan Province of China: A cross-sectional study. *BMC Int. Health Hum. Rights* **2019**, *19*, 4.
39. Tang, X.; Geater, A.; McNeil, E.; Zhou, H.; Deng, Q.; Dong, A.; Li, Q. Parental migration and children's timely measles vaccination in rural China: A cross-sectional study. *Trop. Med. Int. Health* **2016**, *21*, 886–894.
40. Tao, S.; Yu, L.; Gao, W.; Xue, W. Food preferences, personality and parental rearing styles: Analysis of factors influencing health of left-behind children. *Qual. Life Res.* **2016**, *25*, 2921–2929.
41. Tian, X.; Ding, C.; Shen, C.; Wang, H. Does Parental Migration Have Negative Impact on the Growth of Left-Behind Children? New Evidence from Longitudinal Data in Rural China. *Int. J. Environ. Res. Public Health* **2017**, *14*, 11.
42. Tong, Y.Y.; Luo, W.X.; Piotrowski, M. The Association Between Parental Migration and Childhood Illness in Rural China. *Eur. J. Popul.* **2015**, *31*, 561–586.
43. Wen, M.; Li, K. Parental and Sibling Migration and High Blood Pressure among Rural Children in China. *J. Biosoc. Sci.* **2016**, *48*, 129–142.
44. Yang, T.; Li, C.; Zhou, C.; Jiang, S.; Chu, J.; Medina, A.; Rozelle, S. Parental migration and smoking behavior of left-behind children: Evidence from a survey in rural Anhui, China. *Int. J. Equity Health* **2016**, *15*, 127.
45. Yue, A.; Bai, Y.; Shi, Y.; Luo, R.; Rozelle, S.; Medina, A.; Sylvia, S. Parental Migration and Early Childhood Development in Rural China. *Demography* **2020**, *57*, 403–422.
46. Zhang, N.; Becares, L.; Chandola, T. Does the timing of parental migration matter for child growth? A life course study on left-behind children in rural China. *BMC Public Health* **2015**, *15*, 966.
47. Zhang, N.; Becares, L.; Chandola, T. A multilevel analysis of the relationship between parental migration and left-behind children's macronutrient intakes in rural China. *Public Health Nutr.* **2015**, *19*, 1913–1927.

48. Zhou, C.; Sylvia, S.; Zhang, L.; Luo, R.; Yi, H.; Liu, C.; Shi, Y.; Loyalka, P.; Chu, J.; Medina, A.; et al. China's Left-Behind Children: Impact of Parental Migration on Health, Nutrition, And Educational Outcomes. *Health Aff.* **2015**, *34*, 1964–1971.
49. Griffiths, S.M.; Dong, D.; Chung, R.Y. Forgotten need of children left behind by migration. *Lancet* **2018**, *392*, 2518–2519.
50. Gassmann, F.; Siegel, M.; Vanore, M.; Waidler, J. Unpacking the Relationship between Parental Migration and Child Well-Being: Evidence from Moldova and Georgia. *Child. Indic. Res.* **2018**, *11*, 423–440.

MDPI

St. Alban-Anlage 66

4052 Basel

Switzerland

Tel. +41 61 683 77 34

Fax +41 61 302 89 18

www.mdpi.com

International Journal of Environmental Research and Public Health Editorial Office

E-mail: ijerph@mdpi.com

www.mdpi.com/journal/ijerph